An Introduction to QUEUEING NETWORKS

An Introduction to QUEUEING NETWORKS

Jean Walrand

University of California
Berkeley

PRENTICE HALL, Englewood Cliffs, New Jersey 07632

Library of Congress Cataloging-in-Publication Data

WALRAND, JEAN. (date)
An introduction to queueing networks/Jean Walrand.
p. cm.
Bibliography: p.
Includes index.
ISBN 0-13-474487-X
1. Queueing theory. I. Title.
T57.9.W36 1988 519.8'2—dc19 87-32613 CIP

Editorial/production supervision: *Gloria L. Jordan*
Cover design: *Marcel Maessen*
Manufacturing buyer: *Lorraine Fumoso*

Printed in the United States of America

10 9 8 7 6 5 4 3 2 1

ISBN 0-13-474487-X

Prentice-Hall International (UK) Limited, *London*
Prentice-Hall of Australia Pty. Limited, *Sydney*
Prentice-Hall Canada Inc., *Toronto*
Prentice-Hall Hispanoamericana, S.A., *Mexico*
Prentice-Hall of India Private Limited, *New Delhi*
Prentice-Hall of Japan, Inc., *Tokyo*
Simon & Schuster Asia Pte. Ltd., *Singapore*
Editora Prentice-Hall do Brasil, Ltda., *Rio de Janeiro*

Le livre de la vie est le livre suprême . . .

Alphonse de Lamartine

To my mother and Annie

In memory of my father

CONTENTS

THREE. PRODUCT-FORM NETWORKS

FOUR. SOJOURN TIMES AND FLOWS

FIVE. NUMERICAL METHODS

SIX. NONPRODUCT-FORM NETWORKS

SEVEN. STATIONARY QUEUEING SYSTEMS

EIGHT. DYNAMIC CONTROL OF QUEUEING SYSTEMS

NINE. MULTIARMED BANDIT PROBLEMS

TEN. STATISTICAL ISSUES

APPENDIX A: PROBABILITY

APPENDIX B: QUEUES

PREFACE

This book is a self-contained introduction to the theory of queueing networks. It is based on a course taught by the author to first-year graduate students in electrical engineering, computer science, and operations research at the University of California in Berkeley. Some familiarity with basic probability theory, at the level of an upper division course, is assumed. The probability concepts reviewed in **Appendix A** should be understood. No prior exposure to queueing theory is necessary.

Traditionally, communication and control engineering have been concerned with analog signals and processes. A number of issues are now shifting to a higher level where operations can be viewed as being atomic and characterized by their functionality and their duration. Thus, many system level issues involve discrete actions and their scheduling, and the resulting measures of performance. This explains the increased interest in the analysis and the design of "discrete event dynamical systems."

The book is devoted to a subset of the relevant models, namely queueing networks. While many discrete event systems cannot be suitably modeled by queueing networks, a significant fraction of the intuition and techniques is useful for the other models.

The text is organized so that it can be covered in a one-term course. Many important results and methods could not be included in this format.

Chapter One is a general overview of the concepts, methods and results of the theory. On a first reading, one may want to skip the mathematical details of that chapter. **Chapter Two** is a presentation of the theory of *Markov chains.* That theory is illustrated by simple queueing systems *(Jackson networks).* Everything is standard, except possibly for the emphasis on *time reversal* and on *embedding* at selected jump times. In **Chapter Three** we discuss the results on *product-form networks.* The main results are that a network of quasi-reversible queues with Markov routing has the same invariant distribution as if all the flows in the network were independent Poisson processes, and that a customer who jumps finds the others with their invariant distribution. **Chapter Four** presents the results about *sojourn times along overtake-free paths* in networks of quasi-reversible queues and the *loop criterion* for Poisson flows in such networks, as well as a few related results. **Chapter Five** is a short discussion of *numerical* issues and methods, such as the *matrix-geometric solutions, asymptotic expansions,* and *fixed-population mean. In* **Chapter Six** we explain a few *approximation methods* for *non-product-form networks,* such as the *heavy and light traffic approximations, the renewal approximations,* and some *comparison results.* The main subject of **Chapter Seven** is the *stability* of queueing systems with *stationary and ergodic arrivals.* **Chapter Eight** explains some methods that have been used for solving *control problems* in simple queueing networks. *Multiarmed bandit problems* and the control of a *single server queueing network* are discussed in **Chapter Nine.** Finally, **Chapter Ten** deals with some *statistical issues,* such as *nonlinear filtering, hypothesis testing,* and *quick simulations* based on *large deviations.* **Appendix B** discusses a few basic queueing concepts.

The precedence relations among the chapters and appendices are as indicated below:

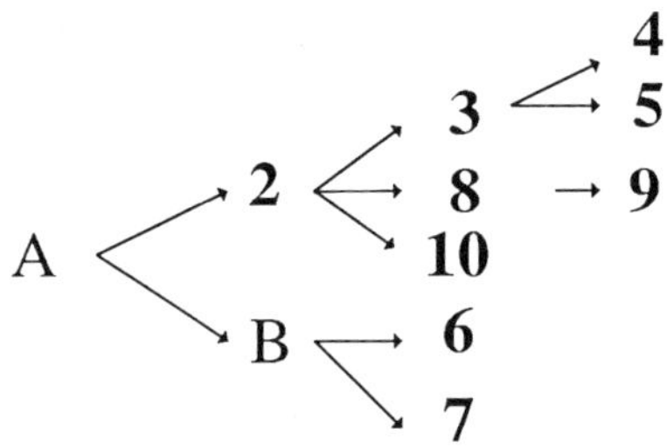

I have been fortunate to receive helpful suggestions from many colleagues and students. They have my deepest gratitude. Special thanks to Carol Block for her careful typesetting job.

J. Walrand
Berkeley

An Introduction to QUEUEING NETWORKS

CHAPTER ONE

AN OVERVIEW

1.1. INTRODUCTION

This chapter discusses a very simple example of network. The objective is to give the readers an idea of the type of questions, methods, and results of the theory of queueing networks.

The main points are the following. It is possible to approximate the evolution of most queueing systems by a **Markov chain** $\mathbf{x} = \{x_t, t \geq 0\}$ (see section 1.3). The advantage that results from using a Markov chain model is that it is possible to calculate the **invariant distribution** of $\mathbf{x}$ by solving a linear system of equations (1.2) called the **balance equations.** Also, under conditions that are easily verified (**irreducibility**) the invariant distribution is also the limiting distribution of $\mathbf{x}$. That is, $\mathbf{x}$ is **asymptotically stationary.** A very useful and informative method for verifying the balance equations is to guess the evolution of $\mathbf{x}$ **reversed in time.** (See section 1.5.) The invariant distribution can be used to calculate the **average sojourn time** of a customer in a network or between two points of the network, as explained in section 1.6. In order to determine the **distribution of the sojourn times** in section 1.11, one has to calculate the **distribution** of the state of the network **at the arrival times** of customers in a given queue. This is done in section 1.7. This result is also useful in characterizing the **flows in the network.** It is explained in section 1.9 that these flows are generally **not Poisson.** The number of possible states in a closed network is often very large. This implies the need for **efficient numerical methods** for normalizing the invariant distributions of these networks. A few words about those issues are found in section 1.8. A **probabilistic** discussion of the role of **quasi-reversibility** can be found in section 1.10, where it is explained that the **product form** is a direct consequence of that property of the queues. The **parametric optimization** of a network is illustrated in sections 1.12 and 1.13. Some examples of **dynamic optimizations** are given in sections 1.14 and 1.15. The **nonlinear filtering problem** of section 1.16 is that of extracting useful information from observations. An application is given in section 1.17. Finally, section 1.18 introduces some simple **comparison** results.

A few topics discussed in the book are not covered by this chapter. That is the case for the results on **nonproduct-form** networks of Chapter Six, the **stationary models** of Chapter Seven, and some **hypothesis testing** and **large deviations** results of Chapter Ten.

A good strategy is to attempt to get the main points of this chapter on a first reading and to come back for the finer points as one progresses through the subsequent chapters.

1.2. THE MODEL

Consider the system illustrated in Figure 1.1.

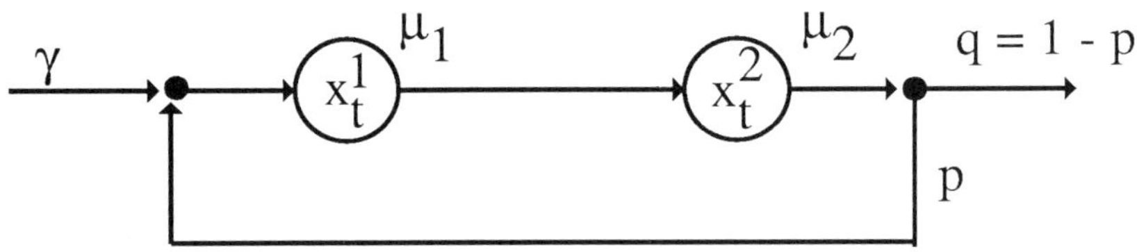

Figure 1.1. Two queues in tandem with feedback.

Customers arrive from outside into the first queue as a **Poisson process with rate** $\gamma \geq 0$. This means that the probability of an arrival in a short time interval $(t, t + \epsilon)$ is roughly equal to $\gamma \epsilon$ and that the occurrence of such an arrival is independent of what happened before time t. Thus, the Poisson process is **memoryless**: future arrival times are independent of past arrival times.

The customers are served on a first-come–first-served (FCFS) basis in the first queue and then immediately join queue 2, where they are also served on a FCFS basis. Upon each service completion in the second queue, a coin is flipped. With probability $p \in [0, 1]$ the outcome is "head," and the customer is immediately sent back to the first queue. With probability $q := 1 - p$, the outcome is "tail," and the customer leaves the system. It is assumed that the service times, the outcomes of the coin flips, the initial queue lengths (at time $t = 0$), and the arrival process are all independent. Moreover, the service times in queue i are **exponentially distributed** with rate μ_i for $i = 1, 2$. This can be shown to imply that if a customer is being served in queue i at time t, then the probability that the service will be completed during $(t, t + \epsilon)$ is approximately equal to $\mu_i \epsilon$, independently of what happened before. This is the **memoryless property** of the exponential distribution. (See section 2.2.)

Denote by x_t^i the queue length (that is the total number of customers, including the one being served if any) in queue i at time t ($t \geq 0$ and $i = 1, 2$).

Analysis issues for this system concern the **evolution** of the vector $x_t = (x_t^1, x_t^2)$ and the **time spent** by customers in the two queues. One is also interested in the **flows** of customers. The motivation is that the queues interact through customer flows, so that we may hope that knowledge about those flows could help us understand the behavior of the network.

Design issues range from the **optimization** of the **parameters** (μ_1, μ_2) to the **control** of the operations of the network. For instance, instead of assuming that the servers work at constant rates, one could let the rates depend on the queue lengths. A related situation is when there is a single server in the network who has to decide which queue to serve at any given time. It might also be possible and desirable to control the arrivals into the network.

The network is said to be **open** if all the customers can leave. This is the case if $p < 1$. If no customer can leave, as happens if $p = 1$, then the network is said to be **closed.** Of course, if $p = 1$, then it is necessary to assume that $\gamma = 0$ for the number of customers not to go to infinity.

1.3. MARKOV CHAIN

In order to analyze the evolution of $\mathbf{x} = \{x_t,\ t \in \mathbb{R}\}$, one first observes that this process is a **Markov chain** in $\mathbf{X} = \{0, 1, 2, \ldots\}^2$. This means that the *evolution of* x_t *starts afresh from* x_t *at time* t. In other words, x_t contains all the information about the past evolution that is relevant for predicting the future evolution of x_t. This follows from the memoryless property of the exponential service times and of the Poisson arrival process, and the independence of the coin flipping experiments that determine the routing of the customers. (See Chapter Two for more details.)

The **rate matrix** of $\mathbf{x}$ is the matrix $Q = [q(x, y),\ x, y \in \mathbf{X}]$ such that, for $0 < \epsilon \ll 1$,

$$q(x, y)\epsilon \approx P[x_{t+\epsilon} = y \mid x_t = x], \text{ for } x \neq y$$

$$q(x, x)\epsilon = -q(x)\epsilon \approx -P[x_{t+\epsilon} \neq x \mid x_t = x] = -\sum_{y \neq x} q(x, y)\epsilon\,.$$

Notice that this definition implies that $\sum_y q(x, y) = 0$ for all $x \in \mathbf{X}$;

i.e., the rows of a rate matrix sum to zero.

This rate matrix will play an important role in determining the *fraction of time* $\pi(x)$ *spent by* x_t *in a given state* $x \in \mathbf{X}$, as we will see below. The rate matrix can be written down by looking at Figure 1.1.

One finds that the rate $q(x, y)$ for $x, y \in \mathbf{X}$ is given by the equations (1.1).

$$q(x, y) = \begin{cases} \gamma, \text{ if } y = A_1 x := (x^1 + 1, x^2) \\ \mu_1 1\{x^1 > 0\}, \text{ if } y = T_{12}x := (x^1 - 1, x^2 + 1) \\ \mu_2 p 1\{x^2 > 0\}, \text{ if } y = T_{21}x := (x^1 + 1, x^2 - 1) \\ \mu_2 q 1\{x^2 > 0\}, \text{ if } y = D_2 x := (x^1, x^2 - 1) \\ -\gamma - \mu_1 1\{x^1 > 0\} - \mu_2 1\{x^2 > 0\}, \text{ if } y = x \\ 0, \text{ otherwise}. \end{cases} \tag{1.1}$$

In these expressions, for any condition C, the term $1\{C\}$ takes the value 1 if the condition is satisfied and takes the value 0 otherwise.

For instance, a transition from x to $y = T_{21}x$ corresponds to a customer being served in queue 2 (at rate μ_2) and then being sent to queue 1 (with probability p). The probability that this occurs in $0 < \epsilon \ll 1$ time unit is then approximated by $\mu_2 1\{x^2 > 0\} p\, \epsilon$. Hence $q(x, T_{21}x)$. The other values can be derived similarly.

A **probability distribution** $\pi = \{\pi(x), x \in \mathbf{X}\}$, i.e., a set of non-negative numbers that sum to one, is said to be **invariant** for **x** if it is such that the probability that $\{x_t = x\}$ is always given by $\pi(x)$ for all $x \in \mathbf{X}$ and all $t > 0$ provided that that is the case for $t = 0$. Thus, π is invariant for **x** if the distribution of x_t does not depend on $t > 0$ when the distribution of x_0 is π.

It turns out that finding an invariant probability distribution $\pi = \{\pi(x), x \in \mathbf{X}\}$ amounts to solving the **balance equations**

$$\pi(x)\, q(x) = \sum_{y \neq x} \pi(y)\, q(y, x), \; x \in \mathbf{X} \tag{1.2}$$

together with the constraint that π be a probability distribution, i.e., that $\pi(x) \geq 0$ for all x and that $\sum_{x \in \mathbf{X}} \pi(x) = 1$.

The intuitive justification of this fact (see Theorem 2.5.2) is that those equations express that the Markov chain x_t observed over a small interval $(t, t + \epsilon)$ is equally likely to make a transition out of state x (this has probability $\epsilon\, \pi(x)\, q(x)$ if the distribution of

x at time t is π) as to make a transition into state x (the probability of such a transition is the right-hand side of (1.2) times ϵ). This condition is an equilibrium condition. It is *necessary and sufficient* for π to be invariant.

1.4. PRODUCT-FORM INVARIANT DISTRIBUTION

Tedious but straightforward algebraic manipulations allow us to verify the following result.

Theorem 1.4.1 [Product-form]

(a) If $p < 1$, then **x** admits an invariant distribution if and only if $\rho_i := \dfrac{\gamma}{q\,\mu_i} < 1$ for $i = 1, 2$. The invariant distribution is then **unique** and is given by

$$\pi(x) = \pi_1(x^1)\,\pi_2(x^2), \quad x = (x^1, x^2) \in \mathbf{X} \tag{1.3}$$

where

$$\pi_i(n) = (1 - \rho_i)\rho_i^n, \qquad n \geq 0, \quad i = 1, 2\,. \tag{1.4}$$

Thus the network admits a **product-form** invariant distribution.

(b) If $p = 1$, then $\mathbf{X}_N := \{x \in \mathbf{X} \mid x^1 + x^2 = N\}$ is a **closed irreducible set** for **x**, for each $N \geq 0$. That is, **x** cannot enter or leave that set, and it can go from any $x \in \mathbf{X}_N$ to any other $y \in \mathbf{X}_N$. As a consequence, **x** admits a **unique** invariant distribution **on each set** $\mathbf{X}_N$. It is given by $\pi_{\mathbf{X}_N}(x)$, where

$$\pi_{\mathbf{X}_N}(n, N-n) = a^n[\sum_{m=0}^{N} a^m]^{-1}, \; 0 \leq n \leq N, \tag{1.5}$$

$$\textit{with } a := \frac{\mu_2}{\mu_1}\,.$$

Remarks 1.4.2

(a) The above result will follow from arguments given in sections 1.5 and 1.10 below.

(b) In the case $p = 1$, let π be given by (1.3)–(1.4) with $\rho_i = \dfrac{\lambda}{\mu_i}$ for $i = 1, 2$ where λ is chosen so that $\rho_i < 1$ for $i = 1, 2$. One can then verify that π is invariant for **x**. In fact, if one extends

$\pi_{\mathbf{X}_N}$ to $\mathbf{X}$ by defining $\pi_{\mathbf{X}_N}(x) = 0$ if $x \in \mathbf{X} - \mathbf{X}_N$, and if $\pi(\mathbf{X}_N) = \sum_{x \in \mathbf{X}_N} \pi(x)$, then one finds that

$$\pi(x) = \sum_{N=0}^{\infty} \pi(\mathbf{X}_N)\, \pi_{\mathbf{X}_N}(x), \quad x \in \mathbf{X}.$$

Thus, π is a convex combination of the invariant distributions $\pi_{\mathbf{X}_N}$ and is therefore an invariant distribution. (For the balance equations are linear.)

(c) Conversely, if one knows that π is an invariant distribution for a Markov chain $\mathbf{x}$ on $\mathbf{X}$ and if $\mathbf{Y} \subset \mathbf{X}$ is a closed set for $\mathbf{x}$ (i.e., if $\mathbf{x}$ cannot enter or leave $\mathbf{Y}$), then

$$\pi_{\mathbf{Y}}(x) := \frac{\pi(x)\, 1\{x \in \mathbf{Y}\}}{\pi(\mathbf{Y})} \tag{1.6}$$

is also invariant for $\mathbf{x}$. See Fact 2.5.7.

(d) The remarks (b) and (c) show that the central result is that (1.3)–(1.4) is invariant for the network of Figure 1.1, *in all cases* ($0 \le p \le 1$). In the closed case $p = 1$, the process $\mathbf{x}$ admits closed irreducible sets $\mathbf{X}_N$. The unique invariant distribution on each of these sets can be found by *normalizing the restriction* of π on $\mathbf{X}_N$, as explained in remark (c).

1.5. TIME REVERSAL

The technique of **time reversal** has proved to be a valuable tool for gaining insight into the behavior of networks. Among the successful applications of that method are simpler proofs and new examples of product-form results, as well as new results on sojourn times in networks. The basic ideas will be explained on the network of Figure 1.1. The key result is the following fact.

Theorem 1.5.1 If $\mathbf{x} = \{x_t, t \in \mathbb{R}\}$ is a stationary (i.e., with an initial distribution that is invariant) Markov chain on $\mathbf{X}$ with rate matrix Q and invariant distribution π, then, for any $T \in \mathbb{R}$, the process $\mathbf{x}$ observed in reversed time, i.e., the process $\tilde{\mathbf{x}} := \{\tilde{x}_t := x_{T-t}, t \in \mathbb{R}\}$ is a stationary Markov chain with invariant distribution π and rate matrix $\tilde{Q}$ given by

$$\pi(x)\, q(x, y) = \pi(y)\, \tilde{q}(y, x), \quad x, y \in \mathbf{X}. \tag{1.7}$$

In particular, if (1.7) is such that $\tilde{Q} = Q$, then $\mathbf{x}$ and $\tilde{\mathbf{x}}$ have the same law. One then says that **x** is **time-reversible.** (See Theorem 2.8.1.)

Corollary 1.5.2 Let Q be the rate matrix of a Markov chain **x**, $\tilde{Q}$ another rate matrix, and π a distribution on **X**. Assume that (1.7) holds. Then π must be invariant for Q and $\tilde{Q}$ must be the rate matrix of $\tilde{\mathbf{x}}$.

Proof: Summing (1.7) over $x \in \mathbf{X}$ shows that $\sum_x \pi(x)\, q(x, y) = 0$ for all y. (For any rate matrix $\tilde{Q}$ must be such that $\sum_x \tilde{q}(y, x) \equiv 0$.)
Hence, π is invariant, and Theorem 1.5.1 implies that $\tilde{Q}$ is the rate matrix of $\tilde{\mathbf{x}}$.

Example 1.5.3 As an application of this idea, let us prove Theorem 1.4.1. To do this, let $\tilde{Q}$ be the rate matrix of the network of Figure 1.2 which is guessed to correspond to the network of Figure 1.1 *reversed in time.*

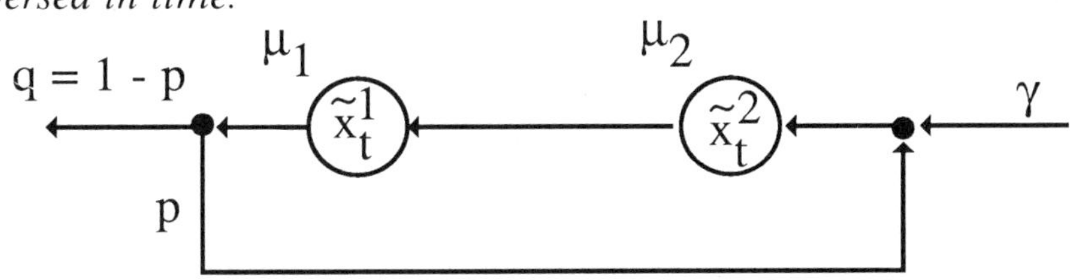

Figure 1.2. A guess for the network reversed in time.

One then checks that this rate matrix and π given in (1.3)–(1.4) satisfy (1.7), where Q is given by (1.1). For instance, let $x \in \mathbf{X}$ with $x^2 > 0$ and let $y = T_{21}x$. One then finds that $q(x, y) = \mu_2 p$ and that $\tilde{q}(y, x) = \mu_1 p$ (see Figure 1.2). It remains to verify that

$$\pi(x)\mu_2 p = \pi(y)\mu_1 p ,$$

which is seen to hold by (1.3)–(1.4). The other possible pairs (x, y) are treated similarly. This proves Theorem 1.4.1.

The essential advantage of the approach is that the verification of (1.7) does not necessitate any summation, contrary to that of the balance equations (1.2). Another important by-product of this method is that it tells us how the network behaves in reversed time. This will be used in section 1.11 to analyze the distribution of sojourn times.

Note that unless $p = 1$ the rate matrices Q and $\tilde{Q}$ do not correspond to the same network. Therefore, the network of Figure 1.1 is *not* time-reversible, unless $p = 1$.

1.6. AVERAGE SOJOURN TIMES

How long does a customer *stay* in a network, *on the average?* What is the *average cycle time* in the case of the closed network? This section addresses these issues for the network of Figure 1.1.

The basic result required to answer these questions is **Little's result** (see section 3.4). That result states that

$$L = \lambda W \tag{1.8}$$

where L is the *average number* of customers in the network, λ is the *average arrival rate* of customers into the network, and W is the *average sojourn time* of a customer in the network. The intuition is that if each customer pays one unit of cost per unit of time while in the system, then (1.8) corresponds to two different evaluations of the average cost paid to the system per unit of time.

Consider first the case $p < 1$, i.e., the open network. Assume that $\rho_i < 1$ for $i = 1, 2$. Then, according to Theorem 1.4.1, the average number of customers in the network will be given by $L = L_1 + L_2$, where L_i, the average number of customers in queue i, is calculated as

$$L_i = \sum_{n=0}^{\infty} n\pi_i(n) = \sum_{n=0}^{\infty} n(1 - \rho_i)\rho_i^n = \frac{\gamma q^{-1}}{\mu_i - \gamma q^{-1}} .$$

Also, the average rate of arrivals into the network is equal to γ. Therefore, (1.8) gives the average time spent by a customer in the network as

$$W = \gamma^{-1} \sum_{i=1,2} \frac{\gamma q^{-1}}{\mu_i - \gamma q^{-1}} = \sum_{i=1,2} \frac{1}{\mu_i q - \gamma} . \tag{1.9}$$

Consider now the case $p = 1$. Assume that there are N customers in the network. The invariant distribution is given by $\pi_{\mathbf{X}_N}$ in (1.5). One would like to calculate the average time spent by a given customer between two successive arrivals into queue 1. That time is the average cycle time of a customer. Let us denote that average time by $C(N)$. One can argue that $N[C(N)]^{-1}$ is the average rate of flow into queue 1, i.e., out of queue 2. Indeed, N customers enter queue 1 every $C(N)$ seconds, on the average.

(Alternatively, one can invoke Little's result applied to the system consisting of the two queues and write that the average number N of customers in the two queues is equal to the average rate into queue 1 times the average time $C(N)$ in the system.) Now, the average rate of flow out of queue 2 is equal to $\mu_2 \pi_{\mathbf{X}_N}(x^2 > 0)$, since the server in queue 2 serves at rate μ_2 whenever the queue is nonempty. Also the probability that the second queue is nonempty when there are N customers in the network is given by $1 - \pi_{\mathbf{X}_N}(N, 0)$. Combining these ideas gives

$$C(N) = \frac{N}{\mu_2 [1 - \pi_{\mathbf{X}_N}(N, 0)]} . \tag{1.10}$$

Algebraic manipulations allow us to rewrite the denominator as

$$\mu_2[1 - \frac{\pi(N, 0)}{\pi(\mathbf{X}_N)}] = \mu_2 \frac{\pi(\mathbf{X}_N) - \pi(N, 0)}{\pi(\mathbf{X}_N)} = \lambda \frac{\pi(\mathbf{X}_{N-1})}{\pi(\mathbf{X}_N)} ,$$

where the last step is obtained by referring to (1.3)–(1.4) where $\rho_i = \lambda / \mu_i$ for $i = 1, 2$. Hence, (1.10) can be written as

$$C(N) = \frac{N}{\lambda} \frac{\pi(\mathbf{X}_N)}{\pi(\mathbf{X}_{N-1})} . \tag{1.11}$$

1.7. THE ARRIVAL THEOREM

In order to determine the *distribution* of the time spent by a customer in given parts of the network, it is necessary to know the distribution of the state of the network when the customer arrives. For instance, the time spent by a customer C in a FCFS queue in the network is the time required to serve the customers who were in the queue when C arrived, plus the time to serve C.

This section will show, on the example of Figure 1.1, that in certain networks, when jumping, a typical customer sees the others with their invariant distribution.

The first step is to define what is meant by a **typical customer.** Consider a given link of the network of Figure 1.1, say the link from queue 2 to queue 1. Denote by $\{T_n, n \in \mathbf{Z}\}$ the successive times when a customer jumps on that link. (By definition, $\mathbf{Z} := \{\ldots, -2, -2, 0, 1, 2, \ldots\}$ is the set of integers.)

Let $z_n := (x^1_{T_n-}, x^2_{T_n})$. That is, z_n is the state of the *customers who do not jump* at time T_n. (The customer who jumped is no longer in queue 2 after the jump, i.e., is no longer in $x^2_{T_n}$, and is not yet in queue 1 just before the jump, i.e., is not in $x^1_{T_n-}$.)

Theorem 1.7.1 [Arrival Theorem] The random sequence $\mathbf{z} = \{z_n, n \in \mathbf{Z}\}$ is a Markov chain with invariant distribution π.

If $p < 1$, then π is the unique invariant distribution of $\mathbf{z}$ on $\mathbf{X}$. Moreover, the distribution of z_n converges to π as $n \to \infty$, for any value of z_0.

If $p = 1$ and $x_0 \in \mathbf{X}_N$, then $\pi_{\mathbf{X}_{N-1}}$ is the unique invariant distribution of $\mathbf{z}$ on $\mathbf{X}_{N-1}$. Moreover, the distribution of z_n converges to $\pi_{\mathbf{X}_{N-1}}$ as $n \to \infty$.

Both results can be stated by saying that **the customer who jumps sees the others with their invariant distribution when that customer is not in the network.**

Proof: [Outline] The fact that $\mathbf{z}$ is a Markov chain means, as in continuous time, that the evolution of $\mathbf{z}$ starts afresh from z_n at time n, for each time n. (See Chapter Two.)

The invariant probability that $\{z_n = x\}$ turns out to be equal to the ratio of the rate of jumps T_n such that $z_n = x$ over the total rate of jumps T_n. This is the *ratio of rates* formula. In other words, the fraction of jumps T_n such that $z_n = x$ is the invariant probability of $\{z_n = x\}$. This intuitive fact is proved in Theorem 2.10.6.

Assume that $\mathbf{x}$ has its invariant distribution π. The ratio of rates can be written as

$$\frac{\pi(x^1, x^2 + 1)\mu_2 p}{\lambda p}, \tag{1.12}$$

where $\lambda = \mu_i \rho_i$ is the average rate of flow through the queues. Indeed, the numerator of (1.12) is the rate of a jump from queue 2 to queue 1 such that $x_{T_n-} = (x^1, x^2 + 1)$, i.e., such that $z_n = (x^1_{T_n-}, x^2_{T_n}) = x$. Also, the denominator of (1.12) is the average rate of flow from queue 2 to queue 1.

It follows from (1.3)–(1.4) that the value of (1.12) is equal to $\pi(x)$, as claimed.

1.8. NUMERICAL CONSIDERATIONS

Formula (1.11) shows that it may be necessary to calculate normalizing constants when the network is not open. These constants are necessary to obtain the normalized invariant distribution, and therefore are needed to calculate average values of functions of the state of the network.

In this simple example, it is not difficult to obtain $\pi(\mathbf{X}_N)$. Indeed, (1.3)–(1.4) show that

$$\pi(\mathbf{X}_N) = \sum_{n=0}^{N} \pi(n, N-n) = (1-\rho_1)(1-\rho_2)\rho_2^N \frac{1-a^{N+1}}{1-a},$$

$$\text{with } a = \frac{\mu_2}{\mu_1}.$$

In more complicated systems, a closed-form expression is not available and efficient algorithms are required in view of the large number of terms involved in the summation.

For instance, in a network with J M/M/1 queues and N customers, the set of possible vectors of queue lengths

$$\mathbf{X}_N = \{x \in \{0, 1, 2, \ldots\}^J \mid x^1 + \cdots + x^J = N\}$$

can be seen to have $\binom{N+J-1}{J-1}$ elements.

A first algorithm is based on the observation that $\pi(\mathbf{X}_N) = P\{x_t^1 + x_t^2 = N\}$, where x_t^1 and x_t^2 are independent and have the geometric distributions given in (1.4). Thus,

$$\pi(\mathbf{X}_N) = \sum_{n=0}^{N} \pi_1(n)\,\pi_2(N-n) = \sum_{n=0}^{N} (1-\rho_1)\rho_1^n \pi_2(N-n)$$

$$= (1-\rho_1)\pi_2(N) + \rho_1 \sum_{n=0}^{N-1} (1-\rho_1)\rho_1^n \pi_2(N-1-n),$$

so that

$$\pi(\mathbf{X}_N) = (1-\rho_1)\pi_2(N) + \rho_1\, \pi(\mathbf{X}_{N-1}). \tag{1.13}$$

This formula allows us to calculate $\pi(\mathbf{X}_n)$ successively for $n = 1, 2, \ldots, N$ in $O(N)$ steps. (Throughout, $O(N)$ denotes a function of N such that $O(N)/N$ converges to a positive number as $N \to \infty$.) This idea is called the **convolution algorithm.** A straightforward calculation of each $\pi(\mathbf{X}_n)$ would require $O(n)$ steps for each n, so that the calculation of $\pi(\mathbf{X}_n)$ for $n = 1, 2, \ldots, N$

would involve $O(N^2)$ steps instead of $O(N)$ by the above approach.

Another algorithm can be derived from Little's result. Denote by $W_i(N)$, $L_i(N)$, and $\lambda_i(N)$ the average time spent by a customer, the average number of customers, and the average rate of flow, at queue i, respectively, when there are N customers in the network. Then Little's result implies that

$$\begin{cases} W_i(N) = \mu_i^{-1}[1 + L_i(N-1)] \\ L_i(N-1) = \lambda_i(N-1)W_i(N-1) \\ \lambda_i(N-1)[W_1(N-1) + W_2(N-1)] = N-1 . \end{cases} \tag{1.14}$$

The first formula is based on the Arrival Theorem 1.7.1 that implies that the average number of customers found in queue i by an arriving customer is $L_i(N-1)$, so that the average time spent in the queue by that customer is $L_i(N-1) + 1$ times the average service time μ_i^{-1} per customer.

These formulas allow to calculate recursively $(W_1(n), W_2(n))$ for $n = 1, 2, \ldots, N$ also in $O(N)$ steps. From these quantities one can then obtain the average cycle times and the average queue lengths. This procedure is known as the **mean value analysis.**

For complicated networks, with many queues and many different classes of customers (a concept to be discussed in section 3.2), the convolution and the mean value analysis algorithms may prove prohibitive in terms of time and memory requirements. A number of approximation methods have been proposed and validated by examples. Computable error bounds are not yet available for these methods. Let us illustrate one such idea on the example.

The basic objective is to avoid the recursion in (1.14). These equations can be used to calculate $L_i(N)$ as a function of $\{L_1(N-1), L_2(N-1)\}$. One finds that

$$L_i(N) = N\frac{\mu_i^{-1}[1 + L_i(N-1)]}{\mu_1^{-1}[1 + L_1(N-1)] + \mu_2^{-1}[1 + L_2(N-1)]} . \tag{1.15}$$

These formulas are then calculated recursively. To avoid the recursion, one can postulate a simple relationship between those quantities. For instance, one can try

$$\frac{L_i(N)}{N} \approx \frac{L_i(N-1)}{N-1}, \tag{1.16}$$

the motivation being that the fraction of the customers being in a given queue might not depend too strongly on the total number of customers in the network. Assuming that this relation is correct, one can replace the $L_i(N-1)$ in (1.15) by their values in terms of the $L_i(N)$. This allows us then to obtain $L_i(N)$ for $i = 1, 2$ by solving two coupled nonlinear equations instead of performing a recursion.

Refined but numerically more complex versions of this approximation have been proposed.

See Chapter Five for more details and for some other methods.

1.9. NON-POISSON FLOWS

The result of Theorem 1.4.1 shows that the invariant distribution of the network of Figure 1.1 is the same *as if* the arrival processes into the two queues were independent Poisson processes with rates given by a solution of the **flow conservation equations**

$$\lambda_1 = \gamma + \lambda_2 p, \quad \lambda_2 = \lambda_1 .$$

Indeed, each π_i is the invariant distribution of a queue with a Poisson arrival process of rate λ_i and with independent exponentially distributed service times μ_i, as is easily verified. (See Exercise 1.2.)

However, it will be shown below that the arrivals into the queues cannot have independent increments, and therefore cannot be Poisson. As may be expected, past arrivals into any of the queues will affect successive arrivals into that queue.

Consider queue 2 in Figure 1.1, where it is assumed that $p < 1$ and that the system has its invariant distribution π. Denote by D_t^1 the number of departures from queue 1 in $[0, t]$, for $t \geq 0$. Choose $0 < \epsilon \ll 1$ and assume that a jump of D_t^1 was observed in $[0, \epsilon]$. The conditional distribution of $(x_\epsilon^1, x_\epsilon^2 - 1)$ is then given by π, by the Arrival Theorem. This shows that given $\{D_\epsilon^1 = 1\}$ the state x_ϵ has the distribution π to which a customer is added in queue 2. It should be clear that adding a customer in queue 2 at time $t = 0$ can only increase D_t^1 for any given $t \geq 0$. (More precisely, one should say that it is possible to define $\mathbf{x}$ so that the last statement is valid.) Also, there is a positive probability that adding that customer will increase D_t^1 by one. This argument, which can be made rigorous, shows that the evolution of $\{D_t^1 - D_\epsilon^1, t \geq \epsilon\}$, i.e., of the jump times of D_t^1 after time ϵ, depends on $\{D_s^1, s \leq \epsilon\}$.

More general arguments can be given that extend to a large class of networks to show that a flow on a link that can be visited more than once by the same customer cannot be Poisson. (See Chapter Four.)

For simple networks, it is also possible to calculate the interarrival time distributions at the queue and show that they are not exponentially distributed when the same customer can visit the queue more than once.

1.10. THE ROLE OF QUASI-REVERSIBILITY

Let us pause a moment to try to identify the properties of the queues that not only made the analysis of the network of Figure 1.1 tractable but also led to surprisingly simple results such as the product form and the Arrival Theorem.

As discussed in section 1.9, the invariant distribution of the network of Figure 1.1 is the same as if the arrival processes into the queues were independent Poisson processes, even though they are not. Also, the two queue length processes $\{x_t^1, t \in \mathbb{R}\}$ and $\{x_t^2, t \in \mathbb{R}\}$ interact when customers jump from one queue to the other. This interaction is such that at any time $t \in \mathbb{R}$ the random variables x_t^1 and x_t^2 are independent.

The key property at work here is the **quasi-reversibility** of the two queues. This result states that the departures up to any time t of a stationary FCFS queue with Poisson arrivals and i.i.d exponentially distributed service times (called an M/M/1 queue) form a Poisson process that is independent of the queue length at time t. (This property is an immediate consequence of the time-reversibility of the stationary queue length process. See Chapter Two.)

To see why the quasi-reversibility implies the product-form of the network of Figure 1.1, let us modify slightly the network by inserting delays of an arbitrarily chosen duration d just before the queues, as indicated on Figure 1.3. The arguments given below apply both to the open and closed networks.

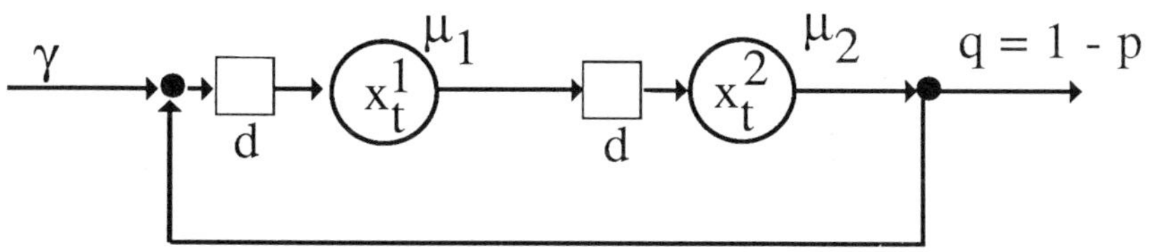

Figure 1.3. The modified network

Assume that at time $t = 0$, x_0 has the distribution π given by (1.3)–(1.4). That is, the two queue lengths are independent and have the invariant distributions for the queues when their arrival processes are Poisson. Assume also that, at $t = 0$, there are two independent Poisson processes in transit in the delays, each with rate $\lambda_1 = \lambda_2 = \lambda$.

It follows that for $t \in [0, d]$ the arrivals in the two queues are independent Poisson processes with rate λ. Thus, for $t \in [0, d]$, the queues behave independently and are stationary. In particular, during $[0, d]$, x_t has distribution π and the two departure processes from the queues are independent Poisson processes with rate λ that are independent of x_d, by the quasi-reversibility of the queues.

The feedback process during $[0, d]$ is obtained by sampling the departures from the second queue with probability p. It will be shown in section 2.11, that this sampling procedure produces two independent Poisson processes: the departures and the process that enters the delay line before queue 1 in $[0, d]$. Moreover, by quasi-reversibility again, these processes are independent of x_d.

Thus, the situation at time d is identical to that at time $t = 0$. Repeating the argument shows that x_t has distribution π for all $t \geq 0$. Hence, the product-form result is an immediate consequence of the quasi-reversibility if there is an arbitrarily small transit time on the links of the network (this is the case for most real systems).

One can show formally that the result is continuous at $d = 0$. This is done in 3.5.6 by constructing the queue length process so that x_t converges to its value for $d = 0$ as $d \downarrow 0$. Since x_t has distribution π for all $t \geq 0$ and all $d > 0$, this implies that x_t has distribution π for all t when $d = 0$.

The Arrival Theorem can be justified in a similar way. Consider once again the modified network of Figure 1.3 with the invariant initial distribution used in the preceding argument.

For $0 < \epsilon < d$, the fact that a customer jumped from queue 2 to queue 1 during $[t - \epsilon, t]$ is independent of x_t, since the flows in transit on the links at time t are independent of x_t. Letting $\epsilon \downarrow 0$ then shows that the distribution of x_τ just after a jump time τ from queue 2 to queue 1 is the invariant distribution π of x_t.

As the delay $d \downarrow 0$, one sees that only the customer who jumped at time τ will be in transit on the links, so that x_τ becomes the state of the customers who did not jump at time τ. Hence, the distribution of the customers who do not jump at the jump time τ is equal to π.

The arguments of this section can be used to show that *a network of quasi-reversible queues with independent routing admits a product-form invariant distribution, and the Arrival Theorem holds.* These results will be the explored further in Chapters Three and Four.

1.11. DISTRIBUTION OF PASSAGE TIMES

What is the *distribution* of the time taken by a typical customer to go once through the two queues of Figure 1.1? The average time was calculated in section 1.6.

The concept of typical customer is as before. A **typical** customer is one who *finds the other customers with their invariant distribution in the network* when entering queue 2, say. This definition is justified by the Arrival Theorem, which states that this is the invariant distribution at the arrival times into queue 2 and that it is also the asymptotic distribution of that random sequence. (The latter statement is shown in section 2.10.)

First consider the case $p < 1$, i.e., the open network. Say that at time 0 a customer C enters queue 2 and that

$$P\{x_0^1 = x^1, x_0^2 - 1 = x^2\}$$
$$= P\{(x_0^1,\ x_{0-}^2) = x\} = \pi(x),\ x \in \mathbf{X}, \tag{1.17}$$

where π is given by (1.3)–(1.4). Thus, C is typical.

Let T_1 be the duration of the last visit (before time 0) of customer C in queue 1. Let T_2 be the duration of the next visit of C in queue 2.

The *key observation* is the following. Since the customer C **cannot be overtaken,** the time T_2 is a function of x_{0-}^2 and of future service times in queue 2. Similarly, by reversing the time, one can show that T_1 can be viewed as a function of x_0^1 and of independent service times in queue 1. Since x_0^1 and x_{0-}^2 are independent by (1.17), this implies that T_1 and T_2 are independent. The details of this argument will be provided in section 4.2.

The marginal distributions of T_1 and T_2 are the same as those of typical sojourn times in M/M/1 queues, since the distributions at the arrival times are the same. That is, T_i is exponentially distributed with mean $[\mu_i - \lambda_i]^{-1}$ (see section 4.2.1).

The case $p = 1$ is quite similar. The distribution of (x_0^1, x_{0-}^2) seen by a typical customer C who enters queue 2 is given by the Arrival Theorem, as discussed in section 1.7. That is,

$$P\{x_0^1 = x^1,\ x_{0-}^2 = x^2\} = \pi_{\mathbf{X}_{N-1}}(x^1,\ x^2), \quad x \in \mathbf{X}_{N-1}\,.$$

One then argues, as in the open case, that given $\{x_0^1 = m, x_{0-}^2 = n\}$, T_1 is distributed as the sum of $m + 1$ exponentially distributed random variables with rate μ_1 and T_2 as the sum of $n + 1$ exponentially distributed random variables with rate μ_2 and that all these random variables are independent. This gives, for complex numbers s_1 and s_2 with real parts larger than $-\mu_1$ and $-\mu_2$, respectively,

$$E\{e^{-s_1T_1 - s_2T_2}\} = \sum_{x \in \mathbf{X}_{N-1}} \pi_{X_{N-1}}(x) \prod_{i=1,2} \left[\frac{\mu_i}{s_i + \mu_i}\right]^{x_i+1} \tag{1.18}$$

where one used the fact that if σ is exponentially distributed with rate μ, then

$$E\{e^{-s\sigma}\} = \frac{\mu}{s + \mu}$$

if the complex number s has a real part larger than $-\mu$. The formula (1.18) is the **Laplace transform** of the joint distribution of (T_1, T_2). It can be used to calculate joint moments. For instance, choosing $s_1 = s_2 = s$ gives the Laplace transform of the distribution of $T_1 + T_2$.

It is important to realize that the above arguments, and the corresponding results, cease to be valid if a customer *can be overtaken* while in a given path. To see this, consider the network of three queues in Figure 1.4.

The queues are M/M/1. The claim is that the successive sojourn times (T_1, T_2, T_3) in the three queues of a customer who happens to go through those three queues *are not independent.*

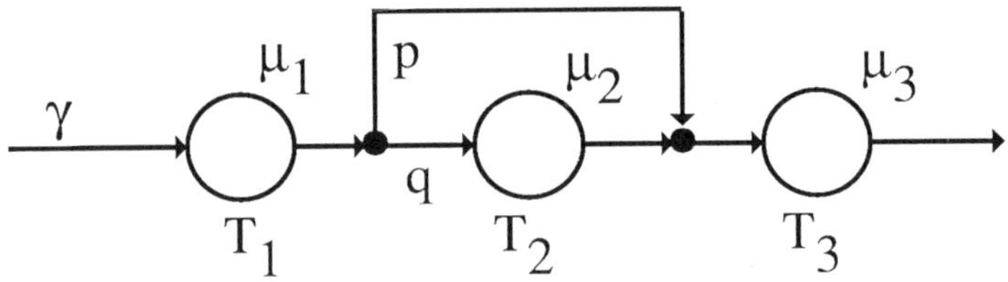

Figure 1.4. A network with overtaking.

The intuition is that if T_1 is very large, then the customer is likely to leave behind a large number of customers in queue 1 when jumping to queue 2. It is therefore likely that he will find a large number of those customers when entering queue 3.

More precisely, by the Arrival Theorem, the queue lengths when the typical customer enters queue 2 are independent. Conditioning on the lengths of the queues at that time, a pathwise coupling argument shows that the number of customers found by the typical customer when entering queue 3 is increasing in the number of customers left behind by that customer in queue 1. This shows that T_1 and T_3 are positively correlated.

Simulation experiments suggest that the dependence between T_1 and T_3 is very mild, for a large range of network parameters.

These issues will be explored further in Chapter Four.

1.12. OPTIMIZATION OF THE SERVICE RATES

Consider once again the network of Figure 1.1. The physical system modeled by this network may be such that the service rates in the queues can be chosen. The objective may be to minimize the average sojourn time or cycle time.

For simplicity, let us start by discussing the case $p < 1$ illustrated on Figure 1.5. As seen in (1.9), the average sojourn time in the network is given by

$$W = \sum_{i=1,2} \frac{1}{\mu_i q - \gamma} .$$

Say that the service requirements in the queues are independent and exponentially distributed with rates r_1 and r_2 in the two queues, respectively. One can choose servers that work at rates α_1 and α_2 in the two queues, respectively. The interpretation is that in one unit of time the server in queue i will accomplish α_i units of work. With that choice, the actual service rates become

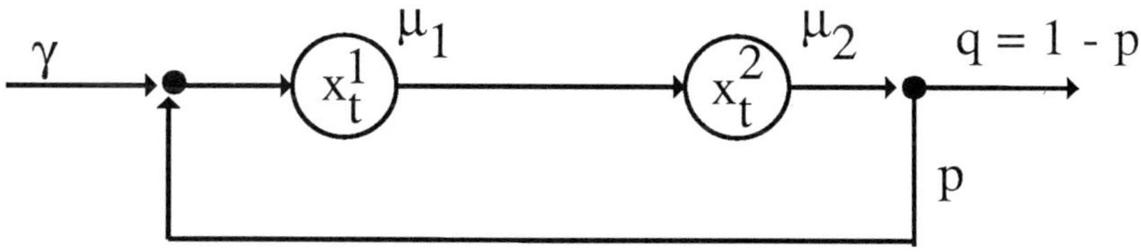

Figure 1.5. Optimization of service rates.

$\mu_i = r_i \alpha_i$. Say that the choice $\alpha = (\alpha_1, \alpha_2)$ costs

$$K(\alpha) = c_1 \alpha_1 + c_2 \alpha_2 .$$

The problem is then

$$\text{minimize } W(\alpha) = \sum_{i=1,2} \frac{1}{q\,\alpha_i\, r_i - \gamma}$$

$$\text{over } \alpha = (\alpha_1, \alpha_2)$$

$$\text{subject to } K(\alpha) = c_1 \alpha_1 + c_2 \alpha_2 \leq K_{\max}$$

$$\text{and } \alpha_i r_i > q^{-1} \gamma, \quad i = 1, 2 .$$

By considering the minimum admissible values of α_1 and α_2, one finds that for the problem to have a solution one must have

$$K_{\max} > \frac{\gamma}{q}\left(\frac{c_1}{r_1} + \frac{c_2}{r_2}\right) =: K_{\min} .$$

A solution is then obtained by observing that if $\alpha_1 > q^{-1} \gamma r_1^{-1}$ and $\alpha_2 > q^{-1} \gamma r_2^{-1}$ solve the problem, then they must be a stationary point of

$$W(\alpha) + \beta K(\alpha) \tag{1.19}$$

for some β that admits the interpretation of the optimal marginal reduction of waiting time per unit of extra cost invested in α_1 or in α_2. This is the standard *Lagrange multiplier* method.

Writing that the derivatives with respect to α_1 and with respect to α_2 of (1.19) are zero, one then finds the α_i in terms of β, and, substituting these expressions in $K(\alpha) = K_{\max}$, then determines β, and therefore the optimal value of α. The algebra gives

$$\alpha_i = \frac{\gamma}{q r_i} + (K_{\max} - K_{\min})\,(c_i r_i)^{-\frac{1}{2}} \left[\left(\frac{c_1}{r_1}\right)^{\frac{1}{2}} + \left(\frac{c_2}{r_2}\right)^{\frac{1}{2}}\right]^{-1} . \tag{1.20}$$

The case $p = 1$ is similar. Say that the problem is to minimize the average cycle time $C(N)$ given by (1.11) as a function of α where $\mu_i = r_i \alpha_i$, and subject to the constraint $K(\alpha) \le K_{max}$. Proceeding as in the preceding example would give the optimal choice.

1.13. FLOW CONTROL IN A CLOSED NETWORK

A more interesting problem than that discussed in the last section is when the service rates are allowed to depend on the queue lengths.

We will consider the case $p = 1$. Queue 1 will be viewed as an inexpensive "storage room" that is used to limit the congestion in queue 2. (See Figure 1.6.)

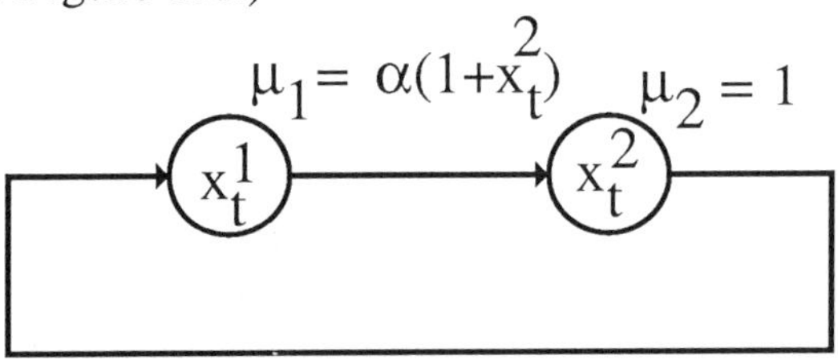

Figure 1.6. The controlled system.

The objective will be to maximize the average rate of flow through queue 2 while keeping the average number in that queue below a given value. This will be done by controlling the rate μ_1 as a function of the number of customers in that queue. Thus, queue 1 controls the flow into queue 2.

Formally, let $N = x_t^1 + x_t^2$ be the number of customers in the network. Assume that $\mu_1 = \alpha(1 + x_t^2)$, for some function $\alpha(.)$, and that $\mu_2 \equiv 1$. This latter assumption amounts to assuming that μ_2 is constant, since by scaling the time one can then assume that $\mu_2 \equiv 1$. It will be required that $\alpha(n) \in [0, a]$ for $n \ge 1$.

The problem is then

$$\begin{cases} \text{minimize } P\{x_t^2 = 0\} \\ \text{over } \alpha = (\alpha(1), \ldots, \alpha(N)) \in [0, a]^N , \\ \text{subject to } E\{x_t^2\} \le A \end{cases}$$

where the probability and the expectation correspond to the stationary distribution of the system for the specific choice of α. Indeed, the average rate of flow through the queues is equal to $\mu_2 = 1$ times the probability that the second queue is not empty.

Now, given α, the process x_t^2 is a Markov chain with the transition diagram in Figure 1.7.

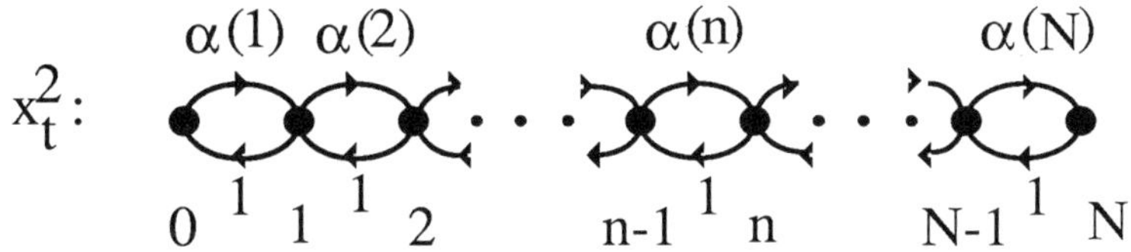

Figure 1.7. Transition diagram of controlled network.

Thus, the invariant distribution of x_t^1 is such that

$$P\{x_t^2 = n\} = \pi(n) = \pi(0)\,\alpha(1) \ldots \alpha(n) =: \pi(0)\,\beta(n), \quad n \ge 1 .$$

Some algebra then shows that the problem becomes

$$\begin{cases} \text{maximize } \dfrac{1}{\pi(0)} - 1 = \sum_{m=1}^{N} \beta(m) \\ \text{subject to } \sum_{m=1}^{N} m\beta(m) \le A[1 + \sum_{n=1}^{N} \beta(n)] . \end{cases}$$

Say that the solution of the problem is such that $\sum_{m=1}^{N} \beta(m) = B$. One then has

$$B = \max \{ \sum_{m=1}^{N} \beta(m) \mid \sum_{n=1}^{N} n\beta(n) \le A[1+B] \text{ and } \beta \text{ is possible} \} . \tag{1.21}$$

By "β is possible," we mean that one can have $\beta(m) = \alpha(1) \ldots \alpha(m)$ for $m \ge 1$ where $\alpha(m) \in [0, a]$ for all $m \ge 1$.

Assume that there is some $n \in \{1, \ldots, N-1\}$ such that $\beta(n-1) > 0$ and $\alpha(n) < a$. (The convention is $\beta(0) = 1$.) The claim is then that $\alpha(n+1) = 0$, so that $\pi(m) = 0$ for $m \ge n+1$ and one may then assume that $\alpha(m) = 0$ for $m \ge n+1$. Indeed, if $\alpha(n+1) > 0$, then replacing $\alpha(n+1)$ by $\alpha(n+1)y < \alpha(n+1)$ and $\alpha(n)$ by $\alpha(n)x > \alpha(n)$ will replace $S := \sum_{m=n}^{N} \beta(m)$ by

$$S' = xyS + x(1-y)\beta(n)$$

and

$$W := \sum_{m=n}^{N} m\beta(m)$$

by $W' = xyW + x(1-y)n\beta(n)$. The last relation shows that there is $0 < y < 1 < x \le a[\alpha(n)]^{-1}$ such that $W' = W$. Since $W > nS$ and $xy < 1$, this is seen to imply that $S' > S$, which contradicts (1.21).

It follows that the optimal solution must be

$$\begin{aligned} a &= \alpha(1) = \ldots = \alpha(n) \\ &> \alpha(n+1) \ge \alpha(n+2) = \ldots = \alpha(N) = 0 \end{aligned} \tag{1.22}$$

for some $n \le N$. It is now easy to determine the values of n and $\alpha(n+1)$ in (1.22). Indeed, these values are those that achieve $E\{x_t^2\} = A$. The details are left to the reader.

In conclusion, the control of the rate of queue 1 that maximizes the rate of flow through the queues while maintaining the average length of the second queue less than A corresponds to

$$\mu_1 = \begin{cases} a, & \text{if } x_t^2 < m \\ \gamma \in [0, a], & \text{if } x_t^2 = m \\ 0, & \text{if } x_t^2 > m, \end{cases}$$

where m and γ are chosen to achieve the largest possible value less than or equal to A of $E\{x_t^2\}$. Such a policy is called a **window flow control.**

1.14. FLOW CONTROL IN AN OPEN NETWORK

As a variation of the problem of section 1.13, consider the case $p = 0$ and say that the objective is to

$$\begin{cases} \text{minimize } E\{x_t^1 + cx_t^2\} \\ \text{over } \mu_1 = \mu_1(x_t^1, x_t^2) \\ \text{subject to } \mu_1 \in [0, a]. \end{cases}$$

In this expression, the expectation is over the stationary distribution that corresponds to the specific choice of the function $\mu_1 : \{0, 1, 2, \ldots\}^2 \rightarrow [0, a]$.
It is assumed that the service rate μ_2 in the second queue is constant. The parameter c is a given positive number. (See Figure 1.8.)

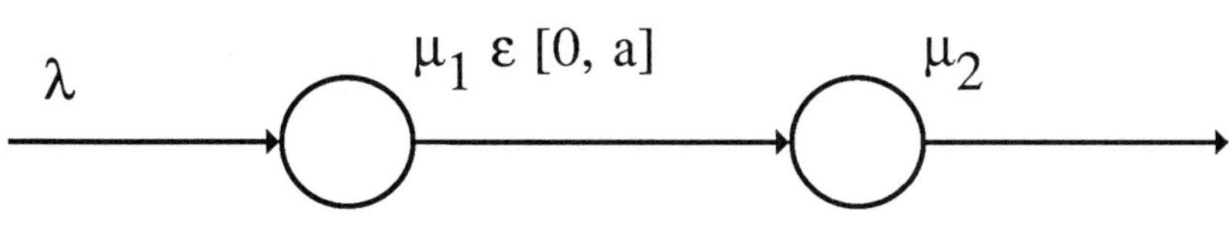

Figure 1.8. The controlled open network.

It should be intuitively clear that if $c \leq 1$, then the optimal strategy is to choose $\mu_1 \equiv a$.

It can be shown, by rather involved dynamic programming arguments, that if $c > 1$, then the optimal strategy is of the **switch-over** type. That is, there is a nondecreasing function $s : \{0, 1, 2, \ldots\} \rightarrow \{0, 1, 2, \ldots\}$ such that the optimal strategy is given by $\mu_1(x^1, x^2) = a. 1\{x^2 \leq s(x^1)\}$. (See Figure 1.9.)

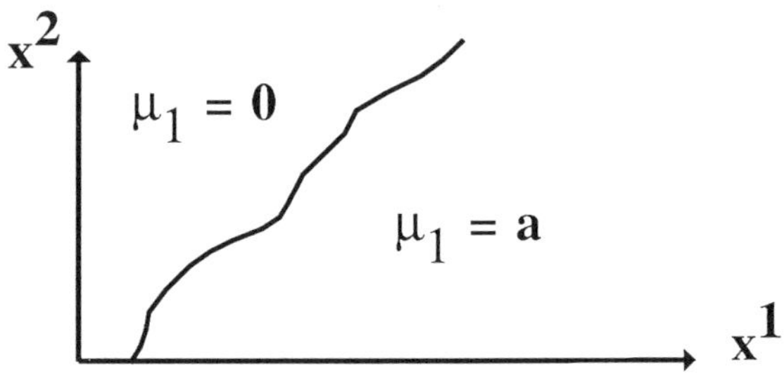

Figure 1.9. Switch-over strategy.

The intuitive justification is that if it is optimal to serve at the maximum rate in queue 1 for a given value of (x^1, x^2), then that should also be optimal if x^1 increases, and also if x^2 decreases. The optimal function $s(.)$ is not known in closed form. It can be approximated by dynamic programming algorithms. (See Chapter Eight for a discussion of this type of results.)

Notice that if the value of μ_1 at time $t \geq 0$ was allowed to depend on x_t^1 alone, then the optimal choice would be $\mu_1 \equiv a$. The reason is that any function $\mu_1 = \mu_1(x_t^1)$ would correspond to a product-form invariant distribution where the marginal distributions of the two queues would be the same as if the arrival processes in the two queues were independent and Poisson with rates γ. This is similar to Theorem 1.4.1 and follows from analogous arguments. (See Chapter Three.) As a consequence, the selection of $\mu_1(x_t^1)$ cannot affect the value of $E\{x_t^2\}$ (provided that the network has an invariant distribution). Hence, the optimal $\mu_1(x_t^1)$ must minimize $E\{x_t^1\}$ and must therefore correspond to serving queue 1 always at the maximum rate a.

1.15. OPTIMAL SERVER ALLOCATION

Consider once again the network of Figure 1.1 but assume that there is a *single server* who must serve the two queues. Say that the service must be **nonpreemptive.** That is, once the server starts serving a customer, then he must complete the service before being allowed to start serving another customer.

The problem is to minimize

$$E\{c_1x_t^1 + c_2x_t^2\} \tag{1.23}$$

over all possible nonpreemptive strategies that are **nonanticipative,** for given positive values of c_1 and c_2. By nonanticipative, it is meant that the decision of which customer to start serving must be based on the queue lengths at that time. In other words, the server cannot be told when future arrivals will occur, nor the values of service times that have not been completed.

First, it must be realized that the cost (1.23) cannot be finite unless

$$\frac{\gamma}{q}\left(\frac{1}{\mu_1} + \frac{1}{\mu_2}\right) < 1 .$$

Indeed, the maximum rate at which work is done by the server is equal to one and the left-hand side of the above expression is the average rate at which work enters the system. To see this, observe that each customer will visit the queues an average number of times equal to q^{-1} and that each visit to the two queues requires an average service time equal to $\mu_1^{-1} + \mu_2^{-1}$.

The optimal strategy for this problem turns out to be remarkably simple. At any given time when the server can start a new service, he should start serving the first nonempty queue in the order (1, 2) if

$$\mu_1(c_1 - c_2) \geq \mu_2(c_2 - pc_1) \tag{1.24}$$

and in the order (2, 1) otherwise.

The interpretation of this result, which is a particular case of **Klimov's Theorem** (see Chapter Nine), is that the left-hand side [respectively, the right-hand side] of (1.24) is the average reduction of waiting cost per unit of service time in queue 1 [respectively, in queue 2]. The result then says that the server should start working in the queue where the reduction of waiting cost per unit of time is the largest.

What is remarkable about the result is that the optimal decision *does not depend on the queue lengths,* only on which queues are not empty, and *does not depend on the arrival rate.* Moreover, it is known that the policy remains optimal if the service time distributions in the queues *are not exponential* but have mean values μ_1^{-1} and μ_2^{-1}, respectively, and have finite second moments.

1.16. THE NONLINEAR FILTERING PROBLEM

In many situations it is not possible to observe directly the state of a queueing system. Instead, one may be able to observe only the states of some queues, or the flows on some links of the network. The problem then arises of determining how the partial observations can be used to control the network or to infer some of its properties.

Applications of this type can be found in *hypothesis testing, disorder,* and *control problems.* (See Chapter Ten.) Some basic ideas of the underlying theory (called **nonlinear filtering)** will be discussed on the example of Figure 1.1. We will discuss the case $p = 0$. (See Figure 1.10.)

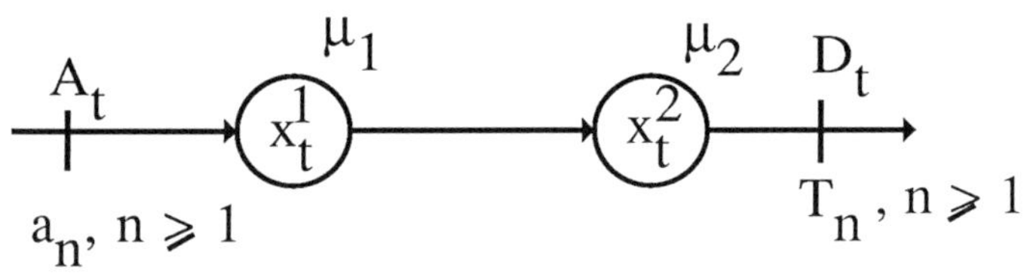

Figure 1.10. The network with $p = 0$.

For $t \geq 0$, let A_t and D_t be the number of arrivals and departures from the network in $[0, t]$, respectively. Assume that one knows $\{A_t, t \geq 0\}$, that one can observe $\{D_t, t \geq 0\}$, and that one knows the probability distribution π_0 of $x_0 = (x_0^1, x_0^2)$. What can one say about x_t for $t \geq 0$ on the basis of that information? Thus, the arrival times $\{a_1, a_2, a_3, \ldots\}$ are known in advance, and the random departure times $\{T_1, T_2, T_3, \ldots\}$ are observed in real time, i.e., as they occur.

For $0 < t_1 < \cdots < t_n < t, x \in \mathbf{X},$ and $0 < \epsilon \ll 1,$ let $\Delta_i := (t_i, t_i + \epsilon)$ and define

$$\nu_{t_n}(t_1, \ldots, t_n, x)\epsilon^n = P\{T_i \in \Delta_i, 1 \le i \le n, x_{T_n} = x\} + o(\epsilon^n) \tag{1.25}$$

$$\nu_t(t_1, \ldots, t_n, x)\epsilon^n = P\{T_i \in \Delta_i, 1 \le i \le n, T_{n+1} > t, x_t = x\} + o(\epsilon^n).$$

(By $o(x)$ one denotes a function of $x > 0$ such that $o(x)/x \to 0$ as $x \to 0$.) It will be shown in Chapter Ten that the following result holds. The arguments $(t_1, \ldots, t_n)$ of $\nu_t(x)$ are omitted for simplicity. Below, the rate matrix Q_0 is the rate matrix of (1.1) with $\gamma = 0$.

Theorem 1.16.1 [Nonlinear Filtering]

(a) The quantity $\nu_t(x)$ can be calculated recursively as follows:

$$\left.\begin{aligned}
\frac{d}{dt}\nu_t(x) &= \sum_y \nu_t(y) q_0(y, x)\, 1\{x \ne D_2 y\}, \\
&\quad \text{if } t \ne T_n \text{ and } t \ne a_n \text{ for all } n \ge 1 \\
\nu_t(x) &= \nu_{t-}(y), \text{ with } x = A_2 y, \\
&\quad \text{if } t = a_n \text{ for some } n \ge 1 \\
\nu_t(x) &= \sum_y \nu_{t-}(y)\, q_0(y, x)\, 1\{x = D_2 y\}, \\
&\quad \text{if } t = T_n \text{ for some } n \ge 1 .
\end{aligned}\right\} \tag{1.26}$$

The initial value of ν_t is given by

$$\nu_0(x) = P\{x_0 = x\} = \pi_0(x), \tag{1.27}$$

as is seen from the definition (1.25).

(b) The finite dimensional distributions of $\{D_s, s \ge 0\}$ are specified by

$$\sum_{x \in \mathbf{X}} \nu_t(x). \tag{1.28}$$

(c) The conditional distribution of x_t given $\{D_s, s \le t\}$ is equal to

$$P[x_t = x \mid D_s, s \le t] = \frac{\nu_t(x)}{\sum_{y \in \mathbf{X}} \nu_t(y)}, \quad x \in \mathbf{X}, \quad t \ge 0. \tag{1.29}$$

Comments:

(a) Formula (1.29) shows that ν_t is an *unnormalized version of the conditional distribution* of x_t given $\{D_s, s \le t\}$. This should be clear from (1.25).

(b) The updating formulas (1.26) are *linear* equations. However, the conditional distribution (1.29) is being updated by nonlinear equations (because of the renormalizing). This justifies the terminology "nonlinear filtering."

(c) The above theorem can be used to calculate the finite dimensional distributions of the departure process from the network, given the initial distribution of **x** and given the arrival times. Similar formulas can be derived for any network modeled by a continuous-time Markov chain. See Chapter Ten for details.

1.17. INTERCHANGEABILITY OF THE QUEUES

This section gives an example of application of the nonlinear filtering approach. It will be shown that if the network of Figure 1.1 with $p = 0$ is initially empty, i.e., if $x_0 = (0, 0)$, then the finite dimensional distributions of $\{D_t, t \ge 0\}$ are symmetric in μ_1 and μ_2. That is, the departure process is not changed if the two queues are interchanged. This is the case for arbitrary arrival times.

As a consequence, the result remains true for the network of Figure 1.1 with $0 < p < 1$.

Similarly, in the case $p = 1$, if one assumes that the N customers are injected into the first queue at time 0, then the departure process of the second queue is not changed if μ_1 and μ_2 are interchanged. In particular, letting the time increase to ∞ shows that the departure process from the first queue has the same finite dimensional distribution in steady state as the departure process from the second queue.

Coming back to the case $p = 0$, Theorem 1.16.1 shows that the result will follow if one can show that $\sum_{x \in \mathbf{X}} \nu_t(x)$ is symmetric in μ_1 and μ_2 for all $t \ge 0$.

To prove this it suffices to show that, for all $0 \le M \le N$,

$$f_t(N, M) := \mu_2^M \sum_{x \in A(N,M)} \nu_t(x) , \tag{1.30}$$

where

$$A(N, M) := \{x \in \mathbf{X} \mid x^1 + x^2 = N, x^2 \geq M\}$$

is symmetric in μ_1 and μ_2. Indeed, $\sum_{x \in \mathbf{X}} \nu_t(x) = \sum_{N=0}^{\infty} f_t(N, 0)$.

At $t = 0$, (1.30) is symmetric since $f_0(N, M) = 1\{N = 0 = M\}$, as follows from $x_0 = (0, 0)$. Now, the nonlinear filtering equations (1.26) show, by summing over $A(N, M)$, that

$$\frac{d}{dt} f_t(N, M) = -(\mu_1 + \mu_2) f_t(N, M) + \mu_1 \mu_2 f_t(N, M-1), \quad (1.31)$$

$$t \neq T_n, \quad t \neq a_n$$

$$f_t(N, M) = f_{t-}(N+1, M+1), \quad t = T_n$$

$$f_t(N, M) = f_{t-}(N-1, M), \quad t = a_n .$$

These formulas prove the symmetry of $f_t(N, M)$ for all $t \geq 0$ and establish the interchangeability of the two queues. See Chapter Ten for more details.

1.18. COMPARISON RESULTS

How does the behavior of the network change when some of its *parameters* are modified? This section discusses some answers to that question.

When a closed-form expression is known for a quantity of interest as a function of the network parameters, then the effect of parameter modifications can be analyzed by discussing the properties of the function. One can then determine whether the quantity of interest is decreasing, convex, symmetric, etc., in some of the network parameters.

As an example, one can check that the throughput $C(N)^{-1}$ (where the average cycle time $C(N)$ is given in (1.11)) is increasing, concave and symmetric in (μ_1, μ_2). Indeed, some algebra shows that

$$C(N)^{-1} N = \frac{\mu_1^{N+1} \mu_2 - \mu_1 \mu_2^{N+1}}{\mu_1^{N+1} - \mu_2^{N+1}}$$

which is seen, after some straightforward analysis, to have the indicated properties. This result is a particular case of the corresponding result for the throughput in Jackson networks. See sections 6.7.1 and 6.7.2.

As another example, let X^1 and X^2 be random variables with the joint distribution $\pi_{\mathbf{X}_N}$. It is intuitively obvious that if one increases μ_1, then X^1 should decrease (in any reasonable sense) and X^2 should increase. For instance, one finds from (1.5) that

$$\frac{P\{X^1 = n-1\}}{P\{X^1 = n\}} = \frac{\mu_1}{\mu_2},$$

which shows that, as μ_1 increases, the distribution of X^1 is pushed down geometrically.

More interesting cases are when closed-form formulas are not available. Simple algebraic arguments are then no longer sufficient.

As a first example, consider the network of Figure 1.9 and assume that it is initially empty and that the arrival times are known in advance. Say that $\mu_1 + \mu_2 = \mu$. It is intuitively to be expected that the customers will go through the network faster if the two queues are equally fast, i.e., if $\mu_1 = \mu_2 = \frac{\mu}{2}$. To state the result precisely, let $\{T_n, n \geq 1\}$ be the successive departure times. The claim is that, for all $n \geq 1$ and $t_1, \ldots, t_n > 0$,

$$P\{T_1 \leq t_1, T_2 \leq t_2, \ldots, T_n \leq t_n\}$$

increases as $|\mu_1 - \mu_2| \to 0$ monotonically.

One way to prove this result is to use the nonlinear filtering method of section 1.16. Other proofs are available. These comparison results will be further discussed in Chapter Six.

As another example, assume that the constant service rate μ_1 is replaced by a controlled rate, possibly random, but that is always at least equal to μ_1. It is then possible to construct the process so that x_t^1 is always smaller than under the constant rate μ_1.

The last example is such that it is possible to compare all realizations of two queueing networks (corresponding to two sets of parameter values). In other situations, such a pathwise comparison is not possible. Such an example is discussed next.

Consider the two networks of Figure 1.11.

It is intuitively clear that if the two systems are initially in the same state, then the second network should be more congested than the first, at all later times. In fact, it can be shown that, for all $t > 0$,

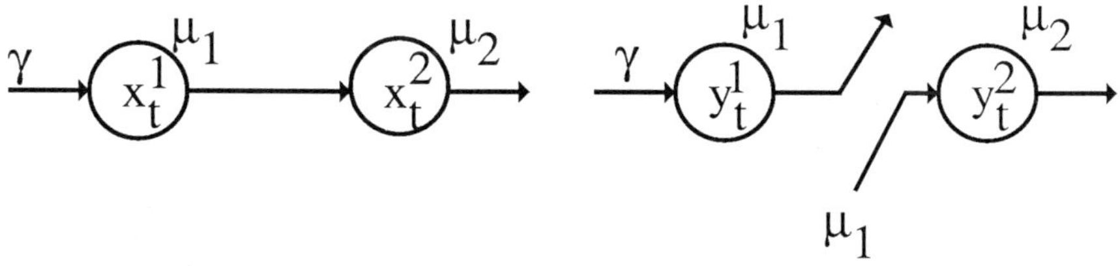

Figure 1.11. Comparing two networks.

$$P\{y_t^1 \geq x^1, y_t^2 \geq x^2\} \geq P\{x_t^1 \geq x^1, x_t^2 \geq x^2\}, \qquad (1.32)$$
$$x^1 \geq 0, \quad x^2 \geq 0\,,$$

provided that this property holds at $t = 0$. This can be done by an induction on the processes observed at their jump times. However, it is not possible to construct the processes so that

$$P\{y_t^1 \geq x_t^1, y_t^2 \geq x_t^2, t \geq 0\} = 1\,, \qquad (1.33)$$

which would have been a sufficient condition for (1.32). Indeed, in the case $\gamma = 0$, if $x_0 = y_0 = (1, 0)$, then $P\{y_\epsilon = (0, 0)\} = \mu_1\epsilon + o(\epsilon)$, while $P\{x_\epsilon = (0, 0)\} = o(\epsilon)$. This would contradict (1.33).

This problem will be discussed again in Chapter Six.

1.19. EXERCISES

1.1. Balance equations Verify directly that π given by (1.3)–(1.4) solves the balance equations (1.2) with Q given in (1.1).

1.2. M/M/1 queue An M/M/1 queue has Poisson arrivals with some rate λ and the service times are independent and exponentially distributed with rate μ. Write the rate matrix for the queue length. Show that the invariant distribution is given by

$$\pi(n) = \rho^n(1 - \rho), \quad n \geq 0$$

if $\rho := \dfrac{\lambda}{\mu} < 1$.

1.3. Reversibility Construct an open network of two M/M/1 queues that is time-reversible.

1.4. Numerical considerations Try to analyze the accuracy of the approximation (1.16).

1.5. Non-Poisson flows Consider the network of Figure 1.1 but with $\mu_2 = \infty$. That is, the system consists of a single queue with feedback. Argue that the arrival process into the queue cannot be Poisson. *Hint:* The rate of the arrival process into queue 1 is equal to γq^{-1}. However, given that an arrival was just observed, the rate is $\gamma + \mu p > \gamma q^{-1}$.

1.6. Sojourn times Consider the formula (1.18) in the case $N = 2$. Calculate the mean value of $T_1 + T_2$ and the variance of that random variable.

1.7. Parameter optimization Verify (1.20).

1.20. REFERENCES AND COMMENTS

Section 1.2 This network is a particular example of Jackson network. These networks will be discussed in Chapter Two.

Section 1.3 Most of the models used in this text, with the exception of Chapter Seven, are Markov chains. The reason is that this set of models is "dense" in the class of models of physical systems to be investigated. Also, the theory of regular Markov chains is fairly simple, as will be seen in Chapter Two. Moreover, a computer investigation of a physical system necessitates a discretization of the state space. This is done from the start with Markov chain models.

Section 1.4 Networks with a product-form invariant distribution constitute the largest class of tractable networks of queues. This class will be investigated in detail in Chapter Three.

Section 1.5 Time reversal is a very important tool in the analysis of product-form networks, as will be made clear in Chapter Three.

Section 1.6 Average sojourn times can be calculated in terms of the invariant distribution (by Little's result). This is not the case for higher moments. For instance, there are many networks with the same invariant distribution but with different values for the variance of the sojourn times.

Section 1.7 The Arrival Theorem is an illustration of the Palm distributions obtained by conditioning on the fact that one jump just took place in a network. A general discussion of that concept will be given in Chapter Seven.

Section 1.8 Chapter Five will be devoted to a more detailed discussion of numerical issues in queueing networks.

Section 1.9 The general discussion of Poisson flows in networks will be given in Chapter Four. The result is that a flow on a link cannot be Poisson if that link can be visited more than once by the same customer.

Section 1.10 This argument will be given for general networks of quasi-reversible queues in Chapter Three.

Section 1.11 Distributions of sojourn times are discussed in Chapter Four.

Sections 1.12 and 1.13 Parameter optimization is discussed extensively in Bertsekas and Gallager (1987). The result in Section 1.13 is from Lazar (1983).

Section 1.14 Dynamic control is discussed in Chapter Eight where the necessary results of dynamic programming are reviewed.

Section 1.15 Klimov's result can be viewed as a limiting case of a multi-armed bandit problem. See Chapter Nine.

Section 1.16 The nonlinear filtering problem for a general Markov chain is discussed in Chapter Ten.

Section 1.17 The result of this section is due to Weber (1979). Our proof is from Anantharam (1985). See also Tsoucas, et al. (1986).

Section 1.18 Further comparison results are discussed in Chapter Six. The monotony result stated about the system of Figure 1.10 is due to Lehtonen (1986). See also Tsoucas, et al. (1986). The result about the networks of Figure 1.11 is due to Massey (1984a-1984b). See also Tsoucas, et al. (1984b).

CHAPTER TWO

MARKOV CHAINS

2.1. INTRODUCTION

This chapter motivates the definition of **Markov chains** by considering a simple queueing system. Markov chains are then defined and their **transition probabilities** are analyzed. We then consider the issues of **stationarity** and **asymptotic stationarity.** Finally, the **time reversal** ideas and the **embedded distributions** are discussed. **Jackson networks** are used to illustrate the ideas.

Let us summarize the essential facts of this chapter. A **Markov chain** $\mathbf{x} = \{x_t, t \geq 0\}$ describes the random evolution of an *amnesic* particle in a countable set. That is, at any given time t, the process $\mathbf{x}$ has forgotten how it got to its present value x_t and its evolution after time t starts afresh from its present value, as if t were equal to zero. (We consider only Markov chains with stationary transition probabilities.) In system-theoretic terminology, $\mathbf{x}$ is a **state process.** In queueing systems, the amnesia comes from the **memoryless** property of the exponential distribution. (See sections 2.2–2.4.) The **regular** Markov chains are defined in 2.4. Regularity means that $\mathbf{x}$ makes finitely many jumps in finite time, which is a natural assumption for models of physical systems. (See section 2.4.5.) The evolution of a regular Markov chain is completely specified by its **initial distribution** and its **rate matrix.** Theorem 2.5.2 contains the key properties that allow us to calculate the distribution of x_t for $t \geq 0$. The existence, uniqueness and convergence to an **invariant distribution** are discussed in sections 2.6–2.7. The main result is that a Markov chain that is **irreducible** and admits an invariant distribution is **asymptotically stationary** and **ergodic.** (See Theorem 2.6.2.) The **time reversal** of a Markov chain is explained in section 2.8. The key observation is that a stationary Markov chain reversed in time behaves as another stationary Markov chain. (See Theorem 2.8.1.) This leads to a very useful method for verifying the invariance of a distribution by guessing the rate matrix of the Markov chain reversed in time, as explained in Lemma 2.8.5. **Discrete-time Markov chains** are discussed in section 2.9. The concept of a **Markov chain watched in a set** (see 2.9.9) is used in section 2.10 to analyze the distribution of $\mathbf{x}$ at certain jump times. The main result on that topic is Theorem 2.10.6, where the distribution at the jumps is shown to be given by a **ratio**

of rates. This fact is a prelude to the concept of Palm distributions of Chapter Seven.

2.2. A SIMPLE QUEUE

Consider the system illustrated in Figure 2.1. Customers arrive at a service facility at the random times $\{S_n, n = 1, 2, 3, \ldots\}$ with $0 < S_1 < S_2 < \cdots$. For simplicity, it will be assumed that the **interarrival times** $S_{n+1} - S_n$ for $n \geq 0$ (with $S_0 = 0$) are **independent** and have probability distribution $F(.)$. Thus, $F(t) = P\{S_{n+1} - S_n \leq t\}$ for $t \geq 0$ and for $n \geq 0$.

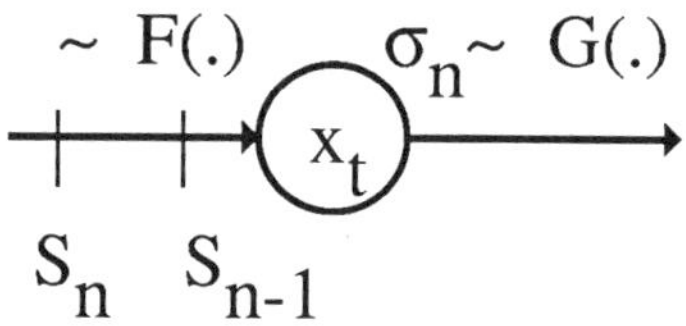

Figure 2.1. The simple queue.

The customers are served on a first-come–first-served (**FCFS**) basis by a single server. The customers require **independent service times** with distribution $G(.)$. The service times are independent of the arrival times and of the number of customers in the facility (called the **queue length**) at time 0. To avoid degenerate cases, it will be assumed that the interarrival times and the service times are not all constant (deterministic).

The questions of interest concern the evolution of the queue length and the time spent by the customers in the facility.

Say that the system is observed from time 0 until some time $t > 0$. What should be remembered about that evolution in order to predict the evolution after time t? That is, what is a suitable **state** description?

The simplest possible state, if one wants to be able to discuss the evolution of the queue length, is the queue length itself. Let x_t denote the queue length at time $t \geq 0$. For x_t to be a state, one must have, for any set **A** of realizations of $\{x_t, t \geq 0\}$,

$$P[(x_s, s \geq t) \in \mathbf{A} \mid x_s, s \leq t] = P[(x_s, s \geq t) \in \mathbf{A} \mid x_t]. \qquad (2.1)$$

This equation, called the **Markov property,** expresses the requirement that x_t must contain all the information about the past evolution of the system that is relevant for predicting the future evolution of the queue length. A process x_t with the Markov property is called a **Markov process.**

The following fact indicates when (2.1) holds for the queueing system described above. It emphasizes the central role played by the exponential distribution. Recall that F is the **exponential distribution with rate** $\lambda \in (0, \infty)$ (i.e., with **mean** λ^{-1}) if

$$F(t) = 1 - e^{-\lambda t}, \quad t \geq 0. \tag{2.2}$$

Fact 2.2.1 The property (2.1) holds if and only if F and G are both exponential distributions.

Proof: We will discuss the case of G. That of F is similar. Consider a situation where $x_t > 0$ and assume that the last time before time t that the queue length decreased is time $t - a > 0$. (See Figure 2.2.)

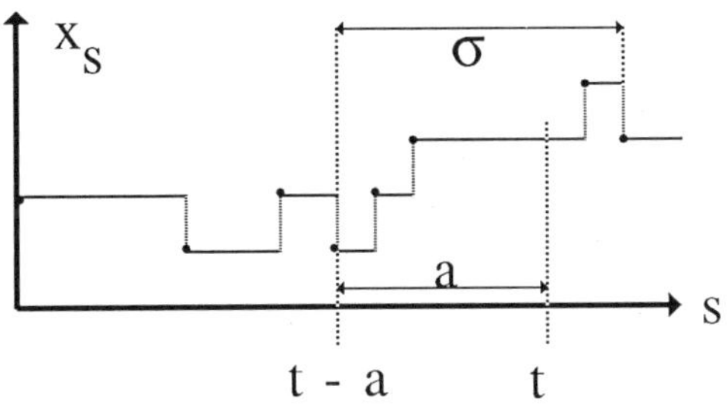

Figure 2.2. A realization of the process.

Denote by σ the service time of the customer being served at time t. By observing x_s for $s \in [0, t]$, one knows that $\sigma > a$. Indeed, the customer being served at time t started service at the last departure time $t - a$. If (2.1) is to hold, then it must be that this information is irrelevant for predicting the amount $\sigma - a$ of service time still required by the customer being served at time t. (For that information is not contained in x_t.)

Hence, (2.1) implies that

$$P[\sigma - a > b \mid \sigma > a] = P\{\sigma > b\}, \quad \text{for all } a, b > 0. \tag{2.3}$$

(Indeed, the value of the left-hand side of (2.3) should not depend on the value of a and must therefore be the same as when $a = 0$, which is the right-hand side.)

Recalling the defining formula $P[A \mid B] = \dfrac{P\{A \cap B\}}{P\{B\}}$ of conditional probability, one finds that (2.3) is equivalent to

$$P\{\sigma > a + b\} = P\{\sigma > a\}P\{\sigma > b\}, \quad \text{for all } a, b > 0 .$$

It now takes only some simple analysis to conclude that

$$P\{\sigma > t\} \equiv e^{-\mu t}, \quad t \geq 0 ,$$

for some μ, which shows that G is indeed an exponential distribution.

Equation (2.3) is the **memoryless property** of the exponential distribution. It says, for instance, that if a light bulb has an exponentially distributed lifetime, then the bulb is as good as new as long as it is still burning.

To see that the conditions are sufficient, notice that the only information in $\{x_s, s \leq t\}$ that is potentially relevant for predicting the future evolution is the time since the last arrival and the amount of service already received by the customer being served. (Because of the assumed independence of the interarrival times and service times.) This information is useless by the memoryless property of the exponential distribution.

2.3. THE M/M/1 QUEUE

The queueing system above with F and G being exponential distributions is called an **M/M/1 queue.** The first M in the notation indicates that the arrival process is memoryless, i.e., that it has **i.i.d.** (independent and identically distributed) and exponentially distributed interarrival times. Such a process is called a **Poisson process.** The second M indicates that the service times are memoryless, i.e., exponentially distributed. The 1 refers to the fact that there is a single server.

The general notation is **A/B/s** and **A/B/s/N** if the maximum number of customers in the queue is N. (The customers arriving when there are N customers in the queue are rejected and do not come back.) The symbol "A" is G if the arrival process is general; it is GI if the interarrival times are i.i.d. with a general distribution, and it is D if they are equal to a common constant (deterministic). Similarly, B characterizes the service times. The value of s is the number of servers. Other possibilities will be discussed in due time. Appendix B contains a few examples.

The following observation helps visualize the evolution of the queue length x_t of an M/M/1 queue.

Fact 2.3.1 Let $\{\sigma_k, 1 \le k \le K\}$ be independent exponentially distributed random variables with respective rates λ_k. Then, for all $t \ge 0$ and $1 \le i \le K$,

$$P\{\sigma_i = \min\{\sigma_k, 1 \le k \le K\} > t\} = \frac{\lambda_i}{\lambda} e^{-\lambda t}, \tag{2.4}$$

with $\lambda := \lambda_1 + \cdots + \lambda_K$.

This fact, the proof of which is left as an exercise, states that the index of the smallest of K independent exponential random variables is independent of the value of that minimum which is also exponentially distributed.

The evolution of the queue length x_t of an M/M/1 queue can therefore be described as follows (the possible values of i and j below are in $\mathbf{X} = \{0, 1, 2, \ldots\}$).

If $x_0 = i > 0$, then x_t keeps that value until an arrival or a service completion occurs. Since the times until these events are two independent exponentially distributed random variables with respective rates λ and μ, we conclude from Fact 2.3.1 that the queue length will keep the value i for a random time τ that is exponentially distributed with rate $q(i) = \lambda + \mu$, after which it jumps to state j with probability $\dfrac{q(i, j)}{q(i)}$ where $q(i, i+1) = \lambda$, $q(i, i-1) = \mu$, and $q(i, j) = 0$ if $|i - j| > 1$. The case $i = 0$ is similar, except that $q(0) = \lambda$ and $q(0, -1) = 0$. The evolution then starts afresh, i.e., independently of how the process got to state j, from x_τ at time τ (in view of the memoryless property of the exponential distribution).

This process x_t is an example of a continuous-time Markov chain.

2.4. MARKOV CHAINS

Recall that a set $\mathbf{X}$ is **countable** if one can enumerate its elements as $\mathbf{X} = \{i_1, i_2, \ldots\}$. Finite sets are also said to be countable. For instance, $\mathbb{Z} := \{\ldots, -2, -1, 0, 1, 2, \ldots\}$ is countable. A countable union $\bigcup_{n=1}^{\infty} \mathbf{X}_n$ of countable sets $\mathbf{X}_n$ is countable. A **finite** product $\times_{n=1}^{N} \mathbf{X}_n := \{i = (i_1, \ldots, i_N) \mid i_n \in \mathbf{X}_n, 1 \le n \le N\}$ of countable sets $\mathbf{X}_n$ is countable. The interval [0, 1] of the real line is not countable.

Let $\mathbf{X}$ be a countable set. One will define a Markov process x_t that takes values in $\mathbf{X}$. In addition, x_t will be such that its law of evolution does not depend explicitly on time. Thus, the probability of going from i at time t to j at time $t + s$ (for $s > 0$) will not depend on t. Such a process is called a **Markov chain** (with stationary transition probabilities). To avoid some (interesting) technical difficulties, one will make sure that x_t makes at most finitely many jumps in finite time. This is called **regularity.**

Definition 2.4.1 [Rate matrix] Let $\mathbf{X}$ be a countable set. A matrix $Q = [q(i, j), i, j \in \mathbf{X}]$ indexed by $\mathbf{X}$ is called a **rate matrix** on $\mathbf{X}$ if, for all $i \neq j \in \mathbf{X}$, one has

$$0 \leq q(i, j) \text{ and } q(i) := -q(i, i) = \sum_{l \neq i} q(i, l) < \infty. \tag{2.5}$$

Notice in particular that the *rows of Q sum to zero,* i.e., $\sum_j q(i, j) = 0$ for all $i \in \mathbf{X}$.

For instance, the rate matrix of the M/M/1 queue was defined below Fact 2.3.1.

Definition 2.4.2 [Markov chain] Let $\mathbf{X}$ be a countable set and Q a rate matrix on $\mathbf{X}$. Let also $\pi_0 = \{\pi_0(i), i \in \mathbf{X}\}$ be a (probability) **distribution** on $\mathbf{X}$ (i.e., a collection of nonnegative numbers that sum to one).

A **Markov chain with rate matrix Q and initial distribution π_0** is a process $\mathbf{x} := \{x_t, \ t \geq 0\}$ defined as follows. (See Figure 2.3.) The value of x_0 is chosen randomly in $\mathbf{X}$ with distribution π_0. If $x_0 = i$, then x_t keeps that value for a random time τ that is exponentially distributed with rate $q(i)$. At time τ, the process jumps to j with probability $\Gamma(i, j) := \dfrac{q(i, j)}{q(i)} 1\{i \neq j\}$ for $j \in \mathbf{X}$, independently of τ. The evolution then starts **afresh** from x_τ at time τ (i.e., as if τ were equal to zero and independently of the evolution prior to time τ). The process x_t has **right-continuous realizations,** i.e., $x_{t+} := \lim_{s \downarrow t} x_s = x_t$ for all t.

For instance, the M/M/1 queue is a Markov chain on $\{0, 1, 2, \ldots\}$.

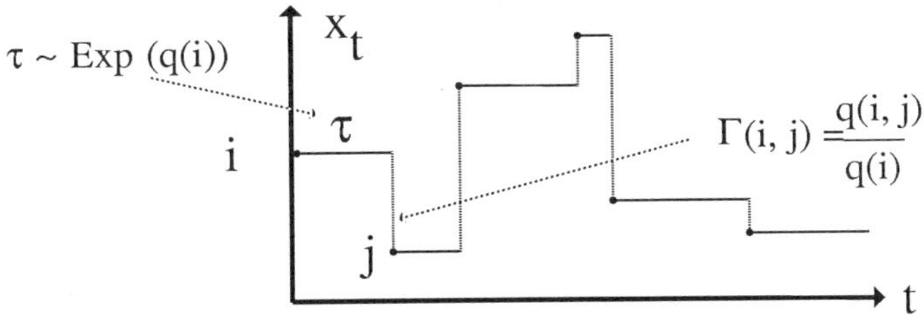

Figure 2.3. Evolution of the Markov chain.

Theorem 2.4.3 [Markov Property] Assume that the above construction defines a process $\mathbf{x} := \{x_t, \quad t \geq 0\}$ (see Remark 2.4.5). Then $\mathbf{x}$ is Markov and has homogeneous transition probabilities. That is, for all sets $\mathbf{A}$ of realizations and for all $i \in \mathbf{X}$,

$$P[(x_s, s \geq t) \in \mathbf{A} \mid x_s, s \leq t; x_t = i] \\ = P[(x_s, s \geq 0) \in \mathbf{A} \mid x_0 = i] \,. \tag{2.6}$$

Indeed, this follows from the memoryless property of the exponential distribution and from the fact that at each jump time the next state is chosen with a distribution that depends only on the previous state. (One has to assume that $\mathbf{A}$ is such that (2.6) is well-defined.)

Example 2.4.4 [Jackson network] Consider the system depicted in Figure 2.4.

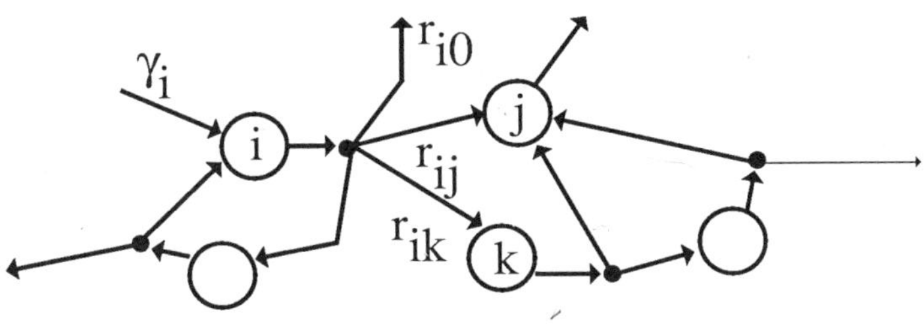

Figure 2.4. Jackson network.

There are $J \geq 1$ queues. Customers arrive from outside as independent Poisson processes, with rate γ_l into queue l (for $1 \leq l \leq J$). They require independent service times at each visit to the queues. The service times in queue l are exponentially distributed with rate μ_l. Also, the service times are independent of the arrival processes. Upon leaving queue l, each customer is sent to queue m with probability $r_{l\,m}$ for $1 \leq m \leq J$ and leaves the network otherwise, i.e., with probability $r_{l0} := 1 - \sum_{m=1}^{J} r_{l\,m}$. This

routing decision is independent of the past evolution of the network. We will call this a **Markov routing** to reflect the fact that the probability of a routing decision depends only on the current position of the customer.

Let $x_t := (x_t^1, \ldots, x_t^J)$, where x_t^l is the queue length (i.e., the number of customers) in queue l at time $t \geq 0$.

The claim is that x_t is a Markov chain on $\mathbf{X} = \{0, 1, 2, \ldots\}^J$. To see this, assume that $x_0 = x = (x^1, \ldots, x^J) \in \mathbf{X}$. The first jump time τ of x_t is the minimum of a finite collection of independent exponentially distributed random variables: the interarrival times and the service times of the customers being served. The rates of those random variables are

$$\gamma_1, \ldots, \gamma_J, \mu_1 1\{x^1 > 0\}, \ldots, \mu_J 1\{x^J > 0\} .$$

In view of Fact 2.3.1, it follows that τ is exponentially distributed with rate

$$q(x) = \sum_{m=1}^{J} [\gamma_m + \mu_m 1\{x^m > 0\}] .$$

Also, the probability that τ corresponds to an arrival in queue m is equal to $\dfrac{\gamma_m}{q(x)}$, again by Fact 2.3.1. Similarly, the probability that τ corresponds to a customer being served in queue m and being sent to queue k is equal to $\dfrac{\mu_m 1\{x^m > 0\}}{q(x)} r_{mk}$. It follows from the memoryless property of the exponential distribution and from the independence of the successive routing decisions that the evolution then starts afresh from x_τ at time τ.

One can write explicitly the rate matrix Q of this queueing network (called a **Jackson network).** One finds

$$\begin{cases} q(x, A_m x) = \gamma_m \\ q(x, T_{mk} x) = \mu_m \, r_{mk} \, 1\{x^m > 0\} \\ q(x, D_m x) = \mu_m \, r_{m0} \, 1\{x^m > 0\} \end{cases} \tag{2.7}$$

where

$$\begin{cases} A_m x &= (x^1, \ldots, x^m+1, \ldots, x^J) \\ T_{mk} x &= (x^1, \ldots, x^m-1, \ldots, x^k+1, \ldots, x^J) \\ D_m x &= (x^1, \ldots, x^m-1, \ldots, x^J). \end{cases} \tag{2.8}$$

The notation is a bit heavy, but everything is rather elementary.

Remark 2.4.5 [Regularity] Let Q be a rate matrix on a countable set $\mathbf{X}$. We have described how a Markov chain x_t with rate matrix Q evolves. There is however the possibility that x_t was not defined for all positive time t. Indeed, it may be that the **jumps will accumulate** in such a way that x_t will make infinitely many jumps in finite time.

Consider the following example. Let $\mathbf{X} = \{0, 1, 2, \ldots\}$ and $q(i) = q(i, i+1)$ for $i \in \mathbf{X}$. Assume that Q is a rate matrix. That is, $q(i, j) = 0$ for $i \neq j \neq i+1$ and $q(i) < \infty$ for all $i \in \mathbf{X}$. Assume that $x_0 = 0$. (See Figure 2.5.)

According to Definition 2.4.2, the process x_t visits successively the states 0, 1, 2, . . . and spends independent exponentially distributed times in those states, with rate $q(i)$ in state i. Denote by τ_i the time spent in state i.

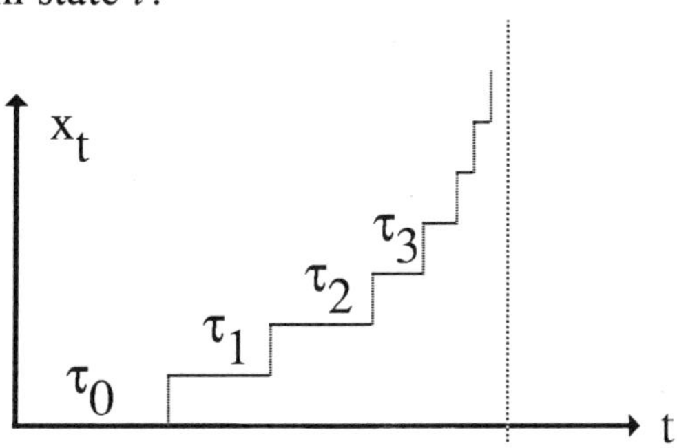

Figure 2.5. A pure-birth process.

One finds that, with $\zeta := \sum_{i=0}^{\infty} \tau_i$,

$$E\{\zeta\} = \sum_{i=0}^{\infty} E\{\tau_i\} = \sum_{i=0}^{\infty} q(i)^{-1}, \text{ and}$$

$$E\{e^{-\zeta}\} = \prod_{i=0}^{\infty} E\{e^{-\tau_i}\} = [\prod_{i=0}^{\infty}(1 + q(i)^{-1})]^{-1}$$
$$\leq [\sum_{i=0}^{\infty} q(i)^{-1}]^{-1} .$$

[We used $E\{e^{-\tau_i}\} = \int_0^{\infty} e^{-t} q(i)\, e^{-q(i)t}\, dt = (1 + q(i)^{-1})^{-1}$. The first equality in the derivation of $E\{\zeta\}$ follows by the Monotone Convergence Theorem (see Appendix A); the first equality in the derivation of $E\{e^{-\zeta}\}$ follows from the Dominated Convergence Theorem (see Appendix A).]

The first relation shows that $E\{\zeta\} < \infty$ if $\sum_{i=0}^{\infty} q(i)^{-1} < \infty$. The second relation shows that $E\{e^{-\zeta}\} = 0$ if $\sum_{i=0}^{\infty} q(i)^{-1} = \infty$. Therefore,

$$P\{\zeta = \infty\} = \begin{cases} 0, \text{ if } \sum_{i=0}^{\infty} q(i)^{-1} < \infty \\ 1, \text{ otherwise} . \end{cases}$$

Thus, if $\sum_{i=0}^{\infty} q(i)^{-1} < \infty$, the process x_t is defined for only a finite time ζ.

There is nothing in the definition that specifies what happens after the **explosion** time ζ. For instance, one could restart the process from $i = 0$ after the explosion, or one could restart it from any other state. All those alternative constructions would be consistent with Definition 2.4.2.

What this shows is that there may be many Markov chains with a given rate matrix, if explosions are possible. Clearly, the behaviors of those different Markov chains are substantially different. For instance, the average fraction of time spent in a given state depends on how the process restarts after explosions. Therefore, this average fraction of time cannot possibly be computed from the rate matrix alone. See also Exercise 2.3 for a related example.

A Markov chain is said to be **regular** if explosions are not possible. Regularity is a property of the rate matrix.

2.5. TRANSITION PROBABILITIES

Let $\mathbf{x} := \{x_t, \ \ t \geq 0\}$ be a regular Markov chain on $\mathbf{X}$ with rate matrix Q and initial distribution π_0. We will learn how to calculate $\pi_t(i) := P\{x_t = i\}$ for $i \in \mathbf{X}$ and for $t > 0$.

Definition 2.5.1 [Transition probabilities] For $t > 0$ and $i, j \in \mathbf{X}$, let

$$P_t(i, j) \ := \ P[x_t = j \mid x_0 = i] \, . \tag{2.9}$$

We will consider $P_t = [P_t(i, j), i, j \in \mathbf{X}]$ as a **matrix.** Also, distributions on $\mathbf{X}$ (such as π_t) will be considered as **row** vectors. Let $P_0 := I$, the identity matrix on $\mathbf{X}$. The P_t are called the **transition probabilities** of x_t.

The following theorem summarizes the essential properties of the transition probabilities.

Theorem 2.5.2 Let $\{P_t, t \geq 0\}$ be the transition probabilities of a regular Markov chain $\mathbf{x} := \{x_t, \ \ t \geq 0\}$ with rate matrix Q.

(a) The transition probabilities satisfy the **Chapman-Kolmogorov equations:**

$$P_{t+s} \ = \ P_t P_s, \quad \text{for all } s, t \geq 0 \, . \tag{2.10}$$

(b) The transition probabilities solve the **Kolmogorov equations:**

$$\frac{d}{dt} P_t \ = \ Q P_t \ = \ P_t Q, \quad t \geq 0 \, . \tag{2.11}$$

(c) If the entries of Q are bounded (for instance, if $\mathbf{X}$ is finite), then Q is said to be **uniform** and

$$P_t \ = \ e^{Qt} \ := \ \sum_{n=0}^{\infty} \frac{Q^n t^n}{n!}, \quad t \geq 0 \, . \tag{2.12}$$

(If Q is not uniform, then it is possible for Q to be regular and for the series in (2.12) not to converge, so that (2.12) does not hold in general; see Exercise 2.8.)

(d) The probability distribution π_t of x_t is given by

$$\pi_t \ = \ \pi_0 P_t, \quad t \geq 0 \, . \tag{2.13}$$

(e) The distribution π_t does not depend on t if and only if π_0 is **invariant,** i.e., if π_0 solves the following **balance equations:**

$$\pi Q \ = \ 0 \, . \tag{2.14}$$

These equations can be rewritten as

$$\pi(i)\, q(i) = \sum_{j \neq i} \pi(j)\, q(j, i), \quad i \in \mathbf{X}.$$

They express the equality of the rate of transitions out of any state i and the rate of transitions into that state i. In that case, $\mathbf{x}$ is **stationary,** i.e., such that the distribution of $\{x_{t+s}, s \geq 0\}$ does not depend on $t \geq 0$.

Proof: Let us just sketch the proof when $\mathbf{X}$ is a finite set. The countable case is outlined in Exercises 2.6 and 2.9.

The equation (2.10) can be seen by the calculation below:

$$\begin{aligned} P[x_{t+s} = j \mid x_0 = i] &= \sum_k P[x_{t+s} = j,\ x_t = k \mid x_0 = i] \\ &= \sum_k P[x_{t+s} = j \mid x_t = k,\ x_0 = i]\, P[x_t = k \mid x_0 = i] \\ &= \sum_k P_s(k, j)\, P_t(i, k). \end{aligned}$$

The second equality follows from

$$P[A, B \mid C] = P[A \mid B, C]\, P[B \mid C]$$

and the third one from the property (2.6) of x_t.

In order to justify (2.11) one first shows that it holds at $t = 0$, i.e., that Q is the derivative of P_t at $t = 0$. To do this, choose $j \neq i$ and observe that if τ denotes the first jump time of x_t, then

$$\begin{aligned} P[x_t = j \mid x_0 = i] &= P[\tau \leq t \mid x_0 = i]\, P[x_\tau = j \mid x_0 = i] + o(t) \\ &= (1 - e^{-q(i)\,t})\, \frac{q(i, j)}{q(i)} + o(t) \\ &= q(i, j)\, t + o(t). \end{aligned}$$

(Throughout, $o(t)$ denotes a function of t such that $\frac{o(t)}{t}$ goes to zero with t, such as t^2.) In the second and third terms, $o(t)$ is the probability of going from i to j in t time units by making at least two jumps. (See Exercise 2.5.) In the fourth term, we used the fact that $e^\alpha = 1 + \alpha + o(\alpha)$.

The above calculation shows that $q(i, j)\, t$ is, to the first order in t, the probability of going from i to j in t time units. This justifies the name **rate** given to those quantities. It also shows that $q(i, j) = \frac{d}{dt} P_0(i, j)$ for $j \neq i$. The case $j = i$ follows by summing the above result over $j \neq i$.

To get (2.11) in the finite case, one writes

$$P_{t+\epsilon} = P_t P_\epsilon = P_t[I + Q\epsilon + o(\epsilon)] = P_\epsilon P_t = [I + Q\epsilon + o(\epsilon)]P_t \,.$$

One then subtracts P_t and divides by ϵ to get the result as $\epsilon \downarrow 0$. Of course, you have noticed that this does not quite work in the countable case. For instance, it is not clear that an infinite sum of terms $o(\epsilon)$ is still $o(\epsilon)$, and the limit of an infinite sum may not be the sum of the limits.

To derive (2.12) in the finite case, one can differentiate the right-hand side of (2.12) term by term (this is legitimate for absolutely convergent series) and verify that it solves (2.11). One can also check that (2.11) admits a unique solution in the finite case (with $P_0 = I$).

The equation (2.13) is immediate.

To see the last statement of the theorem (recall that **X** is finite), one first observes that if $\pi_0 Q = 0$, then (2.13) and (2.12) show that $\pi_t = \pi_0$. Next, to see that this condition is necessary, assume that $\pi_t = \pi_0 P_t = \pi_0$ and differentiate at $t = 0$ to get (2.14). The stationarity of x_t can be seen by writing that for any set of realizations **A** one has

$$\begin{aligned} P\{(x_{t+s}, s \geq 0) \in \mathbf{A}\} &= \sum_i P\{x_t = i\} P[(x_{s+t}, s \geq 0) \in \mathbf{A} \mid x_t = i] \\ &= \sum_i \pi_t(i) P[(x_s, s \geq 0) \in \mathbf{A} \mid x_0 = i] \end{aligned}$$

where the last equality follows from Theorem 2.4.3. Hence, the left-hand side does not depend on t if $\pi_t \equiv \pi_0$.

Example 2.5.3 Let $\mathbf{X} = \{0, 1\}$ and $q(0) = q(0, 1) = \lambda$, $q(1) = q(1, 0) = \mu$ for some positive λ and μ. This is summarized by the **state transition diagram** indicated in Figure 2.6.

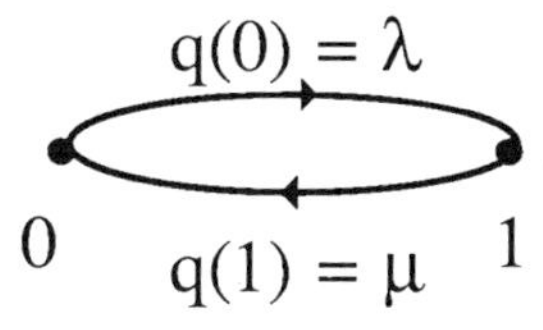

Figure 2.6. A two state Markov chain.

One can solve (2.11) to find that

$$P_t = (\lambda + \mu)^{-1} \begin{bmatrix} \mu + \lambda\gamma_t & \lambda - \lambda\gamma_t \\ \mu - \mu\gamma_t & \lambda + \mu\gamma_t \end{bmatrix}$$

where $\gamma_t = \exp\{-(\lambda + \mu)t\}$.

Solving the balance equations (2.14), together with the requirement that the elements of π sum to one, shows that the unique invariant distribution is given by

$$\pi = (\pi(0), \quad \pi(1)) = (\frac{\mu}{\lambda + \mu}, \frac{\lambda}{\lambda + \mu}) .$$

Notice that $P_t \rightarrow \begin{bmatrix} \pi \\ \pi \end{bmatrix}$ as $t \rightarrow \infty$. As a consequence, for any π_0 one sees from (2.13) that $\pi_t \rightarrow \pi$ as $t \rightarrow \infty$. Thus x_t is asymptotically stationary.
It will be seen in Theorem 2.6.2 that this is not a coincidence.

Example 2.5.4 [M/M/1 queue] The M/M/1 queue corresponds to the transition diagram of Figure 2.7.

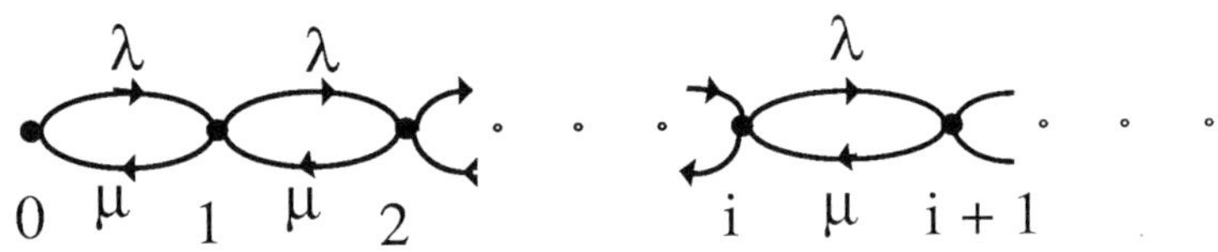

Figure 2.7. Transition diagram of M/M/1 queue.

To simplify the solution of the balance equations, one will use the fact that they imply

$$\pi(i)\, q(i, i + 1) = \pi(i + 1)\, q(i + 1, i), \quad \text{for all } i \geq 0 . \qquad (2.15)$$

Assuming this fact for the time being, we see that it implies that

$$\pi(i)\lambda = \pi(i + 1)\mu, \quad i \geq 0,$$

which shows that $\pi(i)$ must be of the form $\pi(i) = c\rho^i$ for $i \geq 0$ with $\rho := \frac{\lambda}{\mu}$ and where c must be chosen so that $\sum_{i=0}^{\infty} \pi(i) = 1$.
Hence,

$$\pi(i) = (1 - \rho)\rho^i, \quad i \geq 0 \quad \text{if } \rho := \frac{\lambda}{\mu} < 1 . \qquad (2.16)$$

Also, if $\lambda \geq \mu$, then there is no invariant distribution. Intuitively, the queue blows up, i.e., $x_t \rightarrow \infty$. This is the case (almost surely when $\rho > 1$ and in probability when $\rho = 1$) as will be seen in Example 2.7.2.

It remains to justify (2.15). The intuitive meaning is that the "flow of probability" from the left of i to its right must be equal to the flow of probability from the right of i to its left. This idea is made precise next. The equation (2.15) is a particular case of (2.17) below corresponding to $A = \{0, 1, \ldots, i\}$.

Fact 2.5.5 [Probability flows] Assume that π and Q are such that $\pi Q = 0$. Then, for all $\mathbf{A} \subset \mathbf{X}$,

$$\sum_{i \in \mathbf{A}} \sum_{j \in \mathbf{A}^c} \pi(i)\, q(i, j) = \sum_{j \in \mathbf{A}^c} \sum_{i \in \mathbf{A}} \pi(j)\, q(j, i), \qquad (2.17)$$

$$\text{with } \mathbf{A}^c := \mathbf{X} - \mathbf{A}.$$

To see this, let $F(\mathbf{A}, \mathbf{B}) = \sum_{i \in \mathbf{A}} \sum_{j \in \mathbf{B}} \pi(i)\, q(i, j)$. Then $F(\mathbf{A}, \mathbf{X}) = 0$ since $\sum_{j \in \mathbf{X}} q(i, j) = 0$. Also, $F(\mathbf{X}, \mathbf{A}) = 0$ since $\pi Q = 0$. Hence,

$$F(\mathbf{A}, \mathbf{A}^c) = F(\mathbf{A}, \mathbf{X}) - F(\mathbf{A}, \mathbf{A}) = F(\mathbf{X}, \mathbf{A}) - F(\mathbf{A}, \mathbf{A}) = F(\mathbf{A}^c, \mathbf{A})$$

as was to be shown.

Example 2.5.6 [Jackson network] As explained in Example 2.4.4, the vector of queue lengths x_t of a Jackson network is a Markov chain with rate matrix Q given by (2.7)–(2.8).

Consider the following set of **flow conservation equations**

$$\lambda_i = \gamma_i + \sum_j \lambda_j r_{ji}, \quad 1 \leq i \leq J. \qquad (2.18)$$

The interpretation of a solution $\lambda = (\lambda_1, \ldots, \lambda_J)$ is a collection of possible **average rates of flow** of customers through the queues. Indeed, if λ_j is the average rate of customers through queue j, then $\sum_j \lambda_j r_{ji}$ is the average rate of customers joining queue i per unit of time from other queues. Thus, the right-hand side of (2.18) is the total average rate of customers going through queue i per unit of time, i.e., λ_i.

The claim is that x_t has at least one invariant distribution π if and only if there is a nonnegative solution λ of (2.18) with $\lambda_i < \mu_i$ for $1 \leq i \leq J$. An invariant distribution is then given by

$$\pi(x) = \pi_1(x^1) \ldots \pi_J(x^J) \text{ with } \pi_i(n) = (1 - \rho_i)\rho_i^n, \quad n \geq 0 \quad (2.19)$$

where $\rho_i := \dfrac{\lambda_i}{\mu_i}$. This is a **product-form** invariant distribution.

The invariant distribution is unique if there is a unique solution with $\lambda_i < \mu_i$ $(i = 1, \ldots, J)$ of (2.18).

A brute force proof of this claim consists in writing out $\pi Q = 0$. We leave the details to the patient reader. More elegant proofs are available, they will be discussed later. (See Fact 2.8.4.)

Let us look at the three examples illustrated in Figure 2.8.

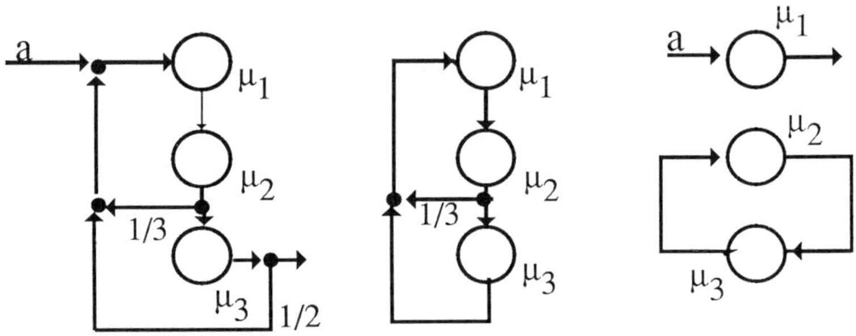

Figure 2.8. Three networks.

The first network is **open** (such that every customer eventually leaves), the second is **closed** (no customer can leave nor enter), and the third is **mixed** (the customers who can enter can leave, the others cannot).
Considering (2.18) for these networks, one finds the following solutions, respectively:

$$\lambda = (3a, 3a, 2a) \text{ is the unique solution;}$$

$$\lambda = (3b, 3b, 2b) \text{ for any } b \geq 0;$$

$$\lambda = (a, b, b) \text{ for any } b \geq 0 .$$

The first network has a unique invariant distribution if $3a < \mu_1$, $3a < \mu_2$, and $2a < \mu_3$. The second has infinitely many invariant distributions. (For one can choose b small enough, and different values of b give different invariant distributions by (2.19).) The third has infinitely many invariant distributions if $a < \mu_1$ and has no invariant distribution if $a \geq \mu_1$.

Let us reflect one instant on the case of the second network. Choosing $b = a$, one finds that the first two networks admit the same invariant distribution. However, these two networks behave quite differently. For instance, the **total population** $|x_t| := x_t^1 + x_t^2 + x_t^3$ in the second network is constant. Say that $|x_0| = N$. Is there an invariant distribution π such that $|x_t| = N$? It is clearly not one of the solutions (2.19). The answer is in the following fact.

Fact 2.5.7 Assume that $\pi Q = 0$ and that there is some set $\mathbf{A} \subset \mathbf{X}$ such that

$$q(i, j) = 0 \text{ if } (i, j) \in \mathbf{A} \times \mathbf{A}^c \text{ or } (i, j) \in \mathbf{A}^c \times \mathbf{A} .$$

That is, it is not possible for x_t to leave or to enter **A**. Such a set is said to be **closed.** Then

$$\pi_A(i) := \frac{\pi(i)\, 1\{i \in \mathbf{A}\}}{\pi(\mathbf{A})}, \quad \text{with } \pi(\mathbf{A}) = \sum_{j \in \mathbf{A}} \pi(j) \qquad (2.20)$$

is invariant for Q.

The proof is straightforward since

$$\sum_{j \in \mathbf{A}} \pi(j)\, q(j, i) = \begin{cases} \sum_{j \in \mathbf{X}} \pi(j)\, q(j, i) = 0, & \text{for } i \in \mathbf{A} \\ 0, \text{ for } i \in \mathbf{A}^c \end{cases}$$

and therefore, $\sum_{j \in X} \pi_A(j)\, q(j, i) = 0$ for all $i \in X$.

Example 2.5.8 [Nonopen Jackson networks] Going back to the second network of Figure 2.8, one finds that for any $N \geq 1$, the set

$$\mathbf{X}_N = \{x = (x^1, x^2, x^3) \mid x^i \geq 0, \quad 1 \leq i \leq 3 \text{ and } |x| = x^1 + x^2 + x^3 = N\}$$

is closed. Therefore, an invariant distribution concentrated on $\mathbf{X}_N$ is given by (2.20) with $\mathbf{A} = \mathbf{X}_N$. It has the form

$$\pi_{X_N}(x) = G(N)(\frac{3b}{\mu_1})^{x^1}(\frac{3b}{\mu_2})^{x^2}(\frac{2b}{\mu_3})^{x^3}, \quad x \in \mathbf{X}_N$$

where $G(N)$ is the normalizing constant such that $\pi_{X_N}(x)$ sums to one over $\mathbf{X}_N$.

At first glance, π_{X_N} depends on b. This is not so, however. Indeed, since $|x| = N$, the factor b^N can be incorporated in $G(N)$, which is then uniquely determined by the normalizing condition. That is, π_{X_N} is given by

$$\pi_{X_N}(x) = G(N)(\frac{3}{\mu_1})^{x^1}(\frac{3}{\mu_2})^{x^2}(\frac{2}{\mu_3})^{x^3}, \quad x \in \mathbf{X}_N$$

where $G(N)$ is the normalizing constant.

In the case of the third network of Figure 2.8, the set

$$\mathbf{Y}_N = \{x = (x^1, x^2, x^3) \mid x^i \geq 0, \quad 1 \leq i \leq 3 \text{ and } x^2 + x^3 = N\}$$

is closed. As in the previous example, one finds that the invariant distribution on $\mathbf{Y}_N$ is given by

$$\pi_{Y_N}(x) = K(N) \left(\frac{a}{\mu_1}\right)^{x^1} \left(\frac{1}{\mu_2}\right)^{x^2} \left(\frac{1}{\mu_3}\right)^{x^3}$$

where $K(N)$ is such that π_{Y_N} sums to one on $\mathbf{Y}_N$.

The next issue is that of **uniqueness** of the stationary distribution, and also the relevance of that distribution.

2.6. IRREDUCIBILITY AND ERGODICITY

Definition 2.6.1 [Irreducibility] Let $\mathbf{x} := \{x_t, \quad t \geq 0\}$ be a regular Markov chain. It is **irreducible** if it is possible for x_t to go from every $i \in \mathbf{X}$ to every other $j \in \mathbf{X}$.

For instance, the M/M/1 queue is irreducible on $\{0, 1, 2, \ldots\}$. The first Jackson network of Example 2.5.6 is irreducible on $\mathbf{X} = \{0, 1, 2, \ldots\}^3$. The second one is irreducible on each $\mathbf{X}_N$. The third is irreducible on each $\mathbf{Y}_N$.

The importance of this concept lies in the next theorem.

Theorem 2.6.2 [Ergodicity] Let $\mathbf{x} := \{x_t, \quad t \geq 0\}$ be an irreducible regular Markov chain.

(a) The Markov chain has **at most one** invariant distribution.

(b) If it has an invariant distribution π (which is certainly the case if X is finite), then $\mathbf{x}$ is **asymptotically stationary,** i.e., such that

$$\lim_{t \to \infty} P_t(i, j) = \pi(j), \quad \text{for all } i, j \in \mathbf{X} \tag{2.21}$$

and, for any initial distribution π_0,

$$\lim_{t \to \infty} \pi_t = \pi . \tag{2.22}$$

Moreover, $\mathbf{x}$ is **ergodic,** i.e., such that, almost surely,

$$\lim_{t \to \infty} \frac{1}{t} \int_0^t f(x_s)\, ds = \sum_i \pi(i) f(i) \tag{2.23}$$

for any $f : \mathbf{X} \to [0, \infty)$ and also for any $f : \mathbf{X} \to (-\infty, +\infty)$

such that $\sum_i \pi(i) \mid f(i) \mid < \infty$.

(c) If the Markov chain does **not** have an invariant distribution, then

$$\lim_{t \to \infty} P_t(i, j) = 0, \quad \text{for all } i, j \in \mathbf{X}, \tag{2.24}$$

and, for any π_0,

$$\lim_{t \to \infty} \frac{1}{t} \int_0^t 1\{x_s = i\}\, ds = 0, \quad \text{for all } i \in \mathbf{X}. \tag{2.25}$$

Examples 2.6.3 Before discussing the meaning of ergodicity, let us comment on the implications of this result on the examples seen so far.

Part (a) of the theorem implies that the stationary distributions obtained in Example 2.5.8 are the only ones for the closed Jackson network on $\mathbf{X}_N$ and for the mixed Jackson network on $\mathbf{Y}_N$.

Part (b) implies that if the closed Jackson network of Example 2.5.6 is started in $\mathbf{X}_N$, then the probability distribution of x_t converges to π_{X_N}. This says that the network approaches "steady state." This asymptotic stationarity result justifies the interest in the invariant distribution. One point should be made, however. The result does not say anything about the rate of convergence to the invariant distribution. It might be that this convergence is slow to the point of being irrelevant in some applications. Similar remarks can be made about the M/M/1 queue and the mixed network. See the references.

Part (c) has implications for the M/M/1 queue and for the open and mixed networks of Example 2.5.6. In the case of the M/M/1 queue, it implies that if $\lambda \geq \mu$, then

$$P\{x_t \leq K\} = \sum_i \sum_{j=0}^{K} \pi_0(i)\, P_t(i, j) \to 0 \text{ as } t \to \infty.$$

That is, the queue length exceeds any given value with a probability that approaches one as t increases. The queue "blows up." This does not quite mean that any realization of $\{x_t, t \geq 0\}$ is such that $x_t \to \infty$, however. In fact, the latter statement is true if $\lambda > \mu$, but it is not if $\lambda = \mu$. Indeed, we will see in Example 2.7.2 that if $\lambda = \mu$, then the queue becomes empty infinitely often. This may seem like a technical point, but it should invite some care in defining the concept of stability of a queueing system. What happens, as suggested by (2.25), is that the fraction of time that the queue length is small is negligible: the average time between two visits to the empty state is infinite. (See Fact 2.7.1.)

Comments 2.6.4 [Ergodicity] The concept of ergodicity is both subtle and essential. Its intuitive meaning is that a process is ergodic if all its possible realizations behave similarly. It is then possible to infer the statistical properties of the "ensemble" of realizations by observing any one of them. This has obvious implications in the context of simulations, for instance.

A stochastic process is **ergodic** if any property that does not depend on the **time origin** is either always true or always false. For instance, if a queue length process is ergodic, then either all its realizations become empty infinitely often, or none of its possible realizations becomes empty infinitely often. As another example, any deterministic process is ergodic (trivially, since it has only one realization).

It is known that if a process $\mathbf{x} := \{x_t, \quad t \geq 0\}$ is **ergodic and stationary,** then

$$\lim_{t \to \infty} \frac{1}{t} \int_0^t f(\mathbf{x}' + s) \; ds \;=\; E\{f(\mathbf{x})\} \text{ for all possible realizations } \mathbf{x}'$$

where $\mathbf{x}' + s := \{x'_{s+u}, \; u \geq 0\}$ and for any nonnegative function f. This is **Birkhoff's individual ergodic theorem.** Those issues will be reexamined in Chapter Seven.

Example 2.6.5 Consider the Markov chain with the transition diagram of Figure 2.9.

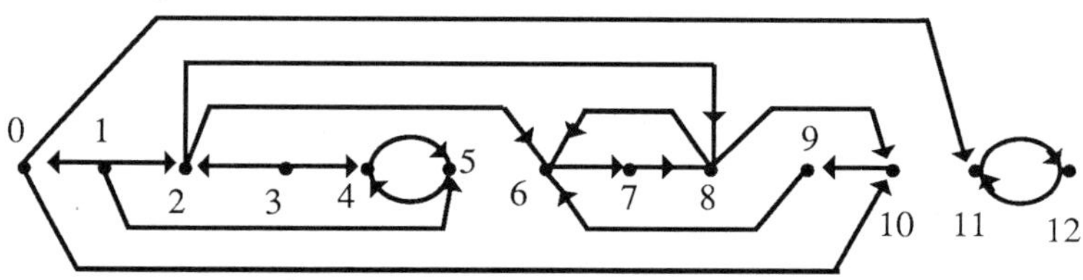

Figure 2.9. Transition diagram.

In this diagram, an arrow from a state to another indicates that the corresponding entry of the rate matrix is positive.

The Markov chain with this transition diagram will spend a finite time in the subset $T = \{0, 1, 2, 3\}$, irrespective of the initial distribution. These states are said to be **transient.** The other states can be partitioned into subsets that do not communicate: $C_1 = \{4, 5\}$, $C_2 = \{6, 7, 8, 9, 10\}$, and $C_3 = \{11, 12\}$. The Markov chain is irreducible on each C_i, so that a unique invariant distribution π_i concentrated on C_i can be found according to Theorem 2.6.2 for $i = 1, 2, 3$. Hence $\pi_i Q = 0$ for $i = 1, 2, 3$. As a consequence, any **convex combination** $\pi = a_1\pi_1 + a_2\pi_2 + a_3\pi_3$ (with $a_i \geq 0$ and $a_1 + a_2 + a_3 = 1$) is again invariant (such that $\pi Q = 0$). One can check that all the invariant distributions are of that form.

This Markov chain has an infinite number of invariant distribution. Given an initial distribution π_0, one can see that π_t will converge to a convex combination $\sum a_i \pi_i$ of the π_i. The coefficients a_is will depend on π_0. For instance, one finds that (see Figure 2.9)

$$a_1 = \pi_0(1)\, q(1, 5)\, q(1)^{-1} + \pi_0(3)\, q(3, 4)\, q(3)^{-1} + \pi_0(4) + \pi_0(5)\,.$$

The Markov chain is stationary and ergodic only if the initial distribution is one of the π_i. Indeed, the property that the Markov chain spends an infinite time in the set C_j does not depend on the time origin. For the probability of that property to be zero or one for each j it is necessary for π_0 to be one of the π_i.

This shows that the invariant distributions under which the Markov chain is ergodic are the **extreme points** of the convex set of stationary distributions. This is a general result about stationary ergodic distributions of stochastic processes.

Notice that if $x_0 \equiv 4$, then the Markov chain is also ergodic, even though it is not stationary.

2.7. TRANSIENCE AND RECURRENCE

How often is a given state visited by a Markov chain? For instance, is a queue empty infinitely often? The answer is discussed in this section.

Fact 2.7.1 Let $\mathbf{x} := \{x_t,\ \ t \geq 0\}$ be a regular Markov chain. One can partition $\mathbf{X}$ as (below, + denotes the union of disjoint sets)

$$\mathbf{X} = T + NR + PR$$

where T [resp. NR, PR] denotes the set of states that are **transient** [resp. **null recurrent, positive recurrent**]. State $i \in \mathbf{X}$ is transient if, when $x_0 = i$, the Markov chain spends at most a finite time in i (i.e., goes back to i at most finitely many times). It is **recurrent** if it is not transient, i.e., if the Markov chain goes back to i with probability 1 after having left that state. It is then **null** [resp. **positive**] **recurrent** if the average return time to i is infinite [resp. finite].

It should be clear that *if a Markov chain is irreducible, then all the states are of the same type* (i.e., transient, null recurrent, or positive recurrent). The **Markov chain is then** said to be **of the corresponding type.**

For instance, let us show that if **x** is irreducible and if a state j is recurrent, then any other state i must also be recurrent. Since j is recurrent and since there is a positive probability α of going from j to i (by irreducibility), it must be that the probability of going from i to j (eventually) is one. (For otherwise, j would not be recurrent.) Say that $x_0 = i$. Eventually, $x_t = j$, after which time j is visited infinitely often, and after each visit there is a probability α of going to i. This implies that the number of visits to i is infinite, as was to be shown.

It can be shown that if an irreducible Markov chain is positive recurrent, with average intervisit times $a(i)$ for state $i \in X$, then its unique invariant distribution is given by

$$\pi(i) = \frac{1}{1 + a(i)\, q(i)}, \quad i \in X . \tag{2.26}$$

One way to see (2.26) is to notice that the Markov chain spends $a(i)$ time units away from i for every $q(i)^{-1}$ time units that it spends in i, on the average. The fraction of time spent in i is then $q(i)^{-1}[a(i) + q(i)^{-1}]^{-1}$, which agrees with (2.26).

Example 2.7.2 [M/M/1 queue] Let us show that the M/M/1 queue is transient for $\lambda > \mu$ and null recurrent for $\lambda = \mu$. The main objective of this example is to introduce the **first-step analysis.**

First, consider the case $\lambda = \mu$. Define

$$\alpha(i) = P[x_t = 0 \text{ for some } t < \infty \mid x_0 = i], \quad i \geq 0 .$$

Then $\alpha(0) = 1$. For $i > 0$, the Markov chain spends a finite time in i and then jumps to $i + 1$ with probability $\dfrac{\lambda}{\lambda + \mu} = \dfrac{1}{2}$ or to $i - 1$ otherwise. Therefore, the probability of going from $i > 0$ to 0 in finite time is equal to $\dfrac{1}{2}$ times the probability of going from $i - 1$ to 0 in finite time, plus $\dfrac{1}{2}$ times the probability of going from $i + 1$ to 0 in finite time. Hence,

$$\alpha(i) = \frac{1}{2}\alpha(i-1) + \frac{1}{2}\alpha(i+1), \quad i \geq 1 . \tag{2.27}$$

The solution of this equation with $\alpha(0) = 1$ is seen to be $\alpha(i) \equiv 1$. (For instance, the above equation shows that $(i, \alpha(i))$ is a straight line. Since it goes through (0, 1) and since the values of $\alpha(i)$ are in [0, 1], it must be that $\alpha(i) \equiv 1$.)

This shows that x_t goes to 0 in finite time from any initial state. Similarly, x_t goes from 0 to any state i in finite time. Hence, any state will be visited infinitely often. Thus, the Markov chain is recurrent. It cannot be positive recurrent, for this would imply the existence of an invariant distribution (2.26), which would contradict the previously made observation that the balance equations are not solved by a probability distribution when $\lambda = \mu$. (See Example 2.5.4.)

The equations (2.27) were derived by considering the first transition of the Markov chain. This procedure is called **first-step analysis** (or **backward method**). The equations are called the **first-step equations,** or the **backward equations.**

Second, consider the case $\lambda > \mu$. Let us show that the Markov chain is transient by first-step analysis. A computationally simpler proof will be given later. Fix $a > 0$ in **X** and define

$$\alpha(i; a) = P[x_t \text{ reaches } 0 \text{ before } a \mid x_0 = i], \quad \text{for } 0 \leq i \leq a .$$

The first-step equations are

$$\alpha(i; a) = q\alpha(i-1; a) + p\alpha(i+1; a), \quad 0 < i < a$$

with $p = 1 - q = \lambda(\lambda + \mu)^{-1}$. The solution, with the boundary conditions $\alpha(0; a) = 1$ and $\alpha(a; a) = 0$, can be checked to be given by

$$\alpha(i; a) = (\rho^{-i} - \rho^{-a})(1 - \rho^{-a})^{-1}, \quad \text{with } \rho = \frac{\lambda}{\mu} .$$

Now,

$$\alpha := \lim_{a \to \infty} \alpha(1; a) = \rho^{-1} < 1$$

is the probability of reaching 0 in finite time, starting from 1. Indeed,

$$\alpha = \lim_{a \to \infty} P[\tau_0 < \tau_a \mid x_0 = 1]$$

where $\tau_j := \min\{n \geq 0 \mid x_n = j\}$ for $j \geq 0$. Also, $\tau_a \to \infty$ as $a \to \infty$. Therefore, by the Dominated Convergence Theorem (Appendix A),

$$\begin{aligned} \alpha &= \lim_{a \to \infty} E[1\{\tau_0 < \tau_a\} \mid x_0 = 1] = E[\lim_{a \to \infty} 1\{\tau_0 < \tau_a\} \mid x_0 = 1] \\ &= E[1\{\tau_0 < \infty\} \mid x_0 = 1] = P[\tau_0 < \infty \mid x_0 = 1] , \end{aligned}$$

as claimed.

Therefore, the probability of returning to 0 after leaving 0 is also α. Hence, the probability of returning to 0 exactly $n \geq 0$ times is equal to $\alpha^n(1 - \alpha)$. This shows that the average number of returns to 0 is given by $\sum n\alpha^n(1 - \alpha) = \alpha(1 - \alpha)^{-1}$. Since this average number is finite, it follows that the number of returns to 0 is finite, with probability one. Hence, 0 is transient. Since the Markov chain is irreducible, it must be that all the states are transient, so that the M/M/1 queue is transient for $\lambda > \mu$.

Finally, let us give a computationally simpler proof of the transience of $\mathbf{x}$ when $\lambda > \mu$. Denote by ξ_n the queue length after the nth transition ($n \geq 1$) and let $\xi_0 = x_0$. One can write

$$\xi_{n+1} = \max\{\xi_n + v_n, 0\}, \quad n \geq 0$$

where the v_n are i.i.d with

$$P\{v_n = +1\} = 1 - q = 1 - P\{v_n = -1\} = p = \lambda(\lambda + \mu)^{-1}.$$

It follows that $\xi_n \geq \xi_0 + v_1 + v_2 + \cdots + v_n$ for $n \geq 1$, so that

$$\frac{\xi_n}{n} \geq \frac{v_1 + v_2 + \cdots + v_n}{n}, \quad n \geq 1.$$

But, by the **strong law of large numbers** (see Appendix A), the right-hand side of the above expression converges to $E\{v_n\} = p - q > 0$. Hence, $\xi_n \to \infty$, so that $x_t \to \infty$, with probability one. This shows that every state will be visited at most finitely often, i.e., that the M/M/1 queue is transient for $\lambda > \mu$.

Example 2.7.3 [Busy cycles of Jackson networks] Consider an open Jackson network (see Example 2.5.6) and assume that it is irreducible, that it has 40 nodes, and that $\rho_i = 0.5$ for every queue. This corresponds to the invariant distribution

$$\pi(x) = \prod_{i=1}^{40} \frac{1}{2}(\frac{1}{2})^{x^i}, \quad x \in \mathbf{X} = \{0, 1, \ldots\}^{40}. \tag{2.28}$$

The average queue length in queue i is given by $\sum n\frac{1}{2}(\frac{1}{2})^n = 1$. The average number of customers in the network is therefore equal to 40.

Let us determine the average return time to the empty state. According to (2.26), this average time is equal to

$$a(\mathbf{0}) = \frac{1 - \pi(\mathbf{0})}{\pi(\mathbf{0})\, q(\mathbf{0})}$$

where $\mathbf{0}$ denotes the empty state, i.e., the vector $(0, 0, \ldots, 0)$. Using (2.28) one finds that $\pi(\mathbf{0}) = 2^{-40}$. Also, $q(\mathbf{0}) = \sum_{i=1}^{40} \gamma_i := \gamma$, the total arrival rate into the network. Hence,

$$a(\mathbf{0}) \approx \gamma^{-1}\, 2^{40} \approx \gamma^{-1} \times 1.1 \times 10^{12} .$$

Consider a **communication network** application (assuming that the model is appropriate, which it almost is), with $\gamma = 200$ arrivals per seconds. Then $a(\mathbf{0}) \approx 5.5 \times 10^9$ seconds, i.e., about **174 years!**

The relevance of this number is that each time the network becomes empty it starts a new "cycle" and all the cycles are i.i.d. It would then be possible to get estimates about functionals of the cycles with known confidence intervals, by using the statistical properties of sequences of i.i.d random variables. This idea is called **regenerative simulation** in the context of simulation experiments. The fact that a network even of a modest size and a relatively light utilization takes so long to regenerate shows that this idea is unlikely to be practical for the investigation of communication networks, at least in its brute force application.

(The estimate of 200 arrivals per second was obtained by assuming that the data packets were of 10^4 bits and arrived into the network on 40 links, each having a transmission rate of 50×10^3 bits per second. These numbers are not unrealistic.)

Example 2.7.4 [Ergodicity of Jackson networks] Consider the Jackson network defined in Example 2.4.4. One might think that the network is irreducible only if it is possible for every customer to visit every queue. This is false, as the example of two queues in parallel shows. In fact, one sees that a Jackson network is irreducible if and only it is open. The formula (2.19) shows that an open network is ergodic if and only if $\lambda_i < \mu_i$ for each queue i, where the λ_i are the solutions of (2.18).

2.8. TIME REVERSAL

Time reversal turns out to be a very fruitful idea in the context of queueing networks. The basic observation is that if a Markov chain is stationary, then it also behaves as a Markov chain when its evolution is observed in **reversed time** (running the movie backward).

This leads to a simple but extremely useful verification lemma (Kelly's Lemma 2.8.5) that permits us to guess and to verify the invariant distribution of certain Markov chains. This idea is one of the major tools of the field. Its importance cannot be overemphasized.

Theorem 2.8.1 [Time Reversal] Let $\mathbf{x} := \{x_t, t \geq 0\}$ be a stationary Markov chain with initial distribution $\pi_0 = \pi$ and with rate matrix Q. We will write this as $\mathbf{x} = (\pi, Q)$. Then, for all $T > 0$, $\{x_{T-t}, 0 \leq t \leq T\}$ has the same law as a Markov chain $(\pi, \tilde{Q})$ on $[0, T]$ where $\tilde{Q} = [\tilde{q}(i, j), i, j \in \mathbf{X}]$ with

$$\tilde{q}(i, j) := \frac{\pi(j)\, q(j, i)}{\pi(i)}, \quad i, j \in \mathbf{X}. \tag{2.29}$$

In particular, the Markov chain has the same law when it is reversed in time if and only if $\tilde{Q} = Q$, i.e., if and only if

$$\pi(i)\, q(i, j) = \pi(j)\, q(j, i), \quad i, j \in \mathbf{X}. \tag{2.30}$$

In the latter case, $\mathbf{x}$ is said to be **time-reversible.**

Before proving this result, let us see how it applies to the M/M/1 queue and to Jackson networks.

Fact 2.8.2 [Time-reversibility of the M/M/1 queue] A stationary M/M/1 queue is time-reversible. In particular, it follows that the *output of a stationary M/M/1 queue is Poisson and that the departure times up to any time t are independent of the queue length at time t*. (For the departure times up to time t are the arrival times after time t for the process reversed in time.) This statement is called the **Output Theorem** for the M/M/1 queue.

To see this, one must check that

$$\begin{aligned} \pi(i) &= \rho^i (1 - \rho) \text{ for } i \geq 0 \text{ with } \rho = \lambda\mu^{-1} \text{ and} \\ q(i, j) &= \lambda 1\{j = i + 1\} + \mu 1\{j = i - 1\} \\ &\qquad - (\lambda + \mu 1\{i > 0\})\, 1\{j = i\} \end{aligned}$$

for $i, j \geq 0$ satisfy (2.30). This is obvious for $|j - i| > 1$ and for $i = j$. It suffices to verify it for $j = i + 1$. That is, one must check that

$$\pi(i)\, \lambda = \pi(i + 1)\mu, \quad i \geq 0.$$

But this follows from the form of π.

Remark 2.8.3 [Trees] Say that **the graph of a Markov chain** (π, Q) **is a tree** if $q(i, j) > 0$ whenever $q(j, i) > 0$ and if the set $G = \{(i, j) \in \mathbf{X}^2 \mid q(i, j) > 0\}$ is a tree, with the obvious meaning of the term. In that case, the Markov chain **is time-reversible when it is stationary.**

Indeed, to check (2.30), observe that to any pair $(i, j) \in G$ corresponds a set $\mathbf{A}$ such that $i \in \mathbf{A}$, $j \in \mathbf{A}^c$, and (i, j) is the only pair (k, l) in $\mathbf{A} \times \mathbf{A}^c$ such that $q(k, l) > 0$. (See Figure 2.10.)

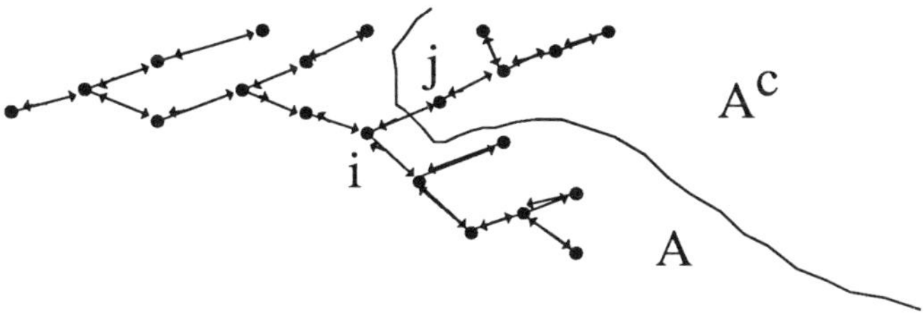

Figure 2.10. Tree criterion.

The equalities (2.30) then follow from Fact 2.5.5.

The time-reversibility of the stationary M/M/1 queue is a particular case of this **tree criterion.** Another example is that of a queue with i.i.d. exponential service times in which the server works at rate $\mu(i)$ when there are i customers in the queue. Such a queue is sometimes denoted by **M/M/(.).** The transition diagram of the queue length is as indicated in Figure 2.11 and the time-reversibility follows from the fact that it is a tree.

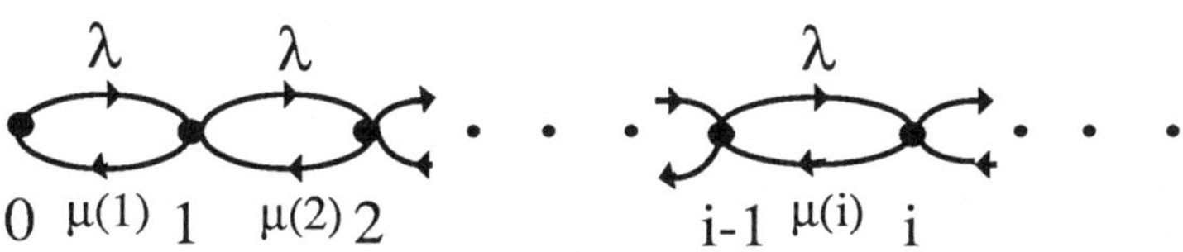

Figure 2.11. Transition diagram of the M/M/(.) queue.

Fact 2.8.4 [Time reversal of Jackson networks] Consider once again the Jackson network of Example 2.4.4. Assume that it has at least one invariant distribution π, which is then of the form (2.19) where the λ_i solve (2.18).

The claim is that *the stationary network reversed in time is again a Jackson network* with the same service rates in the queues. Moreover, its arrival rates γ'_i and routing probabilities r'_{ij} are given by

$$r'_{ij} = \frac{\lambda_j r_{ji}}{\lambda_i}, \; r'_{i0} = \frac{\gamma_i}{\lambda_i}, \; \gamma'_i = \lambda_i r_{i0} \qquad (2.31)$$

for $i \le i, j \le J$. (See Figure 2.12.) In particular, *the departure processes from a stationary Jackson network are independent Poisson processes and their past up to time t is independent of x_t.* This result is the **Output Theorem** for Jackson networks.

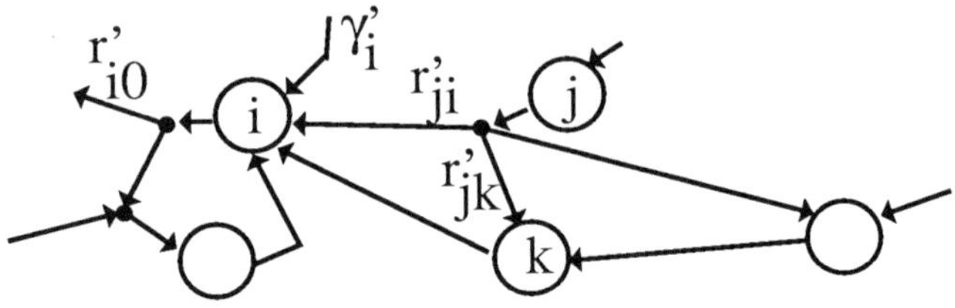

Figure 2.12. The reversed Jackson network.

To verify this claim, denote by Q' the rate matrix of the Jackson network with the above parameters, and by Q the rate matrix of the original Jackson network. We will show that

$$\pi(x)\, q(x, y) \;=\; \pi(y)\, q'(y, x), \quad x, y \in X \tag{2.32}$$

with π given by (2.19).

Using the notation of (2.7)–(2.8), one will show that

$$\pi(x)\, q(x, A_m x) \;=\; \pi(A_m x)\, q'(A_m x, x)\,. \tag{2.33}$$

The other cases of pairs (x, y) with $q(x, y) > 0$ can be checked similarly. The details are left as an exercise.

From (2.7) one has $q(x, A_m x) = \gamma_m$. For the modified network, the transition from $y = A_m x$ to x corresponds to a departure from queue m. That is, $x = D_m y$. Applying (2.7)–(2.8) to this pair, for the modified network, then gives

$$q'(A_m x, x) \;=\; \mu_m r'_{m0}\, 1\{(A_m x)^m > 0\} \;=\; \mu_m \frac{\gamma_m}{\lambda_m}\,.$$

Equation (2.33) then reads

$$\pi(x)\gamma_m \;=\; \pi(A_m x)\mu_m \frac{\gamma_m}{\lambda_m}$$

which is satisfied, in view of the form (2.19) of π.

In fact, the above verification also shows that $\pi Q = 0$. Indeed, this follows from (2.32) by summing that equation over x and using the fact that Q' is a rate matrix, so that $\sum_x q'(y, x) = 0$ for $y \in \mathbf{X}$. We have therefore proved the statements of Example 2.5.6.

The above method for proving the invariance of a distribution is known as Kelly's Lemma.

Lemma 2.8.5 [Kelly's Lemma] Assume that π is a distribution and Q, Q' are two rate matrices on $\mathbf{X}$ such that

$$\pi(x)\, q(x,y) \;=\; \pi(y)\, q'(y,x), \;\; x, y \in \mathbf{X}\,. \tag{2.34}$$

Then $\pi Q = 0$ and $Q' = \tilde{Q}$, the rate matrix of (π, Q) reversed in time.

We now turn to the proof of Theorem 2.8.1.

Proof of Theorem 2.8.1 Denote by P_t the transition probabilities of $\mathbf{x} := \{x_t,\;\; t \geq 0\}$. The claim is that the process $\mathbf{x}$ reversed in time is a Markov chain with transition probabilities $\tilde{P}_t$ given by

$$\tilde{P}_t(i, j) \;=\; \frac{\pi(j)\, P_t(j, i)}{\pi(i)}\,. \tag{2.35}$$

The equation (2.29) then follows by differentiating (2.35) at $t = 0$.

To see why the claim is correct, let $0 < t_1 < \cdots < t_n < T$ and $\tilde{x}_{t_i} = x_{T-t_i}$ for $1 \leq i \leq n$. Then

$$P\{\tilde{x}_{t_1} = i_1, \ldots, \tilde{x}_{t_n} = i_n\} \;=\; P\{x_{T-t_n} = i_n, \ldots, x_{T-t_1} = i_1\}$$

$$= \pi(i_n)\, P_{t_n - t_{n-1}}(i_n, i_{n-1})\, P_{t_{n-1} - t_{n-2}}(i_{n-1}, i_{n-2}) \ldots P_{t_2 - t_1}(i_2, i_1)$$

$$= \tilde{P}_{t_n - t_{n-1}}(i_{n-1}, i_n)\, \pi(i_{n-1})\, P_{t_{n-1} - t_{n-2}}(i_{n-1}, i_{n-2}) \ldots P_{t_2 - t_1}(i_2, i_1)$$

$$= \ldots$$

$$= \tilde{P}_{t_n - t_{n-1}}(i_{n-1}, i_n)\, \tilde{P}_{t_{n-2} - t_{n-1}}(i_{n-2}, i_{n-1}) \ldots \tilde{P}_{t_2 - t_1}(i_1, i_2)\, \pi(i_1)$$

where the last equalities follow from the definition (2.35). This shows that $\tilde{x}_t$ behaves as a Markov chain $(\pi, \tilde{Q})$, as claimed.

Comments 2.8.6 The formula (2.29) expresses that the average rate of transitions from i to j by a Markov chain is equal to the average rate of transitions from j to i by the Markov chain reversed in time. Indeed, if $\mathbf{x}$ is the stationary Markov chain (π, Q), then

$$P\{x_t = i, x_{t+\epsilon} = j\} \;=\; \pi(i)\, P_\epsilon(i, j) \;=\; \pi(i)\, q(i, j)\, \epsilon + o(\epsilon)$$

for $j \neq i$, and similarly for $\mathbf{x}$ reversed in time.

The formulas (2.31) express that the average rate of flow of customers from queue i to queue j in the Jackson network is equal to the rate of flow from queue j to queue i in the reversed network, for $0 \leq i \neq j \leq J$ (with queue 0 being identified with the outside of the network).

For instance, with $1 \le i \ne j \le J$, the rate of customers from i to j in the original network is $\lambda_i r_{ij}$ (the average rate λ_i of flow through queue i times the probability r_{ij} that a customer goes to j when leaving i). For the reversed network, the rate from j to i is $\lambda_j r'_{ji}$.

2.9. DISCRETE-TIME MARKOV CHAINS

Before discussing the jumps of continuous-time Markov chains we need some results about discrete-time Markov chains. The ideas are similar to their continuous-time counterparts. On a first reading, sections 2.9.6–2.9.11 can be skipped.

Definitions 2.9.1 Let $\mathbf{X}$ be a countable set and $P = [P(i,j), i, j \in \mathbf{X}]$ be a matrix with nonnegative entries and such that $\sum_j P(i,j) = 1$ for all $i \in \mathbf{X}$. Such a matrix is called a **stochastic matrix** on $\mathbf{X}$.

The sequence $\{x_n, n \ge 0\}$ is a **Markov chain with transition matrix** P if

$$P[x_{n+1} = j \mid x_m, m \le n-1, x_n = i] = P(i,j), \quad i, j \in \mathbf{X}, \; n \ge 0. \tag{2.36}$$

These equations state that the evolution of x_m starts afresh from x_n at time n. The definition is summarized by the transition diagram of Figure 2.13 where an arrow from i to j is marked with the value of $P(i,j)$.

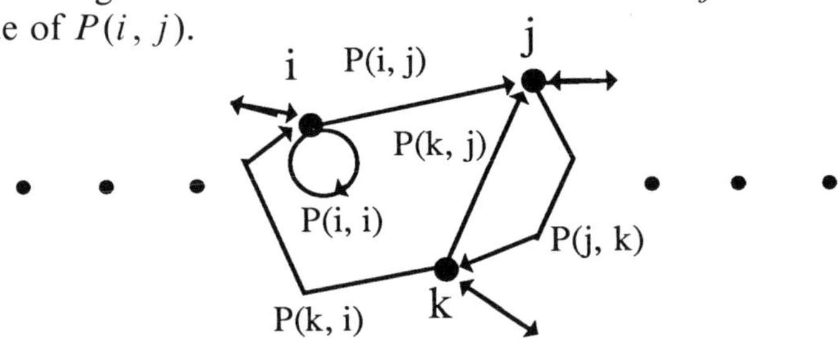

Figure 2.13. State transition diagram of discrete-time chain.

Invariant distribution 2.9.2 A probability distribution π on $\mathbf{X}$ is **invariant** for P if $\pi P = \pi$. It is immediate that $\pi_n(i) := P\{x_n = i\}$ is such that $\pi_n \equiv \pi_0$ if and only if π_0 is invariant.

Irreducibility-ergodicity 2.9.3 The Markov chain $\mathbf{x} := \{x_n, \quad n \geq 0\}$ is **irreducible** if every state can be reached from every other state (not necessarily in one step).

An irreducible Markov chain has **at most one** invariant distribution. It is either **transient, null recurrent,** or **positive recurrent,** defined as in continuous-time. It is positive recurrent if and only if it admits one invariant distribution; this is certainly the case if $\mathbf{X}$ is finite.

If $\mathbf{x} := \{x_n, \quad n \geq 0\}$ is irreducible, then one can show that $d := \gcd\{n > 0 \mid P^n(i, i) > 0\}$ does not depend on $i \in \mathbf{X}$. In this expression, "gcd" denotes the **greatest common divisor** and P^n is the nth power of the matrix P. For instance, $\gcd\{3, 7, 13, 14, 15, 16, \ldots\} = 1$ and $\gcd\{3, 6, 9, 12, \ldots\} = 3$. The value of d is called the **period** of the Markov chain if $d > 1$. The Markov chain is **aperiodic** if $d = 1$. The first two Markov chains in Figure 2.14 are aperiodic; the third has period 3.

Figure 2.14. Three Markov chains.

For an irreducible and aperiodic Markov chain, one has $P^n(i, j) \to \pi(j)$ as $n \to \infty$ for all $i, j \in \mathbf{X}$ where $\pi(j) \equiv 0$ unless the Markov chain is positive recurrent. If the Markov chain is irreducible, aperiodic, and positive recurrent, then it is **ergodic and asymptotically stationary** (defined as in continuous time). Thus, in that case,

$$\lim_{n \to \infty} P^n(i, j) = \pi(j) \quad \text{and}$$

$$\lim_{n \to \infty} \frac{1}{n} \sum_{m=0}^{n-1} f(x_m) = \sum_l \pi(l) f(l) \tag{2.37}$$

for all $i, j \in \mathbf{X}$ and for all f that is either nonnegative or such that $\sum \pi(i) \mid f(i) \mid < \infty$. (Here, π is the unique invariant distribution.)

If the Markov chain $\mathbf{x} := \{x_n, \quad n \geq 0\}$ is irreducible and has period d, then it is possible to decompose $\mathbf{X}$ as $\mathbf{X} = \mathbf{X}_1 + \mathbf{X}_2 + \cdots + \mathbf{X}_d$ in such a way that if $x_n \in \mathbf{X}_i$, then

$x_{n+1} \in \mathbf{X}_j$ with $j = i \bmod d + 1$. The Markov chain $\{x_{nd}, n \geq 0\}$ is then irreducible and aperiodic on each $\mathbf{X}_i$.

Example 2.9.4 [Jumps of $\mathbf{x_t}$] Let $\mathbf{x} := \{x_t, \ t \geq 0\}$ be a continuous-time Markov chain with rate matrix Q and invariant distribution π. Denote by ξ_n the value of x_t after its nth jump and let $\xi_0 = x_0$. We know that $\{\xi_n, n \geq 0\}$ is a Markov chain with transition probabilities $\Gamma(i, j) = q(i, j)\, q(i)^{-1}\, 1\{i \neq j\}$, by definition of x_t. See Definition 2.4.2.

Assume that

$$\beta := \sum_{i \in \mathbf{X}} \pi(i)\, q(i) < \infty . \tag{2.38}$$

The claim is that

$$\nu(i) = \beta^{-1} \pi(i)\, q(i), \; i \in \mathbf{X} \tag{2.39}$$

is an invariant distribution for the **jump chain** $\{\xi_n, n \geq 0\}$. To see this, one must check that $\nu\Gamma = \nu$. This verification is left to the reader.

Example 2.9.5 [Transitions of $\mathbf{x_t}$] Let x_t and ξ_n be as in Example 2.9.4. Define $\mathbf{z} = \{z_n = (\xi_n, \xi_{n+1}), n \geq 0\}$.

A moment of reflection shows that $\mathbf{z}$ is a Markov chain on $\mathbf{Z} = \{(i, j) \in X^2 \mid i \neq j\}$. Its transition matrix is denoted by M and is given by

$$M(ij, kl) = P[z_n = (\xi_{n+1}, \xi_{n+2}) = (k, l) \mid z_n = (\xi_n, \xi_{n+1}) = (i, j)]$$
$$= \Gamma(k, l)\, 1\{j = k\}, \quad \text{for all } (i, j), (k, l) \in \mathbf{Z} .$$

The claim is that if (2.38) holds, then

$$\eta(i, j) := \beta^{-1} \pi(i)\, q(i, j), \quad (i, j) \in \mathbf{Z} \tag{2.40}$$

is an invariant distribution for $\mathbf{z}$. To see this, one must verify that $\eta M = \eta$. This is left to the reader.

Random times 2.9.6 Let $\mathbf{x} := \{x_n, \ n \geq 0\}$ be a Markov chain with transition matrix P. Then, by definition, (2.36) holds for any fixed time n. It is natural to ask whether this property continues to hold when n is replaced by a **random time** (i.e., a random variable that takes values in $\{0, 1, 2, \ldots, \infty\}$).

The answer to this question is **negative,** in general, as the following example shows. (See Figure 2.15.)

Let $\mathbf{X} = \{0, 1\}$ and

$$P(0, 1) = 1 - P(0, 0) = a, \quad P(1, 0) = 1 - P(1, 1) = b$$

with $a, b \in (0, 1)$.

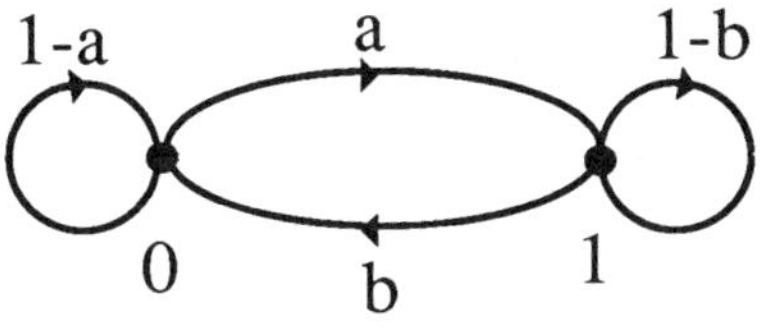

Figure 2.15. The Markov chain of the example.

Let $\tau := \min\{n > 0 \mid x_n = 0\} - 1$. Then, for any initial distribution π_0, one finds

$$P[x_{\tau+1} = 0 \mid x_m, m < \tau; x_\tau = 1] = 1 \neq P(1, 0) = b ,$$

so that (2.36) does not hold when n is replaced by the random time τ. What happens in this example is that one knows already by time τ that $x_{\tau+1} = 0$. Thus, the value of τ is determined by the future evolution of the Markov chain. That is, τ is anticipative.

One suspects that (2.36) will hold if n is replaced by **nonanticipative** random times. That is correct, with a suitable definition of those random times, called **stopping times.**

Definition 2.9.7 [Stopping times] A random time τ is a **stopping time** of $\mathbf{x} := \{x_n, \quad n \geq 0\}$ if

$$P[\tau = n \mid x_m, m \geq 0] = h_n\{x_0, \ldots, x_n\}, \quad n \geq 0, \tag{2.41}$$

for some functions h_n.

For instance, the mth time that $\mathbf{x} := \{x_n, \quad n \geq 0\}$ takes values in a given subset $\mathbf{Y}$ of $\mathbf{X}$ is a stopping time for any $m \geq 1$. In that case, $h_n\{x_0, \ldots, x_n\}$ takes the value 1 if $x_n \in \mathbf{Y}$ and if $m-1$ of the values $x_0, \ldots, x_{n-1}$ belong to $\mathbf{Y}$, and takes the value 0 otherwise.

As another example, define τ as follows. Every time that x_n belongs to a given subset $\mathbf{Y}$ of $\mathbf{X}$, flip a coin and let τ be the first time that the outcome of the coin toss is "head." This random time is seen to be a stopping time.

Strong Markov Property 2.9.8 Let $\mathbf{x} := \{x_n, \quad n \geq 0\}$ be a Markov chain with transition matrix P and let τ be a stopping time. Then, for any initial distribution π_0,

$$P[x_{\tau+1} = j \mid x_m, m < \tau; x_\tau = i] = P(i, j), \quad i, j \in \mathbf{X}. \qquad (2.42)$$

Proof: Using $P\{A, B, C\} = P\{A\} P[B \mid A] P[C \mid A, B]$ one finds that

$$\begin{aligned} P\{x_{\tau+1} = j, x_\tau = i, x_m, m < \tau\} \\ &= \sum_{n=0}^{\infty} P\{x_{n+1} = j, x_n = i, x_m, \ m < n, \tau = n\} \\ &= \sum_{n=0}^{\infty} P\{x_m, m < n, x_n = i\} P(i, j) h_n(x_m, m \le n) \\ &= \sum_{n=0}^{\infty} P\{x_m, m < n, x_n = i, \tau = n\} P(i, j) \\ &= P(i, j) P\{x_\tau = i, x_m, m < \tau\} \end{aligned}$$

which shows (2.42).

The strong Markov property shows that the evolution of a Markov chain $\mathbf{x} := \{x_n, \ n \ge 0\}$ starts afresh from x_τ at time τ, provided that τ is a stopping time.

The next concept will be used to analyze the distribution of a continuous-time Markov chain at selected jump times.

Markov chain watched in a set 2.9.9 Let $\mathbf{x} := \{x_n, \ n \ge 0\}$ be a Markov chain with transition matrix P and invariant distribution π. Let $\mathbf{Y}$ be a proper subset of $\mathbf{X}$ and define the random sequence $\{y_n, n \ge 0\}$ to be the *successive values taken by* $\mathbf{x}$ *in* $\mathbf{Y}$.

Formally, let

$$\begin{aligned} T_0 &= \min\{m \ge 0 \mid x_m \in \mathbf{Y}\} \ \text{ and} \\ T_{n+1} &= \min\{m > T_n \mid x_m \in \mathbf{Y}\} \ \text{ for } n \ge 0 . \end{aligned}$$

Then

$$y_n := x_{T_n} \ \text{ for } n \ge 0 .$$

The sequence $\mathbf{y} := \{y_n, n \ge 0\}$ is said to be **x watched in the set Y**.

Theorem 2.9.10 The sequence **y** defined in 2.9.9 is a Markov chain on **Y** with the invariant distribution π_Y defined by

$$\pi_Y(i) \;=\; \frac{\pi(i)}{\pi(\mathbf{Y})} 1\{i \in \mathbf{Y}\} . \tag{2.43}$$

Before proving this result, let us look at an example.

Example 2.9.11 Consider the Markov chain illustrated in Figure 2.16 with $\mathbf{X} = \{a, b, c, d\}$ and let $\mathbf{Y} = \{a, b\}$. Define **y** as in 2.9.9.

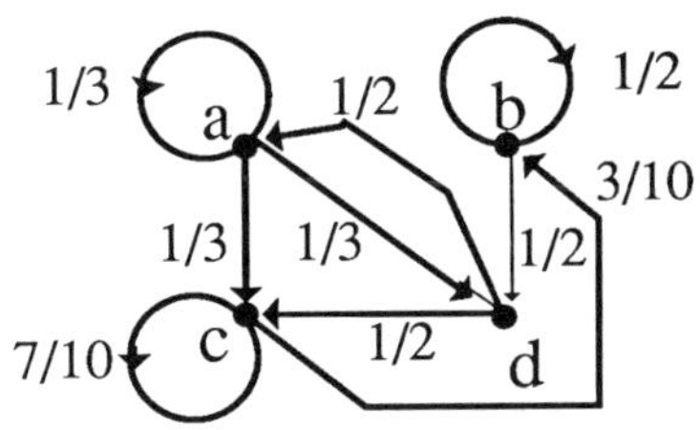

Figure 2.16. A four state Markov chain.

One can check that the invariant distribution of **x** is

$$\pi \;=\; \frac{1}{23}(3, 6, 10, 4) \tag{2.44}$$

The sequence **y** is a Markov chain with the transition diagram indicated in Figure 2.17.

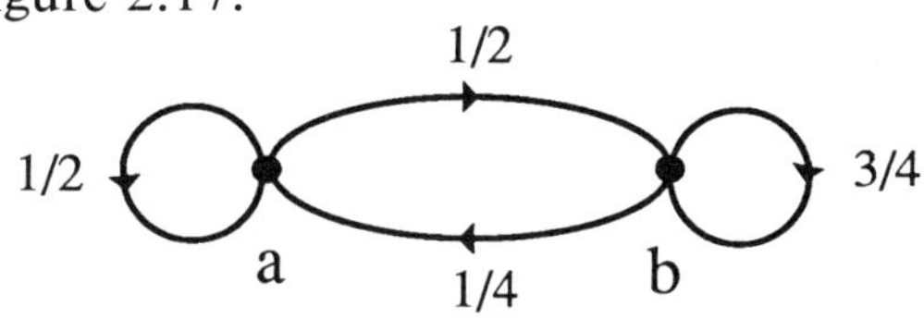

Figure 2.17. The Markov chain watched in **Y**.

Indeed, one has

$$\begin{aligned} P[y_1 = a \mid y_0 = x_0 = a] &= P[x_1 = a \mid x_0 = a] \\ &\quad + P[x_1 = d, x_2 = a \mid x_0 = a] \\ &= \frac{1}{3} + \frac{1}{3} \cdot \frac{1}{2} = \frac{1}{2} \end{aligned}$$

and

$$P[y_1 = a \mid y_0 = x_0 = b] = P[x_1 = d, x_2 = a \mid x_0 = b] = \frac{1}{2} \cdot \frac{1}{2} = \frac{1}{4}.$$

Formulas (2.43)–(2.44) give $\pi_Y = (\frac{1}{3}, \frac{2}{3})$. One can verify that this distribution is indeed invariant for the transition matrix of $\mathbf{y}$, i.e., that

$$(\frac{1}{3}, \frac{2}{3}) = (\frac{1}{3}, \frac{2}{3}) \begin{bmatrix} \frac{1}{2} & \frac{1}{2} \\ \frac{1}{4} & \frac{3}{4} \end{bmatrix}.$$

Proof of Theorem 2.9.10 The first step is to realize that $\mathbf{y}$ is a Markov chain. This follows from the Strong Markov Property and the fact that the T_n are stopping times (see Definition 2.9.7). The fact that the transition probabilities are stationary is easy.

For $n \geq 0$ let y_n [resp. w_n] be the $(n + 1)$st value taken in $\mathbf{Y}$ by $\{x_n, n \geq 0\}$ [resp. by $\{x_n, n \geq 1\}$]. Thus, $w_n = y_m$ with $m = n + 1\{x_0 \in \mathbf{Y}\}$. Now, under the initial distribution π the Markov chain $\mathbf{x}$ is stationary, so that y_n and w_n have the same distribution for all $n \geq 0$. But

$$P\{w_n = i\} = P\{y_{n+1} = i, x_0 \in \mathbf{Y}\} + P\{y_n = i, x_0 \in \mathbf{Y}^c\}$$

and

$$P\{y_n = i\} = P\{y_n = i, x_0 \in \mathbf{Y}\} + P\{y_n = i, x_0 \in \mathbf{Y}^c\}.$$

Comparing these expressions yields, after dividing by $\pi(\mathbf{Y})$,

$$P[y_{n+1} = i \mid x_0 \in \mathbf{Y}] = P[y_n = i \mid x_0 \in \mathbf{Y}].$$

Choosing $n = 0$ shows that this constant distribution is equal to π_Y since $y_0 = x_0$ when $x_0 \in \mathbf{Y}$.

2.10. JUMPS

The results described below are important in the analysis of sojourn times in networks. They will be re-examined in the light of Palm distributions in Chapter Seven. The proof of Theorem 2.10.6 can be skipped on a first reading.

Example 2.10.1 [Arrivals into an M/M/1 queue] Let $\mathbf{x} := \{x_t, \ t \geq 0\}$ be the queue length process of a **stationary** M/M/1 queue with arrival rate λ and service rate μ. Thus, $\rho := \lambda\mu^{-1} < 1$ and the distribution of x_t is given by (2.16).

Denote by S_n the arrival times. We are interested in the sequence $\{x_{S_n-}, n \geq 1\}$ of queue lengths *just before the arrival times.* The motivation is that if $x_{S_n-} = m$, then the customer who arrives at time S_n has to wait for $m + 1$ service times before leaving the queue. That is, the distribution of the sojourn time of the customer who arrives at time S_n is determined by the distribution of x_{S_n-}.

A heuristic way of finding $P\{x_{S_n-} = m\}$ is to argue as follows. The total rate of arrivals is λ. The rate of the arrivals that occur when the queue length is equal to m is $\lambda(m) := \pi(m)\, q(m, m+1) = \pi(m)\lambda$. *Therefore,* the probability that an arrival occurs when the queue length is equal to m must be given by $\frac{\lambda(m)}{\lambda}$. Hence,

$$P\{x_{S_n-} = m\} = \frac{\lambda(m)}{\lambda} = \frac{\pi(m)\lambda}{\lambda} = \pi(m). \tag{2.45}$$

This formula turns out to be correct if the Markov chain $\{x_{S_n-}\}$ is stationary. This will be a particular case of Theorem 2.10.6 below.

Notice that (2.45) states that the invariant distribution of x_{S_n-} is the invariant distribution of x_t. As will be indicated in Theorem 2.10.6, this is a consequence of the fact that the arrivals S_n form a Poisson process with a future that is independent of the present value of x_t. In the calculation above, this translates into $q(m, m+1) \equiv \lambda$.

Example 2.10.2 [Jumps in Jackson networks] Consider the Jackson network of Example 2.5.6 and let S_n be the nth jump time (after $t = 0$) from a given queue i to another given queue j. We want to calculate $P\{x_{S_n-} = x\}$ for $x \in \mathbf{X}$.

Let us first consider the case of an open Jackson network. We will argue as for the M/M/1 queue. The total rate of jumps from queue i to queue j is $\lambda := \lambda_i r_{ij}$. The rate of jumps from i to j that occur when the state of the network is x is equal to $\lambda(x) := \pi(x)\, \mu_i r_{ij} 1\{x^i > 0\}$. Hence, one expects that

$$P\{x_{S_n-} = x\} = \frac{\lambda(x)}{\lambda} = \pi(x)\frac{1\{x^i > 0\}}{\rho_i}$$

$$= \pi(D_i x), \quad x \in \mathbf{X} \tag{2.46}$$

where the last equality follows from (2.19). (The notation $D_i x$ was defined in (2.8).)

It will turn out that the result is correct in the same sense as in the case of the M/M/1 queue. That is, x_{S_n-} is an ergodic and asymptotically stationary Markov chain with invariant distribution given by (2.46).

Notice that $D_i x$ is the configuration of the customers who do not jump at time S_n. Thus, one can describe the result by saying that **the customer who jumps sees the others with their invariant distribution in the network.** Indeed, (2.46) states that the probability that the customers who do not jump at time S_n have the configuration $D_i x$ is equal to $\pi(D_i x)$.

The same result also holds in the case of a mixed or closed network. Let us check it in the case of a mixed network. (See Figure 2.18.) Denote by $\mathbf{X}_v$ a closed irreducible set for x_t. The set $\mathbf{X}_v$ corresponds to given populations in the closed subnetworks.

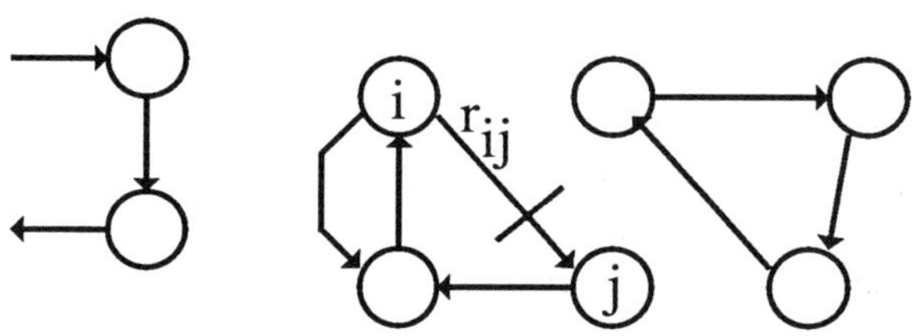

Figure 2.18. A mixed Jackson network.

Let i, j be two queues that belong to the same closed subnetwork. One then finds that the rate of jumps from i to j that occur when the state of the network is x is $\lambda(x) = \pi_{X_v}(x)\, \mu_i 1\{x^i > 0\} r_{ij}$. The total rate of jumps from i to j is $\lambda = \sum_{x \in \mathbf{X}_v} \lambda(x)$. Hence, one expects

$$P\{x_{S_n-} = x\} = \frac{\lambda(x)}{\lambda}.$$

This is correct, in the same sense as (2.45) and (2.46), as will follow from Theorem 2.10.6.

To evaluate this expression one makes the following observations:

$$\pi_{X_v}(x) = \frac{\pi(x)}{\pi(\mathbf{X}_v)}$$

$$\lambda(x) = \frac{\pi(x)\,\mu_i\,1\{x^i > 0\} r_{ij}}{\pi(\mathbf{X}_v)} = \frac{\lambda_i\,\pi(D_i x)\,r_{ij}}{\pi(\mathbf{X}_v)}$$

$$\lambda = \sum_{x \in \mathbf{X}_v} \lambda(x) = \frac{\lambda_i\,r_{ij}}{\pi(\mathbf{X}_v)} \sum_{x \in \mathbf{X}_v} \pi(D_i x) = \frac{\lambda_i\,r_{ij}}{\pi(\mathbf{X}_v)}\,\pi(\mathbf{X}_w)$$

where $\mathbf{X}_w := \{D_i x \mid x \in \mathbf{X}_v\}$. Hence,

$$P\{x_{S_n-} = x\} = \frac{\lambda(x)}{\lambda} = \frac{\pi(D_i x)}{\pi(\mathbf{X}_w)} = \pi_{X_w}(D_i x)\,. \quad (2.47)$$

Now, notice that $\mathbf{X}_w$ is the irreducible set of states that correspond to the configurations of the customers that do not jump at time S_n. Indeed, $\mathbf{X}_w$ consists of all the states in $\mathbf{X}_v$ with one customer removed from the closed subnetwork that contains queues i and j. Therefore, (2.47) shows that the configuration of the customers that do not jump at time S_n have their invariant distribution in the network from which the customer who jumps has been removed.

This result is called the **Arrival Theorem** for Jackson networks.

Remark 2.10.3 At this point, one notices that the algebra is getting tedious and that it results in unexpected simplifications. This should lead one to suspect that there must be a simpler justification of results such as 2.10.2. That is indeed the case.

Let us indicate one possible justification (due to F. Kelly). (See also section 4.4.) Insert an M/M/1 queue with a very large service rate on the link from queue i to queue j. Conditioning on a customer jumping from i to j in the original network is equivalent to conditioning on one customer being in the additional queue of the modified network (for its service rate is very large). Now, the set of possible states of the customers who do not jump when there is one customer in the extra queue is exactly the set of possible states of the other customers in the original network (without the one who jumps). Moreover, the invariant distribution of the modified network has a product form, and the marginal distribution of the original queues agrees with those in the original network. Thus, under the invariant distribution of the modified network, the conditional distribution of the customers who are not in the extra queue, given that there is a customer in the extra queue, is the invariant distribution of the original network. Normalizing as in Fact 2.5.7 allows us to conclude.

Definition 2.10.4 [Jumps of a Markov chain] Let $\mathbf{x} := \{x_t, \quad t \geq 0\}$ be a Markov chain with rate matrix Q. Let $\mathbf{J}$ be a subset of $\{(i, j) \in \mathbf{X}^2 \mid i \neq j\}$. Thus $\mathbf{J}$ selects some of the possible jumps of x_t in $\mathbf{X}$. The corresponding jump times are defined as follows: (see Figure 2.19)

$$S_1 = \min\{t > 0 \mid (x_{t-}, x_t) \in \mathbf{J}\}$$

$$S_{n+1} = \min\{t > S_n \mid (x_{t-}, x_t) \in \mathbf{J}\}, \quad n \geq 1$$

$$J_t = \max\{n \geq 1 \mid S_n \leq t\} \quad \text{with } J_t := 0 \text{ for } t < S_1 .$$

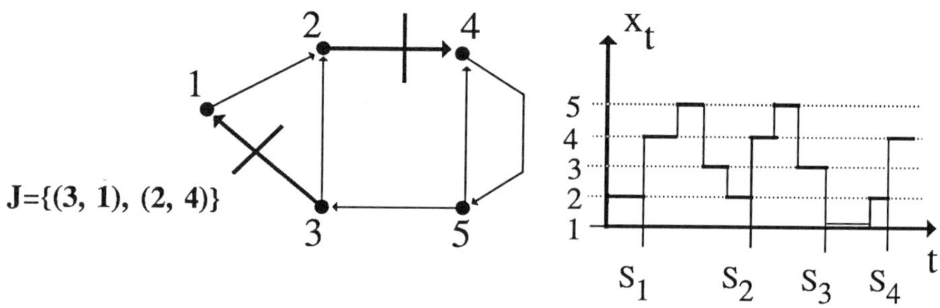

Figure 2.19. The definition of the jump times.

Thus, S_n is the nth time that $(x_{t-}, x_t) \in \mathbf{J}$ for $t > 0$, and J_t is the number of such jumps in $[0, t]$.

Examples 2.10.5

(a) In the case of the arrivals at an M/M/1 queue, one has $\mathbf{X} = \{0, 1, 2, \ldots\}$ and $\mathbf{J} = \{(i, i + 1), i \geq 0\}$.

(b) The jumps from queue i to queue j in a Jackson network with N queues correspond to $\mathbf{X} = \{0, 1, 2, \ldots\}^N$ and $\mathbf{J} = \{(x, T_{ij}x) \mid x \in \mathbf{X} \text{ and } x^i > 0\}$.

(c) The departures from an M/M/1 queue correspond to $\mathbf{X} = \{0, 1, 2, \ldots\}$ and $\mathbf{J} = \{(i + 1, i), i \geq 0\}$.

The evolutions of x_{S_n-} and x_{S_n} are analyzed below.

Theorem 2.10.6 Let $\mathbf{x}$, $\mathbf{J}$, and $\{S_n, n \geq 1\}$ be as in Definition 2.10.4. Assume that $\mathbf{x}$ has an invariant distribution π such that

$$\beta := \sum_i \pi(i)\, q(i) < \infty \quad \text{and} \quad \lambda := \sum_{(i,j) \in \mathbf{J}} \pi(i)\, q(i,j) > 0\,. \tag{2.48}$$

Then

(a) The sequence $\{x_{S_n-}, n \geq 1\}$ is a discrete-time Markov chain with invariant distribution π_{S-} given by

$$\pi_{S-}(i) = \lambda^{-1} \sum_j \pi(i)\, q(i,j)\, 1\{(i,j) \in \mathbf{J}\}, \quad i \in \mathbf{X}\,. \tag{2.49}$$

(b) The sequence $\{x_{S_n}, n \geq 1\}$ is a discrete-time Markov chain with invariant distribution π_S given by

$$\pi_S(j) = \lambda^{-1} \sum_i \pi(i)\, q(i,j)\, 1\{(i,j) \in \mathbf{J}\}, \quad j \in \mathbf{X}\,. \tag{2.50}$$

(c) Under the initial distribution π, the present state x_t and the future jump times $\{J_{t+s} - J_t, s \geq 0\}$ are independent for all $t \geq 0$ if and only if $\pi_{S-} = \pi$. In that case, J_t is a Poisson process with rate λ.

(d) Under the initial distribution π, the present state x_t and the past jump times $\{J_s, s \leq t\}$ are independent for all $t \geq 0$ if and only if $\pi_S = \pi$. In that case, J_t is a Poisson process with rate λ.

Comments 2.10.7 The formulas (2.48)–(2.50) admit the following simple interpretation.

(i) λ is the average rate of transitions (x_{t-}, x_t) in $\mathbf{J}$.

(ii) Formula (2.49) is the ratio of the rate of transitions (x_{t-}, x_t) in $\mathbf{J}$ with $x_{t-} = i$ over the rate of transitions in $\mathbf{J}$.

(iii) Formula (2.50) is the ratio of the rate of transitions (x_{t-}, x_t) in $\mathbf{J}$ with $x_t = j$ over the rate of transitions in $\mathbf{J}$.

The formulas are what should be expected, as discussed in Examples 2.10.1 and 2.10.2. The theorem states that they are invariant probabilities. Ergodicity and asymptotic stationarity require more information about Q and $\mathbf{J}$.

Part (c) of the theorem is often summarized by stating that **Poisson arrivals see time averages (PASTA).** One can also refer to part (d) by saying that **Poisson departures see time averages.** Of course, it should be kept in mind that the statements involve some **independence conditions.**

Notice that, in part (c), the fact that J_t is Poisson is a *consequence* of the assumption $\pi_{S-} = \pi$. This might create some confusion as to the meaning of "Poisson arrivals see time averages." What the result is saying is that the system is such that J_t (which can be thought of an arrival process) is Poisson with a future that is independent of the present state if and only if $\pi_{S-} = \pi$. In a queueing system, the Poisson arrival process may be produced by some mechanism that operates independently (in the intuitive sense) of the queue described by **x**. In that case, the rate $\sum_j q(i, j)\, 1\{(i, j) \in \mathbf{J}\}$ of arrivals when $x_t = i$ must be equal to λ for all i. This then *implies* that $\pi_{S-} = \pi$, by (2.49).

Proof of Theorem 2.10.6

Part (b): Let ξ_n and **z** be as in Examples 2.9.4 and 2.9.5. Observe that $v := \{(x_{S_n-}, x_{S_n}), n \geq 1\}$ is the Markov chain **z** watched in **J**. It follows from (2.40) and Theorem 2.9.10 that it admits the invariant distribution

$$\eta(i, j) = \lambda^{-1}\, \pi(i)\, q(i, j)\, 1\{(i, j) \in \mathbf{J}\} .$$

Assume then that v is given the initial distribution η. Then, x_{S_n} has the invariant distribution π_S where

$$\pi_S(j) = \sum_i \eta(i, j) ,$$

which agrees with (2.50).

It remains to show that x_{S_n} is a Markov chain. To see this, notice that $x_{S_m} = \xi_{T_m}$, where T_m is the mth value of $n \geq 1$ such that $(\xi_{n-1}, \xi_n) \in \mathbf{J}$. Hence, the T_m are stopping times of ξ_n and the fact that $\xi_{T_m} = x_{S_m}$ is a Markov chain then follows from the Strong Markov Property 2.9.8. The fact that the transition probabilities are stationary is easy.

Part (a): The results from part (a) are those of part (b) applied to the Markov chain **x** reversed in time with **J** replaced by $\tilde{\mathbf{J}} = \{(i, j) \mid (j, i) \in \mathbf{J}\}$.

Part (c): Assume that $\pi_{S-} = \pi$. Then, by (2.49),

$$\lambda[i] := \sum_j q(i, j)\, 1\{(i, j) \in \mathbf{J}\} \equiv \lambda, \quad i \in \mathbf{X} . \tag{2.51}$$

Now, it is readily verified that

$$\lambda[i] = \lim_{\epsilon \downarrow 0} \frac{1}{\epsilon} P[J_{t+\epsilon} \geq J_t + 1 \mid x_t = i] . \tag{2.52}$$

To see that (2.51)–(2.52) imply that J_t is a Poisson process with a future that is independent of x_t, one proceeds as follows. For any time u one has

$$P[J_{u+\epsilon} = J_u \mid x_s, s \le u] = 1 - \lambda\epsilon + o(\epsilon), \tag{2.53}$$

by (2.51)–(2.52). Fix $t \ge 0$ and let $t + T_1$ be the first jump time of J_s for $s > t$. Then,

$$P[T_1 > a + \epsilon \mid x_t = i]$$
$$= P[T_1 > a \mid x_t = i]\, P[T_1 > a + \epsilon \mid x_t = i, T_1 > a]$$
$$= P[T_1 > a \mid x_t = i]\,(1 - \lambda\epsilon + o(\epsilon)) \tag{2.54}$$

where the last equality follows from the fact that the information in $\{x_t = i, T_1 > a\}$ is contained in $\{x_s, s \le u = t + a\}$ and from (2.53).

The equation (2.54) is seen to imply that

$$\frac{d}{da} P[T_1 > a \mid x_t = i] = -\lambda P[T_1 > a \mid x_t = i]$$

which shows that

$$P[T_1 > a \mid x_t = i] = e^{-\lambda a}, \quad a \ge 0 .$$

Thus, the first jump time of J_s for $s > t$ is exponentially distributed and is independent of x_t. Repeating the argument after time T_1 (which is legitimate by the Strong Markov Property), one can show that all the subsequent interarrival times of J_s are i.i.d. and independent of x_t.

It remains to show that if J_t is a Poisson process with rate λ and with a future that is independent of x_t, then $\pi_{S-} = \pi$. But this is easy since the assumption implies that $\lambda[i] = \lambda$ and the conclusion then follows from (2.49).

Part (d): The result of part (d) is precisely that of part (c) applied to the Markov chain **x** reversed in time.

Example 2.10.8 Let us go back to Example 2.10.2 and examine the issues of asymptotic stationarity and ergodicity of the mixed Jackson network observed at the jump times S_n from queue i to queue j.

It is known that x_{S_n-} is a Markov chain with the invariant distribution (2.47). This follows from Theorem 2.10.6, part (a). One can verify, independently, that $y_n := D_i(x_{S_n-})$ is a Markov chain that is irreducible and aperiodic on $\mathbf{X}_w$. It then follows from section 2.9.3 that π_{X_w} is the unique invariant distribution of y_n on $\mathbf{X}_w$ and that the Markov chain is ergodic and asymptotically stationary.

Similar considerations apply to the M/M/1 queue or to the open Jackson networks. The readers are invited to check the details.

2.11. POISSON PROCESS

Let us conclude this section by reviewing some basic facts about Poisson processes.

Definition 2.11.1 A **Poisson process with rate** $\lambda > 0$ is a sequence of random times $\{S_n, n \geq 1\}$ with $0 < S_1 < S_2 < \cdots$ and such that the interarrival times $S_{n+1} - S_n$ for $n \geq 0$ (with $S_0 := 0$) are independent and exponentially distributed with rate λ.

The **counting process**

$$N_t = \max\{n \geq 0 \mid S_n \leq t\}, \quad t \geq 0$$

is then also called a Poisson process with rate λ.

The following properties are important.

Theorem 2.11.2

(a) For all $t \geq 0$, $\{N_{t+s} - N_t, s \geq 0\}$ is a Poisson process independent of $\{N_s, s \leq t\}$. This is called the **memoryless property of the Poisson process.**

(b) For all $t \geq 0$ and $n \geq 0$ one has

$$P\{N_t = n\} = \frac{(\lambda t)^n}{n!} e^{-\lambda t}. \tag{2.55}$$

(c) The sum of two independent Poisson processes is a Poisson process.

(d) **Bernoulli sampling** of a Poisson process yields independent Poisson processes. Bernoulli sampling means that each jump S_n is independently considered to be a jump of a process N_t^k with probability p_k for $1 \leq k \leq K$, with $p_k \geq 0$ and $\sum p_k = 1$.

(e) Fix $t > 0$. Given $\{N_t = n\}$, the jump times of the process in $[0, t]$ are independent and uniformly distributed in $[0, t]$.

(f) For all $t > 0$,

$$\lim_{\epsilon \downarrow 0} \frac{1}{\epsilon} P[N_{t+\epsilon} \geq N_t + 1 \mid N_s, s \leq t] = \lambda. \tag{2.56}$$

Moreover, the Poisson process is the only counting process with this property.

Proof: The proofs of these properties can be found in any probability textbook. Let us just indicate that the independence of $S_1, S_2 - S_1, \ldots$ and their exponential distribution imply that, for $n \geq 1$ and $0 < t_1 < \cdots < t_n < t$,

$$P\{S_1 \in (t_1, t_1 + dt_1), \ldots, S_n \in (t_n, t_n + dt_n), S_{n+1} > t\}$$

$$= \lambda e^{-\lambda t_1} dt_1 \ldots \lambda e^{-\lambda(t_n - t_{n-1})} dt_n e^{-\lambda(t - t_n)}$$

$$= \lambda^n e^{-\lambda t} dt_1 \ldots dt_n . \tag{2.57}$$

This shows (2.55), by integrating over $0 < t_1 < \ldots < t_n < t$. (Indeed, the volume of that subset of $\mathbb{R}^n$ is a fraction $\frac{1}{n!}$ of the cube $[0, t]^n$ since the $n!$ permutations of $(t_1, \ldots, t_n)$ have identical volumes and cover the cube.)

Also, dividing (2.57) by (2.55) shows property (e).

The Bernoulli sampling property is also immediate from (2.57). To see this, call S_n^k the positive jump times of process k. Choose a set $\{t_n^k, 1 \leq n \leq N_k, 1 \leq k \leq K\}$ of distinct values in $[0, t]$ and order them as $0 \leq t_1 < \cdots < t_N \leq t$, with $N = \sum N_k$. One then has,

$$P\{S_n^k \in (t_n^k, t_n^k + dt), 1 \leq n \leq N_k, 1 \leq k \leq K\}$$

$$= P\{S_n \in (t_n, t_n + dt), 1 \leq n \leq N\} \prod_{k=1}^{K} (p_k)^{N_k}$$

$$= (\lambda\, dt)^N e^{-\lambda t} \prod_{k=1}^{K} (p_k)^{N_k} = \prod_{k=1}^{K} [(\lambda p_k\, dt)^{N_k} e^{-\lambda p_k t}]$$

which shows that this probability is the product of the probabilities corresponding to Poisson processes with rates λp_k.

The sampling property trivially implies that the sum of independent Poisson processes is Poisson, since one can consider that the independent Poisson processes were obtained by Bernoulli sampling of a Poisson process that is then their sum.

The fact that (2.56) implies that N_t is Poisson can be checked by verifying that this property implies (2.57). For instance, (2.56) gives

$$P\{N_{t+\epsilon} = 0\} = P\{N_t = 0\} [1 - \lambda\epsilon + o(\epsilon)] ,$$

which implies that $P\{N_t = 0\} = P\{S_1 > t\} = e^{-\lambda t}$.

2.12. EXERCISES

2.1. Exponential Distribution Let τ be exponentially distributed with mean λ^{-1}.

(a) Show that $E\{\tau\} = \lambda^{-1}$; calculate $\text{var}(\tau)$.

(b) Show that for $\mu > 0$ the random variable $\sigma = \frac{\lambda}{\mu}\tau$ is exponentially distributed with mean μ^{-1}.

2.2. Exponential Distribution (cont.) Let $\tau_1, \tau_2, \ldots, \tau_n$ be i.i.d. and exponentially distributed with mean 1. Let $\sigma_n = \max\{\tau_1, \ldots, \tau_n\}$. Show that

(a) $E\{\sigma_n\} = 1 + \frac{1}{2} + \cdots + \frac{1}{n}$;

(b) $\text{var}(\sigma_n) = 1 + \frac{1}{2^2} + \cdots + \frac{1}{n^2}$.

Hint: By conditioning on the value of $\xi_n := \min\{\tau_1, \ldots, \tau_n\}$, and by using the memoryless property of the exponential distribution, one finds that $\sigma_n - \xi_n$ and ξ_n are independent, and that $\sigma_n - \xi_n$ has the same distribution as σ_{n-1}.

This shows that $E\{\sigma_n\} = E\{\xi_n\} + E\{\sigma_{n-1}\}$ and that

$$\text{var}(\sigma_n) = \text{var}(\xi_n) + \text{var}(\sigma_{n-1}).$$

2.3. An irregular Markov chain The objective of this exercise is to illustrate the fact that there are Markov processes on countable sets (i.e., Markov chains) that cannot be defined as in Definition 2.4.2.

Let $\mathbf{X} = \{0, 1, 2, \ldots\}$ and let $\{\tau_n, n \geq 0\}$ be independent with $P\{\tau_n \geq t\} = \exp\{-q_n t\}$, where $q_n > 0$ and $\sum_{n=0}^{\infty} q_n^{-1} < \infty$. Define $\mathbf{x} = \{x_t, t \geq 0\}$ by $x_0 = 0$ and, for $t > 0$,

$$x_t = n \text{ if } \sum_{m=n}^{\infty} \tau_m \leq t < \sum_{m=n-1}^{\infty} \tau_m \quad \text{for some } n \geq 1$$

and restart the construction independently from 0 at time $\sum_{m=0}^{\infty} \tau_m$.

Argue that $\mathbf{x}$ is a Markov process. Notice that $\mathbf{x}$ does not have a first jump time.

2.4. Transition probabilities Verify the result of Example 2.5.3.

2.5. Transitions of a Markov chain Let $\mathbf{x}$ be a Markov chain with rate matrix Q (see Definition 2.4.2). Prove that the probability that $\mathbf{x}$ makes at least two jumps in $[0, t]$ given that $x_0 = i$ is $o(t)$ for all $i \in \mathbf{X}$.

2.6. Kolmogorov equations Prove the equations (2.11) for a regular Markov chain $\mathbf{x}$. *Hint:* If x is regular, it makes a last jump τ in $[0, t]$. By construction,

$$P_t(i, j) = 1\{i = j\}e^{-q(i)\,t} + \sum_{k \neq j} \int_0^t P_i\{\tau \in (s, s + ds), x_{\tau-} = k, x_\tau = j\}$$

$$= 1\{i = j\}e^{-q(i)t} + \sum_{k \neq j} \int_0^t P_s(i, k)\, q(k) \frac{q(k, j)}{q(k)} e^{-q(j)(t - s)} ds\,.$$

Differentiating with respect to t gives $\frac{d}{dt}P_t = P_t Q$. By considering the first jump in $[0, t]$, show that the other equation holds.

2.7. Kolmogorov Equations (cont.) Show that the second of the equations (2.11) may fail to hold if Q is not regular. *Hint:* Consider the Markov chain of Remark 2.4.5.

2.8. Nonuniform Markov Chain Show that the series in (2.12) may fail to converge when Q is not uniform. *Hint:* Let $\mathbf{X} = \{0, 1, 2, \ldots\}$ and $q(0,i) = i^{-2}$, $q(i,0) = q(i) = i$ for $i \geq 1$, so that $q(0) = \sum_{i \geq 1} i^{-2} = \frac{\pi^2}{6}$. Then Q is regular since every other state value is 0, and Q^2 is not finite, so that the series in (2.12) is not defined.

2.9. Balance Equations Prove the last statement of Theorem 2.5.2 for a regular Markov chain. *Hint:* For $\lambda > 0$ let $R_\lambda(i) = E[\int_0^\infty e^{-\lambda t} f(x_t)\, dt \mid x_0 = i]$ where $f := \mathbf{X} \to \mathbb{R}$ is a given bounded function. Then,

$$R_\lambda(i) = E[\int_0^\tau e^{-\lambda t} f(i)\, dt \mid x_0 = i]$$

$$+ \sum_{j \neq i} \frac{q(i,j)}{q(i)} E[\int_\tau^\infty e^{-\lambda t} f(x_t)\, dt \mid x_\tau = j]$$

where τ denotes the first jump time of $\mathbf{x}$. This leads to

$$(\lambda I - Q)\, R_\lambda = f$$

where f and R_λ are the column vectors with components $f(i)$ and $R_\lambda(i)$, respectively. Now,

$$R_\lambda(i) = \int_0^\infty e^{-\lambda t} E[f(x_t) \mid x_0 = i]\, dt ,$$

so that $R(\lambda) := \sum_i \pi(i)\, R_\lambda(i)$ is the Laplace transform of $\{E\{f(x_t)\}, t \geq 0\}$ where the expectation corresponds to the initial distribution π for x_0. Assume first that $\pi Q = 0$. One then finds that

$$\pi(\lambda I - Q)\, R_\lambda = \lambda R(\lambda) = \pi f ,$$

which shows that $E\{f(x_t)\}$ is constant (by uniqueness of the inverse Laplace transform). Since f was an arbitrary bounded function, it follows that the distribution of x_t is constant. Second, assume that the distribution of x_t is constant. Then $\lambda R(\lambda) = \pi f$, which implies that $\pi\, Q\, R_\lambda = 0$. This is the Laplace transform of

$$\sum_{i,j} \pi(i)\, Q(i,j)\, E[f(x_t) \mid x_0 = j]$$

which must then be equal to zero for all $t \geq 0$. In particular, $\pi Q f = 0$. Since f is arbitrary, this implies that $\pi Q = 0$, as claimed.

2.10. Balance Equations (cont.) Show that the last part of Theorem 2.5.2 mail fail to hold if Q is not regular. *Hint:* Try the example of Exercise 2.3.

2.11. M/M/s Queue Consider an FCFS queue with Poisson arrivals with rate λ and with $s \geq 1$ servers. The service times are independent and exponentially distributed with rate μ. Explain why the queue length is a Markov chain. Write the balance equations for the invariant queue length distribution. Show that the queue length is ergodic if and only if $\lambda < s\mu$.

2.12. Uniformization Let **x** be a Markov chain with a uniform rate matrix Q. Assume that $|q(i, j)| \leq \lambda < \infty$ for all i, j. Show that $P = I + \lambda^{-1}Q$ is a transition matrix and that

$$P_t = e^{Qt} = \sum_{n=0}^{\infty} \frac{(\lambda t)^n}{n!} e^{-\lambda t} P^n, \quad t \geq 0 .$$

Let $\{y_n, n \geq 0\}$ be a Markov chain with transition matrix P and $\{N_t, t \geq 0\}$ an independent Poisson process with rate λ. Explain that the above formula shows that **x** has the same law as

$$\{y_{N_t}, t \geq 0\} .$$

Thus, the properties of a uniform continuous-time Markov chain can be determined by analyzing a discrete-time Markov chain.

2.13. Ross' Embedding Let **x** be a Markov chain with rate matrix Q. Let $\tau_1, \tau_2, \ldots$ be independent random variables exponentially distributed with rate λ and independent of **x**. Define $y_n = x_{\sigma_n}$ with $\sigma_n = \tau_1 + \tau_2 + \cdots + \tau_n$. Let also $y_0 = x_0$.

Show that $\{y_n, n \geq 0\}$ is a Markov chain with transition matrix P given by

$$P(i, j) = \frac{\lambda 1\{i = j\}}{\lambda + q(i)} + \frac{q(i, j)\, 1\{i \neq j\}}{\lambda + q(i)} .$$

A possible application is for calculating the transition probabilities P_t of **x**. To do this, choose $t > 0$, n moderately large, and $\lambda = nt^{-1}$. Justify that

$$P_t(i, j) := P[x_t = j \mid x_0 = i] \approx P[y_n = j \mid y_0 = i] = P^n(i, j) .$$

This method is due to Ross (1987).

2.14. Discrete-Time Markov Chain Let

$$\mathbf{X} = \{0, 1\}, P(0, 1) = 1 - P(0, 0) = a \in (0, 1) ,$$
$$P(1, 0) = 1 - P(1, 1) = b \in (0, 1) .$$

This is summarized by the **transition diagram** in Figure 2.20.

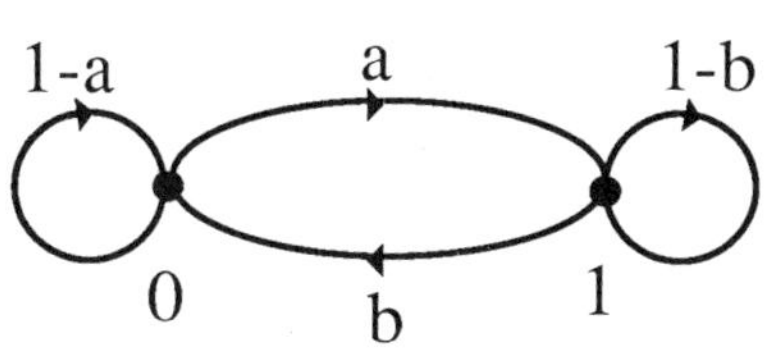

Figure 2.20 Transition Diagram.

Show that this Markov chain is ergodic with invariant probability distribution $\pi = (\pi(0), \pi(1)) = (\frac{b}{a+b}, \frac{a}{a+b})$. Calculate P^n and verify that (2.37) holds. Show that $\mathbf{x}$ is time-reversible when it is stationary.

2.15. Pakes' Lemma Let x be irreducible and aperiodic on $\mathbf{X} = \{0, 1, 2, \ldots\}$. For $i \in \mathbf{X}$ define

$$\gamma(i) = E[x_{n+1} - x_n \mid x_n = i].$$

Show that x is ergodic if

(a) $|\gamma(i)| < \infty, \quad i \in \mathbf{X}$;

(b) $\gamma(i) \leq -a < 0, \quad i \geq i_0$ for some $i_0 \in \mathbf{X}$ and some $a > 0$.

The interpretation is that x_n tends to decrease when it is large. *Hint:* Notice that $\gamma(i) \leq (a+b)\, 1\{i < i_0\} - a$, where $b := \max_{i < i_0} \gamma(i)$. Hence, with $E_i\{.\} := E[. \mid x_0 = i]$,

$$-i \leq E_i\{x_{n+1} - i\} = E_i\{\gamma(x_1) + \cdots + \gamma(x_n)\}$$

$$\leq (a+b) \sum_{m=0}^{n} P_i\{x_m < i_0\} - na.$$

Dividing by n, letting $n \to \infty$, and using the fact that $P_i\{x_m = j\} \to \pi(j)$, one finds that $(a+b) \sum_{j=0}^{i_0} \pi(j) \geq a > 0$. Hence, π cannot be identically 0 and section 2.9.3 then implies that $\mathbf{x}$ must be ergodic.

2.16. Time Reversal Let $\mathbf{X} = \{0, 1, 2\}$ with the transition diagram in Figure 2.21. Find the transition diagram of the Markov chain $\mathbf{x}$ reversed in time (i.e., corresponding to $\tilde{P}$).

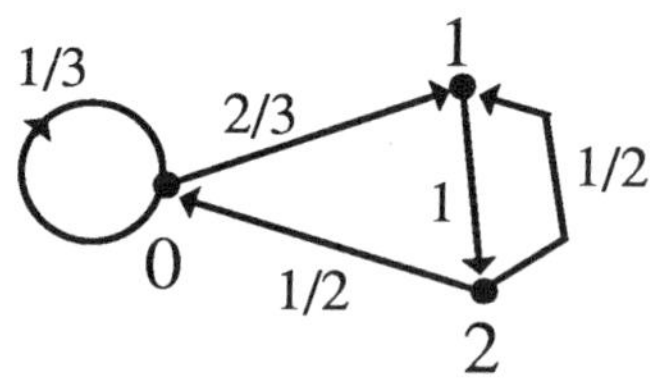

Figure 2.21 Transition Diagram.

2.17. Function of a Markov chain Let $x = \{x_n, n \geq 0\}$ be a Markov chain and $f : \mathbf{X} \to R$ be some function. Show by an example that $\{f(x_n), n \geq 0\}$ need not be a Markov chain.

2.18. Poisson Process Show that (2.55) does not imply that N_t is a Poisson process. *Hint:* Let

$$F_0(t) = e^{-\lambda t}, \quad \text{and}$$

$$F_n(t) = e^{-\lambda t}\{1 - \lambda t - \cdots - \frac{(\lambda t)^{n-1}}{(n-1)!}\}, \quad n \geq 1 .$$

Define $\phi_n : (0, 1) \to (0, +\infty)$ so that $\phi_0(a) + \cdots + \phi_n(a) \leq t$ if and only if $a \leq F_n(t)$. Let then σ be uniformly distributed in (0, 1) and define $\tau_n = \phi_n(\sigma)$ and $N_t = \min\{n \geq 0 \mid \tau_0 + \cdots + \tau_n > t\}$.

Clearly, the τ_ns are not independent (in fact, if one is known, then so are all the others), so that N_t is not Poisson. However,

$$\begin{aligned} P\{N_t = n\} &= P\{\tau_0 + \cdots + \tau_n \leq t < \tau_0 + \cdots + \tau_{n+1}\} \\ &= P\{\phi_0(\sigma) + \cdots + \phi_n(\sigma) \leq t < \phi_0(\sigma) + \cdots + \phi_{n+1}(\sigma)\} \\ &= P\{\phi_0(\sigma) + \cdots + \phi_n(\sigma) \leq t\} - P\{\phi_0(\sigma) + \cdots + \phi_{n+1}(\sigma) \leq t\} \\ &= P\{\sigma \leq F_n(t)\} - P\{\sigma \leq F_{n+1}(t)\} \\ &= F_n(t) - F_{n+1}(t) = \frac{(\lambda t)^n}{n!}e^{-\lambda t} , \end{aligned}$$

so that (2.55) holds for this non-Poisson process.

2.19. Poisson Process (cont.) Let $0 < S_1 < S_2 < S_3 < \cdots$ be a sequence of random variables. For any subset A of $[0, \infty)$, let $N(A) := \sum_{n=1}^{\infty} 1\{S_n \in A\}$ be the number of "points" S_n that belong to A. Assume that for any set A that is the union of a finite number of disjoint intervals and for all $n \geq 0$,

$$P\{N(A) = n\} = \frac{(\lambda \mid A \mid)^n}{n!}e^{-\lambda \mid A \mid}$$

where $\mid A \mid$ indicates the length of A, i.e., the sum of the lengths of the intervals that A is the union of. (In this formula, λ is a positive real number.) Show that $\{N_t, t \geq 0\}$ is Poisson with rate λ.

Hint: It suffices to show that if $A_1, \ldots, A_k$ are disjoint intervals, then the random variables $N(A_1), \ldots, N(A_k)$ are independent. To show this in the case $k = 2$, notice that for any $0 < \epsilon \ll 1$, one can assume that A_1 and A_2 are disjoint unions of intervals of lengths less than ϵ, say $A_1 = \sum_{i=1}^{m} A_{1i}$ and $A_2 = \sum_{j=1}^{l} A_{2j}$. The verification is now easy. (See Rénji (1967).)

2.13. REFERENCES AND COMMENTS

Section 2.2 The concept of **state** is central to the investigation of dynamical systems. As this section explains, the extension of that concept to stochastic systems is the Markov property. If the distributions F and G are not exponential, then the queue length is not Markov.

To obtain a Markov process, one should consider, for instance, (x_t, a_t, s_t) where x_t is the queue length at time t, a_t is the time since the last arrival prior to time t, and s_t is the amount of service already received by the customer in service at time t. The difficulty is now that the state space (the set of possible values o the vector (x_t, a_t, s_t) is not countable. Thus, the process is not a Markov chain. This difficulty can be eliminated if the distributions F and G are approximated by **Phase-type distributions.** That idea is discussed in section 3.2.

Standard references on queueing theory are Beneš (1963), Cohen (1982), Cooper (1981), Gross and Harris (1974), Kleinrock (1975), and Prabhu (1980).

Section 2.4 The essential property of a Markov chain is that its evolution starts afresh from x_t at time t, for all fixed time t. The case of nonregular Markov chains is very interesting because it demonstrates the possible discontinuity in the information contained in the evolution of the process at an explosion time. All this is a bit technical for this book, however. See Williams (1979), section 3.46.

Section 2.5 The usual definition of a stochastic process **x** starts by specifying the finite dimensional distributions. That is, one specifies the probability of $\{x_{t_1} = i_1, \ldots, x_{t_n} = i_n\}$. From this definition, one then attempts to determine the existence of processes with those finite dimensional distributions and with nice trajectories (e.g. right-continuous and piecewise constant).

This program, for Markov chains, proceeds as follows. One specifies the finite dimensional distributions so that (2.6) holds.

This gives, for $0 < t_1 < \cdots < t_n$,

$$P\{x_{t_1} = i_1, x_{t_2} = i_2, \ldots, x_{t_n} = i_n\}$$

$$= \sum_{i_0} \pi(i_0)\, P_{t_1}(i_0, i_1)\, P_{t_2 - t_1}(i_1, i_2) \ldots P_{t_n - t_{n-1}}(i_{n-1}, i_n)\,.$$

In these expressions, the matrices P_t must be such that (2.9) holds. (For the distributions to be consistent.) The sample path (trajectory) properties of processes with those finite dimensional distributions then depend on the derivative Q of P_t at $t = 0$. In particular, one can show that if there is a regular version of the process, then there exists a unique realization of the process with right-continuous and piecewise constant paths and this process agrees with our Definition 2.4.2.

For an exposition of the theory of Markov chains, see Doob (1953), Chung (1967), Feller (1971), Ross (1983a), Breiman (1968), and Freedman (1971). The Jackson networks were discussed in Jackson (1954), Jackson (1957), Gordon and Newell (1967), Whittle (1967), and Kingman (1969).

Sections 2.6 and 2.7 A proof of Theorem 2.6.2 and of the statements of section 2.7 can be found in any of the references of section 2.5 on the theory of Markov chains. A nice introduction to ergodic theory can be found in Breiman (1968), Chapter Six. See Anantharam (1987a) for an analysis of relaxation in Jackson networks and Anantharam (1987b) for the buffer allocation problem in such networks.

Section 2.8 The application of time-reversal ideas in queueing theory was pioneered by Reich (1957). We refer to Lemma 2.8.5 as Kelly's lemma to acknowledge the contribution of Kelly in indicating how this simple result (which was already observed by Kolmogorov in 1936) can be used systematically to derive important properties of stochastic networks. See Kelly (1979) for a superb discussion of those ideas. The Output Theorem for the M/M/1 queue is due to Burke (1956). These ideas will be further explored in Chapters Three and Four.

Section 2.9 It may seem awkward to discuss continuous-time Markov chains before their discrete-time counterparts. This was done for two reasons. First, the continuous-time chains are simpler in the sense that they do not present any periodic behavior. This simplifies the limit properties. Second, the simplest queueing models are continuous-time Markov chains.

The connection between continuous-time and discrete-time is through the uniformization method. See Exercise 2.12.

The strong Markov property is crucial in many applications. It is important for the reader to develop an intuitive understanding of that result.

Section 2.10 The key result of this section is Theorem 2.10.6. It says that the intuitive reasoning of Examples 2.10.1 and 2.10.2 is justified. Another viewpoint will be discussed in Chapter Seven.

The discussion of this section is inspired from Melamed (1982a) and Walrand (1983).

CHAPTER THREE

PRODUCT-FORM NETWORKS

3.1. INTRODUCTION

With few exceptions, networks that are analytically tractable are **product-form** networks. That is, they correspond to a Markov chain $x_t = (x_t^1, \ldots, x_t^J)$ that admits an invariant distribution π of the form $\pi(x^1, \ldots, x^J) = \pi_1(x^1) \ldots \pi_J(x^J)$.

One class of such networks is obtained by interconnecting quasi-reversible queues with a routing that does not depend on the past of x_t. The distribution at the jumps shows that a customer who jumps sees the others in equilibrium. This is the **Arrival Theorem.** (See Theorem 3.3.2.) Also, the network reversed in time is obtained by **reversing the routing and each queue.** (See Theorem 3.5.4.)

Other product-form networks can be constructed. as discussed in section 3.7. The procedure almost always consists in showing that a natural guess $\tilde{Q}$ for the rate matrix $\tilde{Q}$ of the network reversed in time satisfies $\pi(x)\, q(x, y) \equiv \pi(y)\, \tilde{q}(y, x)$ with a product-form π. The definition of the network usually involves some free choices of functions (e.g., state dependent service rates and routing probabilities). One then uses the above equations to determine the form that these functions should have for the network to be product-form. More or less natural and interesting models can thus be derived.

The chapter begins with an informal description of **quasi-reversibility** (section 3.2) which is followed by examples of **product-form networks** in section 3.3. **Little's result** is explained and proved in section 3.4. The proofs of the results on product-form networks are given in section 3.5. The quasi-reversible queues are discussed in detail in section 3.6. Some other product-form networks are covered in section 3.7. A useful observation, called **Norton's Theorem** for product-form networks is explained in section 3.8. The concept of **insensitivity** is related to quasi-reversibility in section 3.9.

3.2. QUASI-REVERSIBILITY

Quasi-reversibility is essentially an input-output property of queues. It says that at any time the past departure times, the present state, and the future arrival times are independent. For instance, the M/M/1 queue is quasi-reversible (when stationary).

Other examples will be given in 3.2.2. An **approximation method** for general distributions will be needed. It is explained in 3.2.3 and 3.2.4.

It is useful to be able to distinguish different **classes** of customers. Customers of different classes will have different characteristics (service time distributions and paths in the networks). The set of classes will be denoted by **C**. It will be assumed to be countable. For each class $c \in \mathbf{C}$ there is an arrival process A_t^c and a departure process D_t^c at each queue.

The evolution of a queue is described by a Markov chain $\mathbf{x} := \{x_t, \ \ t \geq 0\}$. This means that by observing $\mathbf{x}$ one can determine the quantities of interest about the evolution of the queue. In particular, it will always be assumed that arrivals and departures correspond to certain **transitions** of x_t in the state space **X**. That is,

$$A_t^c = \sum_{0<s\leq t} 1\{(x_{s-}, x_s) \in \mathbf{A}^c\}, \quad D_t^c = \sum_{0<s\leq t} 1\{(x_{s-}, x_s) \in \mathbf{D}^c\}$$

where the $\mathbf{A}^c$ and $\mathbf{D}^c$ are disjoint subsets of $\{(i, j) \in \mathbf{X} \mid i \neq j\}$.

Definition 3.2.1 [Quasi-reversibility] The queue is **quasi-reversible** if, when it is stationary,

$$\{D_s^c, s \leq t, c \in \mathbf{C}\}, x_t, \text{ and } \{A_{t+s}^c - A_t^c, \ s \geq 0, \ c \in \mathbf{C}\}$$

are independent (for all $t \geq 0$). It then follows that the **arrival processes** are **independent Poisson processes,** and that so are the **departure processes.** (The proof of this fact is similar to that of Theorem 2.10.6(c)–(d). It is suggested in Exercise 3.1.)

One can show, similarly to the proof of Theorem 2.10.6, that a queue is quasi-reversible if and only if, for all $c \in \mathbf{C}$,

$$\sum_j q(i, j)\, 1\{(i, j) \in \mathbf{A}^c\} = \lambda_c, \quad i \in \mathbf{X}, \tag{3.1}$$

and

$$\sum_i \pi(i)\, q(i, j)\, 1\{(i, j) \in \mathbf{D}^c\} \;=\; \mu_c \pi(j), \quad j \in \mathbf{X}\,, \tag{3.2}$$

for some λ_c and μ_c which are then the rates of the arrival and departure processes, respectively. (In (3.2), π denotes the invariant distribution of the Markov chain $\mathbf{x}$.)

Examples 3.2.2 [Quasi-reversible queues] There are a few quasi-reversible queues that are often encountered in applications. That they are quasi-reversible will be explained in section 3.6.

(1) **M/GI/1-PS queue:** (PS = Processor-Sharing.) This queue is defined as the M/GI/1 queue except that the single server divides his service effort equally among the customers in the queue. (The M/GI/1 queue was defined before Fact 2.3.1; see also Appendix B.) Thus, if in the time interval $[t_0, t_1]$ there are n customers in the queue, then each of them will receive $n^{-1}(t_1 - t_0)$ units of service in that time interval.

(2) **M/GI/1-LCFS queue:** (LCFS = Last-Come–First-Served.) This queue is similar to the M/GI/1 queue except that the customers are served on a Last-Come–First-Served basis. The service of a customer is preempted by subsequent arrivals and is resumed when those later arrivals have been served.

(3) **M/GI/∞ queue:** This queue has infinitely many servers, so that all the customers present in the queue are being served in parallel by different servers. The customers do not interact in the queue. They are simply delayed by their service time before leaving the queue.

Remark 3.2.3 [Approximations by Markov chains] An M/GI/1 queue may not be modeled by a Markov chain $\mathbf{x} := \{x_t, \quad t \geq 0\}$. For instance, assume that the service times are all equal to the constant 1. The minimum information x_t that qualifies as a state, if one wants to be able to analyze the evolution of the queue length, is $x_t = (n_t, a_t)$, where n_t is the queue length at time t and a_t is the amount of service already received by the customer being served (if any). Indeed, this information is available in $\{n_s, s \leq t\}$ and it determines the departure time of the customer being served, i.e., the future jump of the queue length.

In this example, $x_t \in \mathbf{X} = \{0, 1, 2, \ldots\} \times [0, 1]$. The set $\mathbf{X}$ is not countable, so x_t is not a Markov chain. (It is seen to be a Markov process.)

One can approximate a_t by a random variable that takes values in a countable set. One way to do this is to observe that if $\tau_1, \ldots, \tau_n$ are independent and exponentially distributed with rate n, then $\tau_1 + \cdots + \tau_n \approx 1$ if $n \gg 1$. (See Exercise 3.2.)

Each service time is approximated by a sum of independent random variables τ_i that are considered as the durations of "phases" of service. One then keeps track of the number of phases that have been completed for the customer being served. The resulting process, together with the queue length, is then a Markov chain, by the memoryless property of the exponential distribution.

This idea is generalized below to the case of an arbitrary distribution.

Definition 3.2.4 [Phase-type distribution] Let $\phi = \{\phi_t, t \geq 0\}$ be a Markov chain on $\{1, 2, \ldots, m+1\}$ with rate matrix

$$Q = \begin{bmatrix} S & s_0 \\ s_1 & s_2 \end{bmatrix}$$

where $S \in R^{m \times m}$, $s_0 \in R^{m \times 1}$, $s_1 \in R^{1 \times m}$, $s_2 \in R$.

Let $\mu = (\alpha(0), \ldots, \alpha(m), 0) = (\alpha, 0)$ be the distribution of ϕ_0. Then $\tau := \min\{t > 0 \mid \phi_t = m+1\}$ is said to have a **phase-type** distribution with parameters (α, S). This will be indicated by writing that τ is $PH(\alpha, S)$.

The claim is that if τ is $PH(\alpha, S)$, then (below, $\mathbf{1}$ is the column vector in R^m with all components equal to one)

$$P\{\tau > t\} = \alpha e^{St}\mathbf{1}, \quad t \geq 0. \tag{3.3}$$

In particular, $P\{\tau < \infty\} = 1$ if and only if S is nonsingular, and then

$$E\{\tau\} = -\alpha S^{-1}\mathbf{1} < \infty. \tag{3.4}$$

To see this, one shows that $g_i(t) := P[\tau > t \mid \phi_0 = i]$ satisfies $\frac{d}{dt}g(t) = Sg(t)$ and $P\{\tau > t\} = \alpha g(t)$. (See Exercises 3.4 and 3.5 for details.)

The two transition diagrams in Figure 3.1 illustrate the idea. If one chooses $\alpha(1) = p = 1 - \alpha(n+1)$, then the first example approximates a random variable that is equal to a with probability p and to b with probability $1 - p$. The second one is a "Coxian" distribution.

It can be shown that any distribution can be approximated (in any suitable sense) by a phase-type distribution. M/GI/1 queues

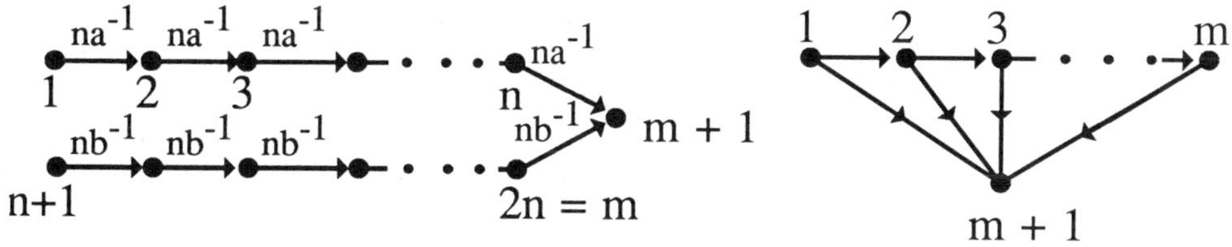

Figure 3.1. Phase-type distributions.

will then be approximated by M/PH/1 queues. This topic will be discussed in section 3.6.

3.3. QUASI-REVERSIBILITY AND PRODUCT-FORM

The basic results about quasi-reversible queues and networks of such queues are described below.

Notation 3.3.1 Let **C** be the countable set of possible customer classes. Assume that customers of classes $c \in \mathbf{C}$ arrive into one of the queues of 3.2.2 as independent Poisson processes with respective rates λ_c and have respective mean service times μ_c^{-1}. Define the **traffic intensity** of the queue as

$$\rho = \sum_{c \in \mathbf{C}} \lambda_c \mu_c^{-1} . \tag{3.5}$$

Thus ρ is the rate at which work enters the system. (With rate λ_c an average service time of μ_c^{-1} enters the queue, for $c \in \mathbf{C}$.)

Theorem 3.3.2

(a) For $n \geq 0$, the **invariant probability** $\pi(n)$ of having n customers in the queue is given by

$$\pi(n) = \begin{cases} \rho^n(1-\rho), \quad \text{if } \rho < 1, \\ \qquad \text{for M/GI/1-PS or -LCFS queues} \\ \dfrac{\rho^n}{n!} e^{-\rho}, \quad n \geq 0 \\ \qquad \text{for M/GI/}\infty \text{ queues} . \end{cases} \tag{3.6}$$

In particular, π depends on the service time distributions only through their means. This is an **insensitivity** property.

(b) Under the invariant probability distribution, *given the number of customers in the queue,* the **classes** of the customers **are**

independent and a customer is of class c with probability

$$p_c = \frac{\lambda_c \mu_c^{-1}}{\rho}.$$

(c) Moreover, when the queue is stationary, *the departure processes of different classes are independent Poisson processes and their past is independent of the present state of the queue.* That is, the queue is **quasi-reversible.**

(d) A **network** of quasi-reversible queues with Markov routing (see below) has a **product-form** invariant distribution that is the same *as if* all the flows in the network were independent Poisson processes with rates given by the flow conservation equations. The routing probabilities in the network are allowed to depend on the class of the customers. Equivalently, the class of a customer can also determine its **path** in the network (see Example 3.3.3). Also, a **typical** customer who jumps in the network sees the others in equilibrium (this is the Arrival Theorem).

These results will be proved later. (See Theorem 3.5.4 for the product-form and section 4.4 for the Arrival Theorem.) The results of (d) are valid for open, closed, and mixed networks. (See Examples 2.5.6.) **Markov routing** means that the routing decisions only depend on the class of the customer, not on the past evolution of the network. A precise definition will be given in 3.5.2.

It is important to keep in mind that an M/GI/1-FCFS queue is quasi-reversible only if it is an M/M/1 queue and if the customers of all classes have the same service time distribution. (See references.)

As indicated in Fact 2.5.7, when the network is not irreducible, the invariant distribution on a closed irreducible subset is found by normalizing π on that subset. Detailed examples will be worked out below.

Example 3.3.3 This example will indicate how Theorem 3.3.2 is used. Consider the network in Figure 3.2. Customers are of two classes 1 and 2; they arrive as independent Poisson processes with respective rates γ_1 and γ_2. Customers of class 1 are fed back exactly 3 times when they leave queue 3. Customers of class 2 are fed back with probability p, independently of each other, and from one time to the others.

One can view the routing of the customers of class 1 as being Markov. To see this, one may consider that the customers of class 1 are first fed back as customers of class $1b$, and that customers of class $1b$ are fed back as customers of class $1c$, and those of class

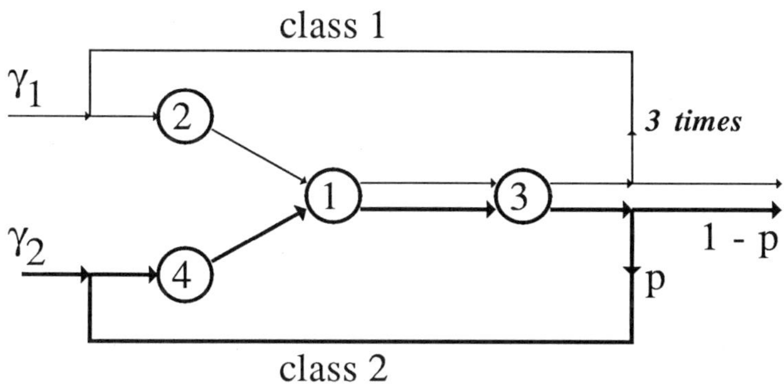

Figure 3.2. A network of quasi-reversible queues.

$1c$ are fed back as customers of class $1d$ who then leave the network. This is a particular case of a model where a customer of class c who leaves queue i is sent to queue j as a customer of class d with probability r_{ij}^{cd}, which is what is meant by Markov routing. The queues are as follows:

1: M/GI/1-PS; $\mu_1^{-1} = 3$; $\mu_2^{-1} = 2$;

2: M/M/1; $\mu_1^{-1} = 1$;

3: M/GI/1-LCFS; $\mu_1^{-1} = 2$; $\mu_2^{-1} = 5$;

4: M/D/∞; $\mu_2^{-1} = 3$.

Let also

$$\gamma_1 = 0.02, \quad \gamma_2 = 0.06, \quad p = 0.5 .$$

Denote by θ the **sojourn time** of a **typical** customer in the network. By typical customer, we mean the following. If the arrival times are S_n and if the state of the network at time t is x_t, then one can show that $\{x_{S_n -}\}$ is an ergodic Markov chain (see 2.9.3). A typical customer is one who arrives when the state of the network has the unique invariant distribution of the Markov chain $\{x_{S_n -}\}$.

It will be shown how to derive the following fact.

Fact: The mean sojourn time in the network is given by

$$E\{\theta\} = 56.7 .$$

The **procedure** for calculating $E\{\theta\}$ is typical of this class of problems:

(a) Find $\pi(n^1, n^2, n^3, n^4)$, the **invariant probability** of having n^i customers in queue i ($i = 1, 2, 3, 4$);

(b) Calculate $E\{N\} = E\{N^1\} + \cdots + E\{N^4\}$, where $E\{N^i\}$ is the **mean number of customers** in queue i;

(c) Let λ be the **total arrival rate** into the network;

(d) Little's result (see Theorem 3.4.1) implies that

$$E\{\theta\} = \lambda^{-1} E\{N\} .$$

These steps are carried out in detail next.

(a) To find π, one uses the fact that the network is **product-form.** (Indeed, the queues are quasi-reversible and the routing is determined by the classes of the customers.) Thus,

$$\pi(n^1, \ldots, n^4) = \pi_1(n^1) \ldots \pi_4(n^4),$$

where $\pi_i(.)$ is given by (3.6). It suffices then to calculate the rates of flows of the customers of both classes through the queues, and to evaluate the corresponding **traffic intensities** needed in (3.6). For instance, the traffic intensity for queue 1 is given by

$$\rho_1 = 4\gamma_1 \times 3 + \frac{\gamma_2}{1-p} \times 2 = 0.48 .$$

To see this, notice that the customers of class 1 go through queue 1 exactly 4 times, and that the customers of class 2 go through that queue an average of $\frac{1}{1-p}$ times.

Similarly, one finds (with $q = 1 - p$)

$$\rho_2 = 4\gamma_1 \times 1 = 0.08 ,$$

$$\rho_3 = 4\gamma_1 \times 2 + \frac{\gamma_2}{q} \times 5 = 0.76 ,$$

$$\rho_4 = \frac{\gamma_2}{q} \times 3 = 0.36 .$$

(b) Now,

$$E\{N^i\} = \sum_{n=0}^{\infty} n\,\pi_i(n) = \begin{cases} \dfrac{\rho_i}{1-\rho_i}, & 1 \le i \le 3 \\ \rho_4, & i = 4\,. \end{cases}$$

Hence

$$E\{N\} = \frac{0.48}{1-0.48} + \frac{0.08}{1-0.08} + \frac{0.76}{1-0.76} + 0.36 = 4.54\,.$$

(c) The **total arrival rate** in the network is given by $\lambda = \gamma_1 + \gamma_2 = 0.08$.

(d) It follows from **Little's result** that

$$E\{\theta\} = \lambda^{-1} E\{N\} = 56.7\,.$$

Example 3.3.4 This example concerns a **mixed** network. The point is to illustrate the role of the **normalizing constants.**

The network in Figure 3.3 consists of an M/GI/1-LCFS and an M/M/1-FCFS queue. There are two classes (1 and 2) of customers. There is a single customer of class 2. Customers of class 1 enter the network as a Poisson process with rate λ. The mean service times for class 1 and class 2 in queue 1 are a and b, respectively. The mean service time for class 2 in queue 2 is c. It is assumed that $\lambda a < 1$.

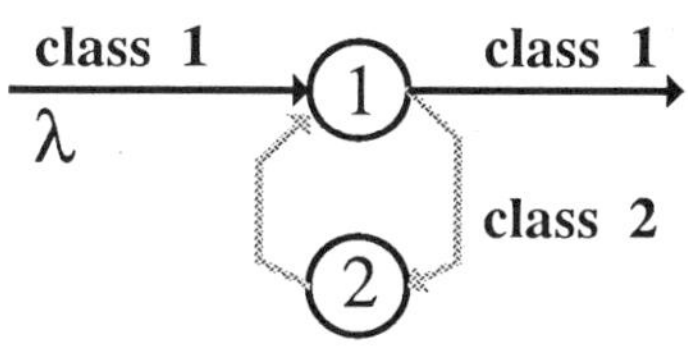

Figure 3.3. A mixed network.

The problem is to determine the average sojourn time for the customers of class 1 in the network. It will be shown that this **average sojourn** time is given by

$$E\{\theta\} = E\{\theta_0\}\frac{1 - \lambda a + 2\beta}{1 - \lambda a + \beta} \tag{3.7}$$

where $\beta := \dfrac{b}{c}$ and $E\{\theta_0\} := \dfrac{a}{1-\lambda a}$ is the average sojourn time when there is no customer of class 2. Notice that $E\{\theta\}$ increases from $E\{\theta_0\}$ to $2E\{\theta_0\}$ when $\beta = \dfrac{b}{c}$ increases from 0 to ∞.

The procedure is the same as in Example 3.3.3:

(a) Find $\pi(m, n)$, the invariant probability of having m customers of class 1 and n customers of class 2 in queue 1 ($m \geq 0$ and $0 \leq n \leq 1$);

(b) Calculate $E\{N\}$, the mean number of customers of class 1 in the network;

(c) The arrival rate of customers of class 1 is λ;

(d) Little's result implies that $E\{\theta\} = \lambda^{-1} E\{N\}$.

(a) To find π, one uses Theorem 3.3.2. To do this, one first has to find a set of rates that solve the flow conservation equations. For this network, it is clear that one can choose an arbitrary rate $\alpha > 0$ for the rate of the customers of class 2 through queue 1 and queue 2.

Next, one uses the theorem to calculate the probability of having m customers of class 1 and n customers of class 2 in queue 1, assuming that the arrival processes into that queue are independent and Poisson, with respective rate λ and α. One finds that this probability is

$$K\binom{m+n}{n}(\lambda a)^m (\alpha b)^n$$

where K is the normalizing constant that exists since $\lambda a < 1$.

To see this, notice that the rate ρ at which work enters queue 1 is $\lambda a + \alpha b$. Therefore, the probability of having $m + n$ customers in queue 1 is $K\rho^{m+n}$, by (3.6). Also, given that there are $m + n$ customers in the queue, the probability that m of them are of class 1 and that the others are of class 2 is $\binom{m+n}{n}(\frac{\lambda a}{\rho})^m (\frac{\alpha b}{\rho})^n$. Indeed, there are $\binom{m+n}{n}$ possibilities for n of the $m + n$ customers to be of class 2 and for the others to be of class 1. Also, the probability of any one of those possibilities is given by $(\frac{\lambda a}{\rho})^m (\frac{\alpha b}{\rho})^n$, by part (b) of Theorem 3.3.2.

Now, the probability of having k customers of class 2 in queue 2, assuming that the arrival process into that queue is Poisson with rate α is given by $G(\alpha c)^k$, by (3.6), where G is the normalizing constant.

Finally, by Theorem 3.3.2(d), one concludes that the probability of having m customers of class 1 in queue 1, n customers of class 2 in queue 1, and $1 - n$ customers of class 2 in queue 2 is

$$\pi(m, n) = G\binom{m+n}{n}(\lambda a)^m (\alpha b)^n (\alpha c)^{1-n}, \quad m \geq 0, \quad n \in \{0, 1\}$$

where G is the normalizing constant.

To determine G, one calculates

$$\sum_{m=0}^{\infty} \sum_{n=0}^{1} \pi(m, n) = G \sum_{m=0}^{\infty} (\lambda a)^m (\alpha c) + G \sum_{m=0}^{\infty} (m+1)(\lambda a)^m (\alpha b)$$

$$= G\left[\frac{\alpha c}{1 - \lambda a} + \frac{\alpha b}{(1 - \lambda a)^2}\right]$$

Hence,

$$G = \left[\frac{\alpha c}{1 - \lambda a} + \frac{\alpha b}{(1 - \lambda a)^2}\right]^{-1}.$$

(b) One then calculates $E\{N\} = \sum_{m=0}^{\infty} \sum_{n=0}^{1} m\, \pi(m, n)$. One finds

$$E\{N\} = G \sum_{m=0}^{\infty} m(\lambda a)^m (\alpha c) + G \sum_{m=0}^{\infty} m(m+1)(\lambda a)^m (\alpha b)$$

$$= G(\alpha c)(\lambda a)\frac{1}{(1 - \lambda a)^2} + G(\alpha b)(\lambda a)\frac{2}{(1 - \lambda a)^3}.$$

Hence, substituting the value of G, and after some simplifications,

$$E\{N\} = \frac{\lambda a(1 - \lambda a + 2\beta)}{(1 - \lambda a)(1 - \lambda a + \beta)}, \quad \text{with } \beta := \frac{b}{c}.$$

(d) This then allows us to calculate $E\{\theta\} = \lambda^{-1} E\{N\}$ and to verify (3.7).

3.4. LITTLE'S RESULT

Consider a service system that is *not necessarily* FCFS and that may have more than one server. Assume that (in a sense to be made precise below) the arrival rate is λ.

Say that the customers pay for using the service system. If C denotes the average cost paid by a customer, and if S is the average cost received per unit of time by the system, then one expects that

$$S = \lambda C . \tag{3.8}$$

For instance, if the average sojourn time per customer in the system is W, and the average number of customers in the system is L, then one expects that

$$L = \lambda W, \tag{3.9}$$

if one considers that each customer pays one unit of cost per unit of time spent in the system.

Little's result is a precise statement of those relations. Its importance stems from its generality.

We will prove (3.9). The derivation of the more general result (3.8) is suggested in Exercise 3.7.

Theorem 3.4.1 [Little's result] Let $\{\tau_n, d_n, \theta_n, n \geq 1\}$ be the arrival times, the departure times, and the sojourn times of customers in an *initially empty* service system. Define, for $t \geq 0$, $A_t = \max\{n \mid \tau_n \leq t\}$ and $D_t = \max\{n \mid d_n \leq t\}$, with the convention that $A_t = 0$ if $t < \tau_1$, and similarly for D_t. Thus A_t and D_t are the number of arrivals and departures in $[0, t]$, so that $A_t - D_t$ is the number of customers in the system at time t.

Assume that the limits

$$\lambda := \lim_{t \to \infty} \frac{A_t}{t} \tag{3.10}$$

and

$$W := \lim_{n \to \infty} \frac{1}{n} \sum_{m=1}^{n} \theta_m \tag{3.11}$$

exist and are finite. Then the limit

$$L := \lim_{t \to \infty} \frac{1}{t} \int_0^t [A_s - D_s]\, ds \tag{3.12}$$

exists and (3.9) holds.

Notice that these statements are for every realization of the queueing process, if it is random. In many random systems, it turns out that the limit λ in (3.10) is the same for all realizations and also that

$$\frac{1}{n} \sum_{m=1}^{n} \theta_m \to \lim_{k \to \infty} E\{\theta_k\} =: E\{\theta\} \text{ a.s. as } n \to \infty , \tag{3.13}$$

and, with $N_t = A_t - D_t$,

$$\frac{1}{t}\int_0^t N_s ds \rightarrow \lim_{u\rightarrow\infty} E\{N_u\} =: E\{N\} \text{ a.s. as } t \rightarrow \infty . \qquad (3.14)$$

It then follows from Little's result (3.9) that

$$E\{N\} = \lambda E\{\theta\} . \qquad (3.15)$$

This result is also called Little's result.

Conditions are known that imply (3.13)–(3.14). In most of the applications, these properties follow from the ergodicity of the queue length process.

Examples 3.4.2

(a) Consider an M/GI/1 queue with arrival rate λ and average service time μ^{-1}. Assume that $\lambda < \mu$. This can be shown to imply the ergodicity of the queue length process. (These issues will be analyzed in Chapter Seven.)

The server is a service system with arrival rate λ and an average sojourn time per customer equal to μ^{-1}. (For the sojourn times with the server are the service times.) As a consequence, the average number of customers with the server is equal to $\lambda\mu^{-1}$, by (3.15). This average number is the fraction of time (by ergodicity) that the queue is not empty.

One can show that this fraction of time is also the invariant **probability** $1 - \pi(0)$ **that the queue is not empty.** Thus,

$$1 - \pi(0) = \frac{\lambda}{\mu} . \qquad (3.16)$$

This gives an important interpretation of the **traffic intensity** $\rho = \lambda\mu^{-1}$.

Also, using the fact that Poisson arrivals see time averages (see 2.10.6), we find that this is the invariant probability that a customer finds the queue busy upon arrival.

This result can be extended to the case of multiple classes.

(b) Let $\mathbf{x} := \{x_t, \quad t \geq 0\}$ be the queue length process of an M/M/1 queue with arrival rate λ and service rate $\mu > \lambda$ (see 2.5.4). The ergodicity of x_t implies that (3.14) holds. Also, the ergodicity of the queue observed at the arrival times can be used to show that (3.13) also holds. The average queue length is given by (see (2.5.8))

$$E\{N\} = \sum_{n=1}^{\infty} n\,\pi(n) = \sum_{n=1}^{\infty} n\,\rho^n(1-\rho) = \frac{\lambda}{\mu - \lambda}. \tag{3.17}$$

It then follows from (3.15) that the **average sojourn time in the queue** is given by

$$E\{\theta\} = \lambda^{-1} E\{N\} = \frac{1}{\mu - \lambda}. \tag{3.18}$$

(c) Consider an open Jackson network, as in 2.4.4. Equation (3.15) implies that the **average sojourn time** $E\{\theta\}$ **in the network** is given by $(\sum_i \gamma_i)^{-1} E\{N\}$, where $E\{N\}$ is the average number of customers in the network. (Indeed, the arrival rate in the network is $\sum_i \gamma_i$.) The value of $E\{N\}$ is the sum of the average numbers of customers in the J queues. The latter can be found as in (3.17) since the marginal distributions of the queue lengths in the queues are the same as for an M/M/1 queue (see (2.5.11)). Hence,

$$E\{\theta\} = (\sum_{i=1}^{J} \gamma_i)^{-1} \sum_{i=1}^{J} \frac{\lambda_i}{\mu_i - \lambda_i}. \tag{3.19}$$

Proof of Little's result It will be assumed that $\lambda > 0$. The steps of the proof are as follows. Say that the customers leave in the order $n_1, n_2, \ldots$. (Recall that the system is not necessarily FCFS.)
(i) First observe that

$$\sum_{i=1}^{D_t} \theta_{n_i} \leq \int_0^t [A_s - D_s]\,ds \leq \sum_{n=1}^{A_t} \theta_n. \tag{3.20}$$

Indeed, the extreme terms are the sums of the total sojourn times of the customers that left [resp. arrived] before time t, while the integral is the sum of the sojourn times up to time t. Indeed, the integral can be written as

$$\int_0^t \sum_{n=0}^{\infty} 1\{\tau_n \leq s < d_n\}\,ds \tag{3.21}$$

which is seen to be the sum (over n) of the time spent by customer n in the system up to time t.

(ii) Assume for the time being that

$$\frac{\theta_n}{\tau_n} \to 0 \text{ as } n \to \infty . \tag{3.22}$$

This then implies that for any $\epsilon > 0$ there is some n_ϵ such that $\theta_n \leq \epsilon\tau_n$, and therefore $d_n \leq (1 + \epsilon)\tau_n$, for $n \geq n_\epsilon$. As a consequence, all the customers that arrive before time t, except possibly for the first n_ϵ, must leave before time $(1 + \epsilon)t$. This implies that $\frac{\tau_n}{n} \to \lambda^{-1}$, so that

$$\sum_{n=n_\epsilon+1}^{A_t} \theta_n \leq \sum_{i=1}^{D_{(1+\epsilon)t}} \theta_{n_i} \leq \int_0^{(1+\epsilon)t} [A_s - D_s]\,ds \leq \sum_{n=1}^{A_{(1+\epsilon)t}} \theta_n .$$

Dividing this relation by $(1 + \epsilon)t$, one finds that

$$\frac{1}{1+\epsilon}\frac{A_t}{t}\frac{1}{A_t}\sum_{n=n_\epsilon+1}^{A_t} \theta_n \leq \frac{1}{(1+\epsilon)t}\int_0^{(1+\epsilon)t} [A_s - D_s]\,ds$$

$$\leq \frac{A_{(1+\epsilon)t}}{(1+\epsilon)t}\frac{1}{A_{(1+\epsilon)t}}\sum_{n=1}^{A_{(1+\epsilon)t}} \theta_n .$$

Letting $t \to \infty$ and using (3.10)–(3.11) and the fact that ϵ is arbitrary then shows (3.9).

(iv) It remains to show (3.22). To do this, first notice that $\lim\frac{1}{n}\sum_{m=1}^{n-1}\theta_m = \lim\frac{1}{n}\sum_{m=1}^{n}\theta_m$, so that $\frac{\theta_n}{n} \to 0$. Also, for $t \geq t_\epsilon$ (some t_ϵ), $A_t \in (\lambda t(1 - \epsilon), \lambda t(1 + \epsilon))$, and $\tau_{A_t} \leq t < \tau_{A_t+1}$.

Therefore, $A_t[\lambda(1 + \epsilon)]^{-1} \leq \tau_{A_t+1}$ and $\tau_{A_t} < A_t[\lambda(1 - \epsilon)]^{-1}$. Hence, $\frac{1}{A_t}\tau_{A_t} \to \lambda^{-1}$ as $t \to \infty$, so that

$$\lim\frac{\theta_n}{\tau_n} = \frac{\lim\frac{\theta_n}{n}}{\lim\frac{\tau_n}{n}} = \frac{0}{\lambda^{-1}} = 0.$$

This concludes the proof.

3.5. PROOF OF PRODUCT-FORM RESULTS

This section is concerned with the proof of Theorem 3.3.2. A proof based on **time reversal** will be given first. The results will then be explained by a **forward time argument.**

In Fact 2.8.4 it was shown that the invariant distribution of a Jackson network can be verified by guessing that, when it is reversed in time, the state process corresponds to another Jackson network (with the reversed routing). This section shows that the same idea extends to networks of quasi-reversible queues with Markov routing.

We first define the queues, then the network.

Notation 3.5.1 [Quasi-reversible queues] For $1 \le i \le J$, $\mathbf{x}^i$ is a Markov chain on $\mathbf{X}^i$ with rate matrix Q_i that describes the evolution of one queue with arrivals $A_t^{ic} = \sum_{0<s\le t} 1\{(x_{s-}, x_s) \in \mathbf{A}_i^c\}$ and departures $D_t^{ic} = \sum_{0<s\le t} 1\{(x_{s-}, x_s) \in \mathbf{D}_i^c\}$ for customers of class $c \in \mathbf{C}$ (a given countable set). It is assumed that $\mathbf{x}^i$ admits the invariant distribution π_i, and that it is quasi-reversible with arrival and departure rate λ_i^c for each class c. Thus, the equalities (3.1)–(3.2) hold with the index i for $1 \le i \le J$.

Notice that it is assumed that the queues are **customer-conserving** in that the average arrival and departure rates of all classes are equal. This assumption can be relaxed.

Say that an arrival of class d occurs into queue j when its state is x^j. The state then jumps to some new value. To find the distribution of that value, observe that

$$
\begin{aligned}
P[x_t^j &= y^j \mid x_{t-}^j = x^j, t = \text{arrival time of class } d] \\
&= \lim_{\epsilon \downarrow 0} P[x_\epsilon^j = y^j \mid x_0^j = x^j, A_\epsilon^{jd} \ge 1] \\
&= \frac{q_j(x^j, y^j)\, 1\{(x^j, y^j) = \in \mathbf{A}_j^d\}}{\lambda_j^d}
\end{aligned}
\tag{3.23}
$$

where one used the fact that the rate λ_j^d of arrivals of class c is independent of the state x^j, by quasi-reversibility.

This observation will be used to specify the rate matrix of the network.

Definition 3.5.2 [Network of quasi-reversible queues] With probability r_{ij}^{cd}, a customer of class c who leaves queue i is sent immediately to queue j and the customer class switches to d; with probability $r_{i0}^{c} := 1 - \sum_{j,d} r_{ij}^{cd}$ that customer leaves the network. Also, an arriving customer is of class c and enters the network into queue i with probability r_{0i}^{c}. Customers arrive from outside the network as a Poisson process with rate γ. (See Figure 3.4.) All the routing decisions are independent. This will be called a **Markov routing.**

The network will be defined by deciding that if a departure from one queue becomes an arrival of class d in queue j, then the state of queue j will jump according to the distribution (3.23).

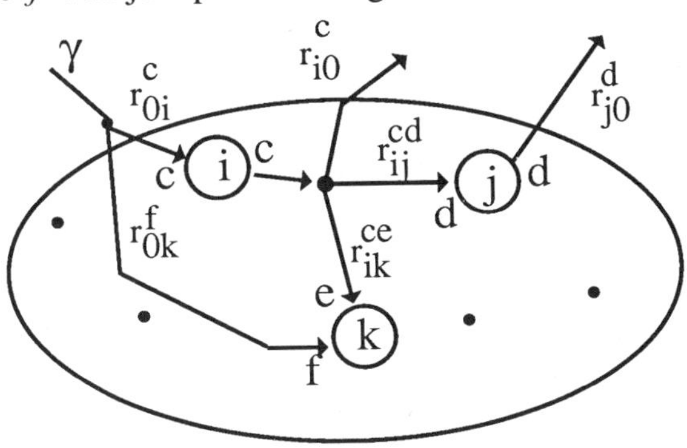

Figure 3.4. The network.

Let then $x = (x^1, \ldots, x^J)$ and $y = (y^1, \ldots, y^J)$ in $\mathbf{X} = \mathbf{X}^1 \times \cdots \times \mathbf{X}^J$ be two possible vectors of states of queues. The network is defined as the Markov chain x_t on $\mathbf{X}$ with the following rate matrix.

$$
q(x, y) = \begin{cases} q_i(x^i, y^i), \text{ if } (x, y) \in \mathbf{I}_i \\ \dfrac{q_i(x^i, y^i)\, q_j(x^j, y^j)}{\lambda_j^d} r_{ij}^{cd}, \text{ if } (x, y) \in \mathbf{T}_{ij}^{cd} \\ \dfrac{q_i(x^i, y^i)}{\lambda_i^c} \gamma r_{0i}^c, \qquad \text{if } (x, y) \in \mathbf{T}_{0i}^c \\ q_i(x^i, y^i) r_{i0}^c, \qquad \text{if } (x, y) \in \mathbf{T}_{i0}^c . \end{cases} \tag{3.24}
$$

In the expressions above, $\mathbf{I}_i$ is the set of pairs (x, y) such that $x^j = y^j$ for $j \neq i$ and such that a transition (x^i, y^i) is **internal** to node i, i.e., does not correspond to an arrival nor to a departure. Also, $\mathbf{T}_{ij}^{cd}$ denotes the set of transitions (x, y) corresponding to a customer of class c **moving** from i to j where it becomes a customer of class d. That is, for $i \neq l \neq j$

$$
(x, y) \in \mathbf{T}_{ij}^{cd} \equiv (x^i, y^i) \in \mathbf{D}_i^c,\ (x^j, y^j) \in \mathbf{A}_i^d, \quad x^l = y^l . \tag{3.25}
$$

Similarly, $(x, y) \in \mathbf{T}_{0i}^c$ if and only if $(x^i, y^i) \in \mathbf{A}_i^c$ and $x^l = y^l$ for $l \neq i$; and $(x, y) \in \mathbf{T}_{i0}^c$ if and only if $(x^i, y^i) \in \mathbf{D}_i^c$ and $x^l = y^l$ for $l \neq i$. Thus, the transition from x to y corresponds to an **arrival** of a customer of class c into queue i if $(x, y) \in \mathbf{T}_{0i}^c$ and to a departure of a customer of class c from queue i if $(x, y) \in \mathbf{T}_{i0}^c$.

For instance, consider a pair $(x, y) \in \mathbf{T}_{ij}^{cd}$. The rate of this transition is that of a departure of a customer of class c from i, times the probability that this customer will jump to j and change class to d, times the probability that an arrival of class d into j would correspond to a transition from x^j to y^j in queue j. This last probability is obtained as in (3.23).

Of course, some routing matrices $R = [r_{ij}^{cd}]$ are such that no customer ever leaves the network; others are such that customers of certain classes can leave while the other customers cannot; also, for some matrices all the customers can leave. The networks corresponding to those matrices are then respectively said to be **closed, mixed,** and **open.** This is similar to the discussion of Example 2.5.6.

For $(i, c) \in \{1, \ldots, J\} \times \mathbf{C}$, define λ_i^c to be the average rate of flow of customers of class c through queue i.

Those rates must satisfy the following **flow conservation equations**:

$$\lambda_i^c = \gamma r_{0i}^c + \sum_{jd} \lambda_j^d r_{ji}^{dc} . \tag{3.26}$$

Indeed, the average rate of flow (i, c) of customers of class c through i must be the sum of the average exogenous rate and of the "feedback" rates from the queues of the network.

The equations (3.26) may have a unique or infinitely many solutions. It will be assumed that the rates λ_i^c are **a** solution.

Definition 3.5.3 [Guess for reversed network] We will define the network obtained by **reversing the routing and by reversing each queue** of the network of Definition 3.5.2. It will be verified below that this is the Markov chain that describes the network of Definition 3.5.2 reversed in time. As a by-product, the invariant distribution of the network will be derived.

First, one defines the queues reversed in time. The queues of Definition 3.5.1, when considered in isolation, can be reversed in time. Queue i reversed in time is described by a Markov chain with rate matrix

$$\tilde{q}_i(x^i, y^i) = \frac{\pi_i(y^i)\, q_i(y^i, x^i)}{\pi_i(x^i)} . \tag{3.27}$$

The discussion of 2.8.6 can be reproduced to justify the fact that the time-reversed routing corresponds to the routing matrix $\tilde{R}$ given by

$$\tilde{r}_{ji}^{dc} = \frac{\lambda_i^c r_{ij}^{cd}}{\lambda_j^d}, \quad \tilde{r}_{0i}^c = \frac{\lambda_i^c r_{i0}^c}{\gamma}, \quad r_{i0}^c = \frac{\gamma r_{0i}^c}{\lambda_i^c} . \tag{3.28}$$

The guess for the network reversed in time is that it should correspond to the reversed queues with the reversed routing. This network has rate matrix $\tilde{Q}$ given by

$$\tilde{q}(y, x) = \begin{cases} \tilde{q}_i(y^i, x^i), & \text{if } (y, x) \in \mathbf{I}_i \\ \dfrac{\tilde{q}_j(y^j, x^j)\, \tilde{q}_i(y^i, x^i)}{\lambda_i^c} \tilde{r}_{ji}^{dc}, & \text{if } (y, x) \in \mathbf{T}_{ji}^{dc}, \\ \dfrac{\tilde{q}_i(y^i, x^i)}{\lambda_i^c} \gamma\, \tilde{r}_{0i}^{c}, & \text{if } (y, x) \in \mathbf{T}_{0i}^{c}, \\ \tilde{q}_i(y^i, x^i)\, \tilde{r}_{i0}^{c}, & \text{if } (y, x) \in \mathbf{T}_{i0}^{c}. \end{cases} \tag{3.29}$$

Theorem 3.5.4 [Product-form and network quasi-reversibility] The network of Definition 3.5.2 admits the **invariant distribution**

$$\pi(x) = \pi_1(x^1) \ldots \pi_J(x^J), \quad x = (x^1, \ldots, x^J) \in \mathbf{X}. \tag{3.30}$$

Moreover, the **network reversed in time** has rate matrix (3.29). That is, the network reversed in time is obtained by reversing the routing and each queue. In particular, in the original network, the *departure processes of the various classes from the J queues are independent Poisson processes and their past up to any time $t > 0$ is independent of x_t*. (For the future arrival times in the original network are independent Poisson processes that are independent of the present state.)

Thus, **a network of quasi-reversible queues with Markov routing has a product-form invariant distribution and is itself quasi-reversible.**

Proof: To prove the Theorem, it suffices to show that

$$\pi(x)\, q(x, y) = \pi(y)\, \tilde{q}(y, x), \quad x, y \in X. \tag{3.31}$$

(See Lemma 2.8.5.)

To verify this equation, one has to consider the cases where (x, y) is in $\mathbf{T}_{ij}^{cd}$, $\mathbf{I}_i$, $\mathbf{T}_{0i}^{c}$, or $\mathbf{T}_{i0}^{c}$. The details are given for the first case; the verification of the others is left to the reader.

For $(x, y) \in \mathbf{T}_{ij}^{cd}$, (3.31) reads

$$\pi(x) \frac{q_i(x^i, y^i)\, q_j(x^j, y^j)}{\lambda_j^d} r_{ij}^{cd} \tag{3.32}$$

$$= \pi(y) \frac{\tilde{q}_j(y^j, x^j)\, \tilde{q}_i(y^i, x^i)}{\lambda_i^c} \tilde{r}_{ji}^{dc},$$

where we used (3.24) and (3.29). Using (3.30) and the fact that $x^l = y^l$ for l not in $\{i, j\}$ allows us to rewrite (3.32) as

$$\lambda_i^c\, r_{ij}^{cd}\, \pi_i(x^i)\, q_i(x^i, y^i)\, \pi_j(x^j)\, q_j(x^j, y^j)$$
$$= \lambda_j^d\, \tilde{r}_{ji}^{dc}\, \pi_j(y^j)\, \tilde{q}_j(y^j, x^j)\, \pi_i(y^i)\, \tilde{q}_i(y^i, x^i).$$

A glance at (3.27) and (3.28) shows that the above equation indeed holds.

This concludes the proof.

Note that the flow conservation equations (3.26) are necessary for Q and $\tilde{Q}$ defined respectively by (3.24) and (3.29) to be rate matrices.

The proof given above is a nice illustration of the power of time-reversal arguments. For a different look at the result we give next a probabilistic proof based on a forward time argument.

Proposition 3.5.5 [Network with delays] Consider the network of Definition 3.5.2 with propagation delays of duration d introduced on the links between queues. (See Figure 3.5.) The value of d is arbitrarily small.

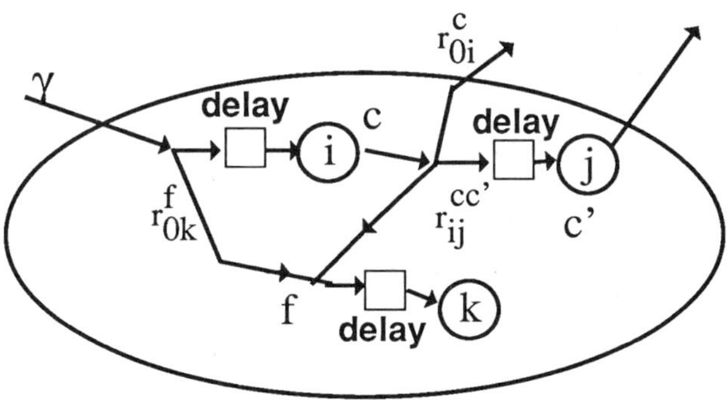

Figure 3.5. The network with delays.

The argument of 1.10 shows that the product-form distribution is invariant. It also shows that the departures from the network in

$[0, t]$ consist of independent pieces of Poisson processes. Therefore, they are Poisson; also, their past up to time t is seen to be independent of x_t.

It will be shown in 3.5.6 that the result continues to hold when $d = 0$ by letting $d \downarrow 0$. If the network models did not have this elementary continuity property they would be of limited practical value, in view of the ever-present modeling approximations.

This argument will also be used later to explain results on the distribution of the state of the network at jump times.

Fact 3.5.6 [Continuity as d ↓ 0] For completeness, we show that the result obtained in 3.5.5 carries to the limit as $d \downarrow 0$.

What is happening as $d \downarrow 0$ is that, for any fixed $t > 0$, each realization of $\{x_t; D_s^{ic},\ s \le t,\ 1 \le i \le J,\ c \in \mathbf{C}\}$ is modified **continuously.** (D_t^{ic} is the number of departures of class c from the network via node i in $[0, t]$.) By "continuously" one means that the realizations converge to their value for $d = 0$ as $d \downarrow 0$ (with probability one).

This is first shown in the case of an M/M/1 queue with feedback. (See Figure 3.6.)

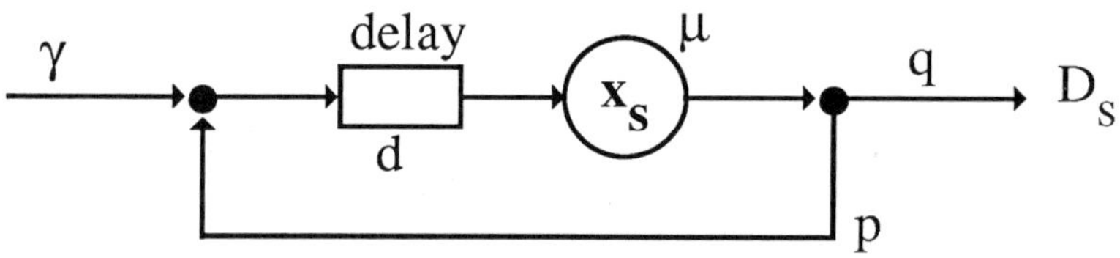

Figure 3.6. M/M/1 queue with feedback.

A realization of the queue length process x_t of the system with an arbitrary delay d is specified by the realizations of the initial queue length x_0 and of four Poisson processes with respective rates γ, λ, μp, and μq. These five random objects are independent. Recall that $\lambda = \gamma q^{-1}$ is the averagc rate of flow through the queue.

The system then evolves as follows: the Poisson process of rate γ is the **external arrival process;** that with rate λ specifies the **customers in transit** (in the delay) at time 0; that with rate μp indicates the **potential service completion times** of customers being fed back; that with rate μq indicates the **potential departure times** from the system. That is, if the queue is nonempty at a given jump time of the process with rate μp, then that time corresponds to a service completion of a customer that is then fed back; if the queue is empty at that jump time, then nothing happens at that time. The situation is similar for the process with rate μq.

This is equivalent to the definition of the Markov chain in view of the memoryless property of the exponential distribution. For instance, if a customer is in the queue at some time t, then his residual service time after time t is the time until the next jump of the process with rate μp or of the process μq. Fact 2.3.1 then shows that that time is exponentially distributed with rate μ and that the customer will be fed back with probability p.

Fix an arbitrary positive time t. Denote by $T_1 < T_2 < \cdots < T_n$ the jump times of the Poisson processes with rates γ, μp, and μq in $[0, t]$. With probability one, t is not a jump time of those processes. Let S_1 be the first jump time of the Poisson process with rate λ. Define

$$\Delta = \min\{S_1, T_2 - T_1, \ldots, T_n - T_{n-1}, t - T_n\}.$$

The claim is that for all $d < \Delta$, the realization of $\{D_s, 0 \le s \le t, x_t\}$ is the same as for $d = 0$. A look at Figure 3.7 explains that fact. In that figure, T_1 and T_5 are jump times of the Poisson process with rate γ, i.e., external arrival times; T_2 and T_3 are jump times of the Poisson process with rate μq, i.e., potential departure times; T_4 is a jump time of the Poisson process with rate μp, i.e., a potential feedback time.

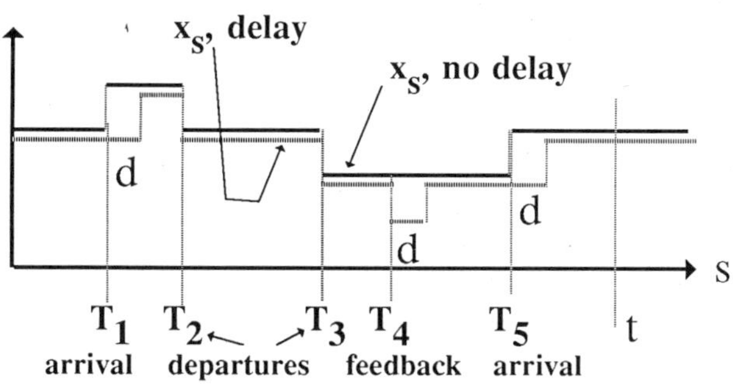

Figure 3.7. Typical realization.

Since $d < S_1$, there is no customer initially in transit. Since $d < T_m - T_{m-1}$ for $2 \le m \le n$, the successive values of the queue length just before the times $\{T_m, 1 \le m \le n\}$ are the same as for $d = 0$. Also, the jump times of x_t that do not correspond to a customer completing his transit in the delay are the same as for $d = 0$. Since $d < t - T_n$, the value of x_t is the same as for $d = 0$.

The argument above shows that $\eta := \{D_s, 0 \le s \le t, x_t\}$ converges with probability one to its value for $d = 0$ as $d \downarrow 0$. Since for all $d > 0$ the process η has the same distribution, it follows that it has the same distribution for $d = 0$. (For convergence with probability one implies convergence of the distribution.)

The only difficulty in generalizing this argument to the case of a general Markov chain model is one of notation. The argument is made possible by the fact that there are only finitely many Poisson processes involved, so that the value of Δ is positive, with probability one. In the general Markov chain model, the state process x_t of the system with delay d takes a finite number of values in $[0, t]$. To each state value i corresponds a collection of independent Poisson processes with rates $\{q(i, j), j \in \mathbf{X}, j \neq i\}$. The interpretation is that if the process with rate $q(i, j)$ is the first one to jump, then the state next jumps to j. By assumption, the sum $\sum_{j \neq i} q(i, j)$ of those rates is finite. It follows that the value of Δ will also be positive with probability one, so that the above argument can again be used.

3.6. QUASI-REVERSIBLE QUEUES

This section justifies the quasi-reversibility properties stated in Examples 3.2.2.

Example 3.6.1 [Multiclass M/M/1 queue] Customers of classes $c \in \mathbf{C}$ arrive at a queue as independent Poisson processes with rates λ_c. They are served on a FCFS basis. The service times are independent and exponentially distributed with the **same** rate μ. Let $\lambda = \sum_c \lambda_c$ and $p_c = \lambda_c \lambda^{-1}$ for $c \in \mathbf{C}$. Assume that $\rho := \lambda \mu^{-1} < 1$.

The **state** of the queue specifies the **classes** of the customers and their order. Thus, the state is

$$x = (c_1, c_2, \ldots, c_n) \tag{3.33}$$

when there are n customers in the queue and when the customer in position i is of class $c_i \in \mathbf{C}$ for $1 \le i \le n$. By convention, position 1 corresponds to the head of the queue (i.e., the customer being served), position 2 is the next one,...,and position n is the tail of the queue.

The claim is that the queue **reversed in time** corresponds to a similar queue except that the **positions are reversed.** That is, when the state of the reversed queue is x, as in (3.33), then c_1 is the class of the customer at the tail of the queue, c_n is the class of the customer at the head of the queue, etc. Also, the **invariant distribution** is claimed to be given by

$$\pi(x) = \pi(c_1, \ldots, c_n) = \rho^n(1-\rho)\, p_{c_1} \cdots p_{c_n}. \qquad (3.34)$$

In particular, the queue is **quasi-reversible** since the past of the departure processes is independent of the present state. (For the future of the arrival processes in the reversed queue is independent of the present state.)

To verify the claim, denote by Q the rate matrix of the multi-class M/M/1 queue and by $\tilde{Q}$ the rate matrix of the claimed reversed queue. It suffices to verify that

$$\pi(x)\, q(x, y) = \pi(y)\, \tilde{q}(y, x), \quad x, y \in \mathbf{X} \qquad (3.35)$$

where $\mathbf{X}$ is the set of possible states of the form (3.33). For instance, let x be as in (3.33) and let $y = (c_1, \ldots, c_n, c)$. That is, the transition from x to y in the original queue corresponds to an arrival of a customer of class c. Thus, $q(x, y) = \lambda_c$. In the claimed reversed queue, the transition from y to x corresponds to the service completion of a customer of class c, so that $\tilde{q}(y, x) = \mu$. Hence, (3.35) becomes

$$\pi(x)\, \lambda_c = \pi(y)\, \mu$$

which is seen to be satisfied by (3.34). The other cases are similar.

Of course, a **forward time** proof is possible by observing that the queue length must behave as that of a single class M/M/1 queue and that the customer classes are independent since the arrival processes can be obtained by Bernoulli sampling of a single Poisson process.

Example 3.6.2 [M/PH/1-LCFS queue] Customers arrive at a queue as independent Poisson processes with rates λ_c, for $c \in \mathbf{C}$. The queue has one server who serves the customers on a LCFS basis, as described in Example 3.2.2(2). The service times are independent and the distribution of the service time of a customer of class c is $PH(\alpha_c, S_c)$ (see Definition 3.2.4). The possible phases will be $\{1, 2, \ldots, m\}$ for all classes, to simplify the notation. (See Figure 3.8.)

The **state** of the queue is

$$x = (c_1, \phi_1; \ldots; c_n, \phi_n) \tag{3.36}$$

when there are n customers and when the customer in position m is of class c_m with the service in phase ϕ_m for $1 \le m \le n$. Position 1 corresponds to the *last* arrived among the present customers, etc. (See Figure 3.8.)

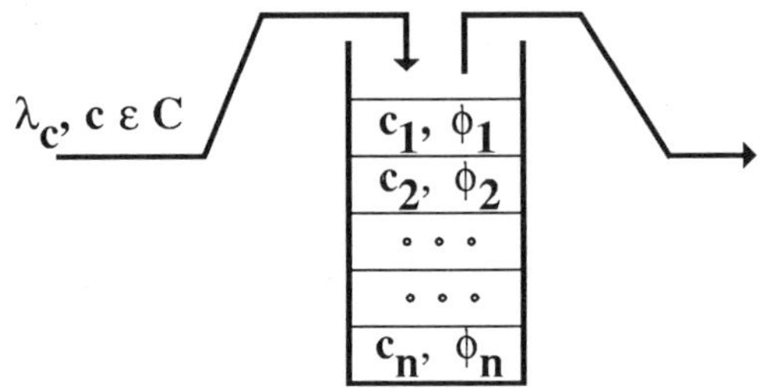

Figure 3.8. The M/PH/1-LCFS queue.

Define the traffic intensity ρ as in (3.5). The claim is that if $\rho < 1$, then the queue is **quasi-reversible** and admits the **invariant distribution** π given by (for x as in (3.36))

$$\pi(x) = \rho^n(1 - \rho) \prod_{m=1}^{n} p(c_m, \phi_m) \tag{3.37}$$

with

$$p(c, \phi) = \frac{\lambda_c \mu_c^{-1}}{\rho} v_c(\phi)$$

where $v_c = (v_c(1), \ldots, v_c(m))$ is the probability distribution that solves

$$v_c[S_c + s_{0c}\alpha_c] = 0 . \tag{3.38}$$

In particular, one can use (3.4) and (3.38) to verify that $\mu_c = v_c s_{0c}$. (See Exercise 3.10.)

Thus, the distribution of the queue length is as claimed in Theorem 3.3.2. Also, given that there are n customers in the queue, their classes are independent and equal to c with probability $\frac{\lambda_c \mu_c^{-1}}{\rho}$ for $c \in \mathbf{C}$. In addition, given that a customer is of class c, the service of that customer is in phase ϕ with probability $v_c(\phi)$.

To prove this result, one verifies that the state process $\mathbf{x}$ **reversed in time** also corresponds to an M/PH/1-LCFS queue with the same arrival rates and with the **reversed phase processes.** The reversed phase process for class c is defined by

$$\left.\begin{aligned} \tilde{S}_c(i, j) &= \frac{v_c(j)\, S_c(j, i)}{v_c(i)}, \quad \tilde{\alpha}_c(i) = \frac{s_{0c}(i)\, v_c(i)}{\mu_c}, \\ \tilde{s}_{0c}(i) &= \frac{\mu_c\, \alpha_c(i)}{v_c(i)} . \end{aligned}\right\} \quad (3.39)$$

To show this, one must verify that

$$\pi(x)\, q(x, y) = \pi(y)\, \tilde{q}(y, x) \quad (3.40)$$

for all x, y where Q is the rate matrix of the original M/PH/1-LCFS queue and $\tilde{Q}$ is the rate matrix of the one that is claimed to correspond to the reversed process.

For instance, if the transition from x to y corresponds to the service phase of a customer of class c in position 1 jumping from the value i to the value j, then

$$q(x, y) = S_c(i, j) \text{ and } \tilde{q}(y, x) = \tilde{S}_c(j, i) .$$

Some algebra allows us to check (3.40).

Similarly, if the transition from x to y corresponds to the arrival of a customer of class c with initial service phase equal to i, then

$$q(x, y) = \lambda_c\, \alpha_c(i) \text{ and } \tilde{q}(y, x) = \tilde{s}_{c0}(i) .$$

One can verify that (3.40) holds.

Example 3.6.3 [M/PH/1-PS] This case is similar. The customers are served on a PS basis as described in Example 3.2.2.(1). The state is defined as in (3.36), except that an arriving customer is equally likely to join the queue in any position. (The motivation is that a customer can leave from any position, so that if one wants the reversed M/PH/1-PS queue to be a queue of the same type, then it is necessary to assume that customers can join the queue in any position.)

The claim is the same as in Example 3.6.2, i.e., that if $\rho < 1$, then the queue is **quasi-reversible** with the **invariant distribution** (3.37).

The proof consists in verifying that the process **reversed in time** corresponds to the M/PH/1-PS queue with the same arrival rates and with the **reversed phase processes** defined by (3.39).

One has to check (3.40). For instance, if (x, y) corresponds to the arrival of customer of class c with initial service phase i in a given position (of the $n + 1$ equally likely positions, if x is as in (3.36)), then

$$q(x, y) = \lambda_c \alpha_c(i)\frac{1}{n+1} \text{ and } \tilde{q}(y, x) = s_{0c}(i)\frac{1}{n+1}$$

where the factor $\frac{1}{n+1}$ in the last term comes from the fact that any given customer in y receives only a fraction $\frac{1}{n+1}$ of the server effort (by processor-sharing). Combining this with (3.39) and (3.37) allows us to verify (3.40).

Example 3.6.4 [M/PH/∞ queue] The arrivals, the service distributions, and the state are in Example 3.6.2. There are infinitely many servers. Customers are equally likely to join the queue in any position.

The claim is that the queue is **quasi-reversible** with the **invariant distribution**

$$\pi(x) = \frac{\rho^n}{n!}e^{-\rho} \prod_{m=1}^{n} p(c_m, \phi_m) \tag{3.41}$$

where $p(c_m, \phi_m)$ is as in Example 3.6.2.

The proof consists in showing that the process **reversed in time** corresponds to the M/PH/∞ with the **reversed phase processes** (3.39).

Once again, one must check (3.40). For instance, if (x, y) corresponds to a departure of a customer of class c from a given position when its service is in phase i, then

$$q(x, y) = s_{0c}(i) \text{ and } \tilde{q}(y, x) = \lambda_c \tilde{\alpha}_c(i)\frac{1}{n}$$

where the factor $\frac{1}{n}$ corresponds to the probability that the arriving customer in the reversed process joining the queue in a given position among the n possible ones. (If there are n customers in state x, then there are n possible positions for a customer to join the queue when its state is y.) Combining this relation with (3.39) and (3.41) allows us to verify (3.40).

Definition 3.6.5 [Symmetric queue] The above examples are particular cases of **symmetric queues.** (See Figure 3.9.)

The arrivals at a symmetric queue are independent Poisson processes with rates λ_c for $c \in \mathbf{C}$. The state and the service times are as in Example 3.6.2. A customer who arrives when the state is (3.36) joins the queue in position l with probability $\gamma(n+1, l)$ for $1 \le l \le n+1$, where $\sum_{l=1}^{n+1} \gamma(n+1, l) = 1$. When the state is

(3.36), the server works at a total rate $\psi(n)$ and a fraction $\gamma(n, l)$ of that effort is allocated to the customer in position l, for $1 \le l \le n$.

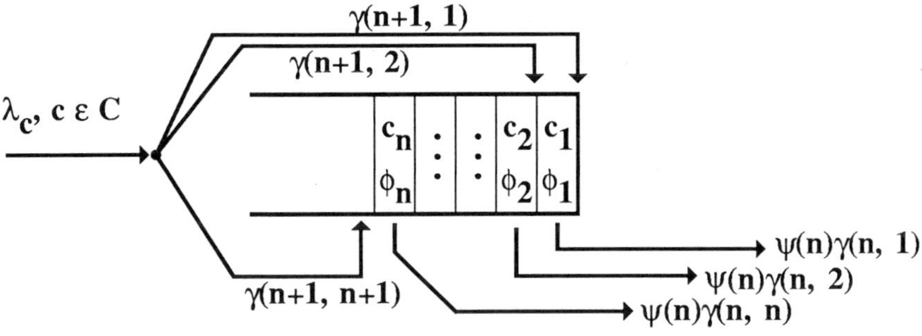

Figure 3.9. The symmetric queue.

The claim is that if there is a constant A that normalizes π given in (3.42) below, then that π is the **invariant distribution** for the symmetric queue and the queue is **quasi-reversible.**

$$\pi(x) = A \frac{\rho^n}{\psi(1)\,\psi(2)\ldots\psi(n)} \prod_{m=1}^{n} p(c_m, \phi_m) . \qquad (3.42)$$

This is shown by proving that the queue **reversed in time** is again a symmetric queue, with the **reversed phase processes** (3.39). One shows that by verifying (3.40). For instance, if (x, y) corresponds to the arrival of a customer of class c in position l with initial service phase i, then

$$q(x, y) = \lambda_c\, \alpha_c(i)\, \gamma(n+1, l) \text{ and}$$

$$\tilde{q}(y, x) = \tilde{s}_{0c}(i)\, \gamma(n+1, l)\, \psi(n+1)$$

and (3.42) and (3.39) can then be used to check (3.40).

To see that examples 3.6.2–3.6.4 are particular cases of symmetric queues, notice that the M/PH/1-PS [resp. LCFS] queue corresponds to $\gamma(n, l) \equiv n^{-1}$ [resp. $\gamma(n, l) = 1\{l = 1\}$] and $\psi(n) \equiv 1$. The M/PH/∞ queue corresponds to $\gamma(n, l) \equiv n^{-1}$ and $\psi(n) = n$.

3.7. OTHER PRODUCT-FORM NETWORKS

So far, the product-form networks that were discussed are constructed by interconnecting quasi-reversible queues with a Markov routing. This section gives some examples of product-form networks that do not fall into this class.

The following simple fact will be needed.

Fact 3.7.1 [Restriction of a time-reversible Markov chain] Let $\mathbf{x}$ be a time-reversible Markov chain on $\mathbf{X}$ with rate matrix Q and invariant distribution π. Let $\mathbf{Y}$ be a given subset of $\mathbf{X}$ and let $\mathbf{y}$ be a Markov chain on $\mathbf{Y}$ with rate matrix Q' defined for $i, j \in \mathbf{Y}$ by

$$q'(i, j) = \begin{cases} q(i, j) & \text{if } j \neq i \\ -\sum_{i \neq k \in \mathbf{Y}} q(i, k) & \text{if } j = i \end{cases}$$

The meaning of this definition is that only the transitions that **do not exit Y** are permitted.

The claim is that $\mathbf{y}$ admits the **invariant distribution**

$$\pi_Y = \{\frac{\pi(i)}{\pi(\mathbf{Y})}, \quad i \in \mathbf{Y}\}$$

under which it is time-reversible. This is immediate since for any $j \neq i$ in $\mathbf{Y}$ one has $\pi_Y(i)\, q'(i, j) = \pi_Y(j)\, q'(j, i)$, from the time-reversibility of $\mathbf{x}$.

Definition 3.7.2 [Restricted reversible network] Consider a network of J symmetric queues with **reversible phase processes** for every customer class, and with **reversible routing.** This network is time-reversible and has a product-form invariant distribution, by Theorem 3.5.4. It can be restricted (in the sense of Fact 3.7.1) by blocking the transitions that would make the state of the network leave a given set. It follows from Fact 3.7.1 that the restricted network still has a **product-form** invariant distribution.

The restriction can correspond to **finite buffers** or to more complicated restrictions (e.g. given **maximum numbers of customers in given subsets** of the network).

Another class of examples is described next.

Definition 3.7.3 [Interfering queues] Consider an open single-class network of J queues. The routing is Markov with routing matrix $R = \{r_{ij},\ 0 \le i,\ j \le J\}$. (As before, r_{0j} is the probability that a customer arriving from outside enters the network in queue j, and $r_{i0} = 1 - \sum_{j=1}^{J} r_{ij}$ is the probability that a customer leaves the network upon leaving queue i.) The customers arrive from outside as a Poisson process with rate γ.

Let $x_t = (x_t^1, \ldots, x_t^J)$ denote the vector of queue lengths at time t. Assume that when $x_t = x$ the server in queue i **serves at rate** $\mu_i(x)$, for $1 \le i \le J$. Thus, *the rate of service in one queue depends on the state of the other queues.*

Fact 3.7.4 Assume that there exists a function ψ such that

$$\begin{aligned} \psi(0) &= 1 \\ \psi(x + e_i) &= \mu_i(x + e_i)\,\psi(x) \end{aligned} \tag{3.43}$$

for all $x \in \mathbf{X} = \{0, 1, \ldots\}^J$. Then the distribution

$$\{\psi(x)\}^{-1}\lambda_1^{x^1} \ldots \lambda_J^{x^J} \tag{3.44}$$

where the λ_is are a solution of the flow conservation equations is invariant for x_t.
Moreover, the network is quasi-reversible if (3.44) can be normalized.

The proof is left as an exercise. (See Exercise 3.12.)

The following two **examples** illustrate the type of networks permitted by this definition.

(a) Choosing $\{\psi(x)\}^{-1} = (x^1 + \cdots + x^J)!$, one finds

$$\mu_i(x) = \frac{\psi(x)}{\psi(x - e_i)} = (x^1 + \cdots + x^J)^{-1}.$$

(b) Choosing

$$\begin{aligned} \{\psi(x)\}^{-1} &= a(1)\,a(2) \ldots a(x^1 + x^2) \\ &\quad \times b(1)\,b(2) \ldots b(x^3 + \cdots + x^J), \end{aligned}$$

one finds

$$\mu_i(x) = \begin{cases} a(x^1 + x^2) & \text{for } i = 1, 2 \\ b(x^3 + \cdots + x^J) & \text{for } 3 \le i \le J \end{cases}$$

Many variations are possible. Notice that the resulting network is not product-form in the strict sense, but its invariant distribution has a closed form. Exercise 3.13 provides another example.

3.8. NORTON'S THEOREM

A result similar to **Norton's Theorem for electrical circuits** is known to hold for networks of quasi-reversible queues. The result states that, under certain conditions, it is possible to **replace a subset** of a network **by a single queue** without modifying the **invariant distribution** of the rest of the network. The *dynamics* of the network are not preserved by this transformation. For instance, the distributions of the sojourn times are modified.

The "equivalent" queue serves customers at a rate that is equal to the **expected rate** of departures from the network subset given the state of the rest of the network.

Algebraic proof are available. (See Exercises 3.16 and 3.17.) Let us attempt to give an **intuitive justification** of the result on an abstract model. This justification is sufficient for most practical cases. It is, however, not the best possible in that it does not cover some situations.

For $i = 1, 2$, $\mathbf{x}^i$ is a Markov chain on $\mathbf{X}^i$ with invariant distribution π_i and with arrivals and departures. Each chain is assumed to be quasi-reversible with arrival and departure rates equal to λ. For simplicity it will be assumed that these arrivals and departures are of a given customer class. One constructs a network by interconnecting the queues so that the departures of one become the arrivals of the other, and vice-versa. (See Figure 3.10.)

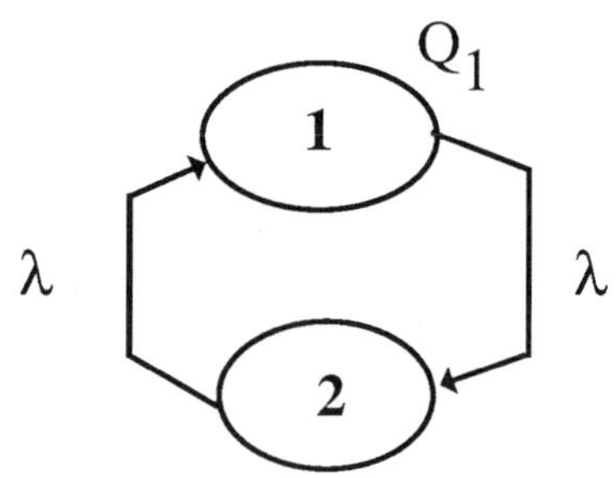

Figure 3.10. The network.

As seen previously, this network admits the invariant distribution $\pi = \pi_1 \pi_2$. Assume that $x_t = (x_t^1, x_t^2)$ belongs to the closed and irreducible set $\mathbf{S} \subset \mathbf{X}^1 \times \mathbf{X}^2$. The interesting case is when $\mathbf{S}$ is not the product of a subset of $\mathbf{X}^1$ and a subset of $\mathbf{X}^2$. Indeed, in that case, x_t^1 and x_t^2 are not independent under the invariant distribution

$$\pi_S(x) = \frac{\pi(x)\, 1\{x \in \mathbf{S}\}}{\pi(\mathbf{S})} = \frac{\pi_1(x^1)\, \pi_2(x^2)\, 1\{(x^1, x^2) \in \mathbf{S}\}}{\pi(\mathbf{S})} . \quad (3.45)$$

Consider the modification of the network illustrated in Figure 3.11.

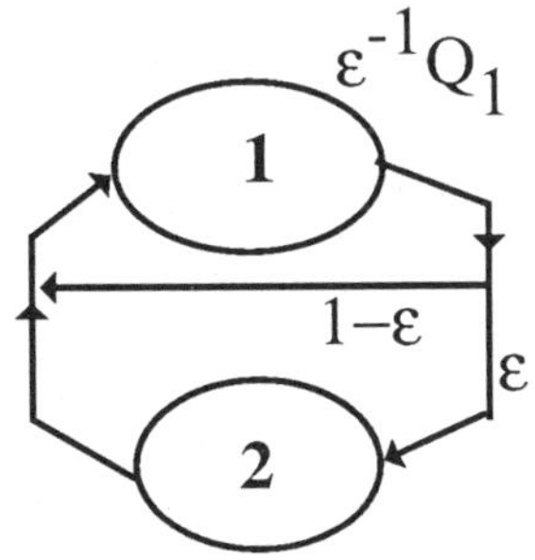

Figure 3.11. The modified network.

In this modified network, the first queue is assumed to evolve ϵ^{-1} times **faster** than in the original network. One can see that this network still admits $\pi = \pi_1 \pi_2$ as an invariant distribution. This can be verified by checking that the network reversed in time is the obvious guess. More simply, one can see that if one inserts just before the queues small delays that are initially filled with independent Poisson processes with rates $\lambda\epsilon^{-1}$ before queue 1 and λ before queue 2, and if x_0 has distribution π, then this situation is invariant.

The basic **assumptions** needed at this point are the following. First, *the modified network is irreducible on the closed set* $\mathbf{S}$. Second, *the "closed-loop" version of queue 1 is irreducible on the closed set* $\mathbf{S}(x^2) = \{x^1 \mid (x^1, x^2) \in \mathbf{S}\}$, for each x^2. This closed-loop queue is as indicated in Figure 3.12.

One can then argue that if ϵ is very small, then queue 1 in Figure 3.11 will make a large number of transitions independently of queue 2 (i.e., without arrivals from or departures into queue 1).

Thus, *between transitions of* x^2, *queue 1 will approach its invariant distribution on* $\mathbf{S}(x_0^2)$. As a consequence, the rate of tran-

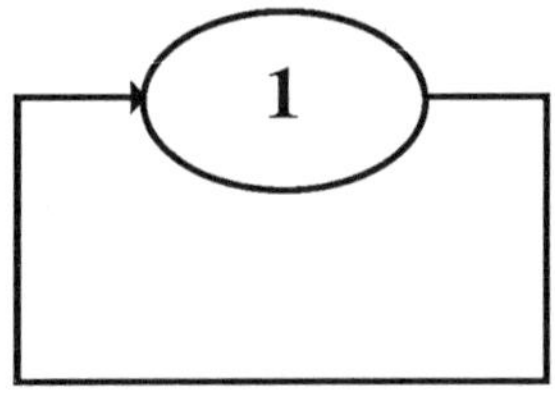

Figure 3.12. The closed-loop version of queue 1.

sitions from queue 1 into queue 2 when queue 2 is in state x^2 is essentially equal to the expected rate of departures from queue 1, where the expectation is with respect to the invariant distribution of queue 1 on $\mathbf{S}(x^2)$. (If x^1 has that invariant distribution, then the rate of a transition into queue 2 is ϵ^{-1} times the expected rate of a departure from queue 1, times the probability ϵ that the customer is routed to queue 2 and not back to queue 1.)

But this invariant distribution is equal to the conditional distribution of x^1 given x^2 for the original network. Indeed, let $X = (X^1, X^2)$ have the distribution $\pi = \pi_1 \pi_2$. Then

$$P[X^1 = x^1 \mid X^1 \in \mathbf{S}(x^2)]$$
$$= P[X^1 = s_1, X^2 = x^2 \mid X \in \mathbf{S}, X^2 = x^2] .$$

The above discussion shows that **one can replace queue 1 by a queue that serves customers at the conditional rate of service in queue 1 given the state of the queue 2.**

Two assumptions were needed for the argument to be justified. They are satisfied when the closed irreducible sets are specified by the **number of customers in the subnetworks.** This is the case for networks of symmetric queues. Notice that the argument extends to the **multiclass case.** Figure 3.13 illustrates the necessary modification.

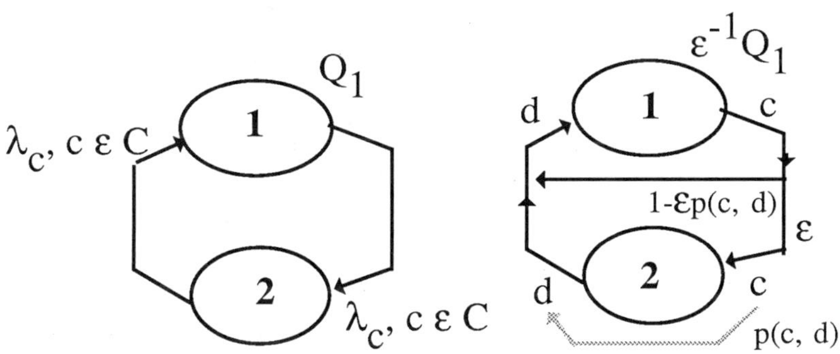

Figure 3.13. The multiclass network and its modification.

3.9. INSENSITIVITY

Insensitivity refers to the property of some stochastic processes that have certain invariant probabilities that depend on a set of probability distributions only through their **mean values.** Our objective in this section is to show that this property can be viewed as a **product-form result** that follows from the now familiar **quasi-reversibility** argument.

Consider the network in Figure 3.14 with two queues and customer classes **C**.

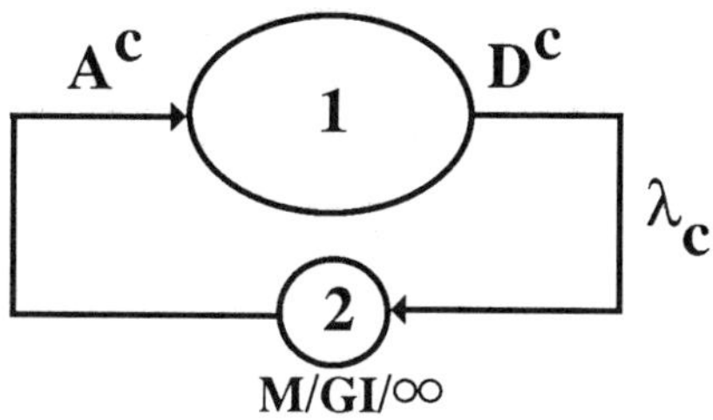

Figure 3.14. The network.

Queue 1 is modeled by a Markov chain **x** with rate matrix Q_1 and invariant distribution π_1. Departures and arrivals of class c correspond to transitions in $\mathbf{D}^c$ and $\mathbf{A}^c$, respectively. Assume that the queue is quasi-reversible with rates λ_c. The queue has the property that it contains **at most one customer of each class** c. That is, if a customer of class c arrives (with Poisson rate λ_c) when there is already a customer of class c in the queue, then the customer "jumps" the queue by immediately joining the departure process. Queue 2 is an M/GI/∞ queue where the customers of class c require a mean service time equal to λ_c^{-1}. (The service time distributions in the M/GI/∞ queue can be approximated by *PH* distributions.)

Assume that the closed network is such that it contains **exactly one customer of class** c for $c \in \mathbf{C}$. This network is **product-form,** since the two queues are quasi-reversible. A possible **state** of the network is $x = (x^1, x^2)$ where $x^2 = \{T^c, c \in \mathbf{C}\}$ is the collection of **residual service times** of the customers. By convention, $T^c = 0$ if the customer of class c is not in the queue. Given the state x^1 of the first queue, there is a subset $\mathbf{C}(x^1)$ of $\mathbf{C}$ that indicates which customer classes are in queue 2. Thus, to find the invariant distribution of the closed network, one must normalize the restriction of the product-form invariant distribution on the set of states x that correspond to having one customer of each class in the network. Denote that set by $\mathbf{Y}$.

Say that queue 2 contains customers of classes $c \in \mathbf{C}(x^1)$. The invariant probability that these customers have residual service times $\{T^c, c \in \mathbf{C}(x^1)\}$ such that $T^c \le t^c$ for $c \in \mathbf{C}(x^1)$ can be verified to be given by

$$\prod_{c \in \mathbf{C}(x^1)} R_c(t^c)$$

where

$$R_c(t) = \lambda_c \int_0^t [1 - F_c(s)]\, ds \tag{3.46}$$

where F_c is the probability distribution of the service times of class c. (See Exercise 3.18.)

This implies that

$$\pi_Y(x) = \pi_1(x^1) \prod_{c \in \mathbf{C}(x^1)} R_c(t^c) \tag{3.47}$$

is the invariant probability that the state of queue 1 is x^1 and that the residual service times in the second queue are not larger than the t^cs.

This example shows that the quasi-reversibility conditions for queue 1 imply that the invariant distribution of x_t^1 in the closed network does not depend on the distributions F_c.

One may view x_t^1 in the closed network as being a process that takes values in a countable set and such that its transitions may start **clocks.** A transition in $\mathbf{D}^c$ starts a clock with an initial value distributed as F_c. Also, when a clock of class c runs out, a transition in $\mathbf{A}^c$ is induced. Such a model is called a **Generalized Semi-Markov Process** (GSMP).

This result extends to the case where the **speeds** of the clocks (the service rates in the M/GI/∞ queue) depend on the state x^1.

Notice also that customers of different classes leaving the M/GI/∞ queue induce different transitions in queue 1. This corresponds to "age-dependent routing" in the GSMP.

3.10. EXERCISES

3.1. Quasi-reversibility Prove the statements of section 3.1.
Hint: First assume (3.2.1). To show that the arrival processes are independent and Poisson, first show that the sum of the arrivals is a Poisson process with rate $\lambda = \sum \lambda_c$.

This follows from Theorem 2.10.6. Then argue that the various arrivals can be viewed as being obtained by Bernoulli sampling of the sum. Conclude with Theorem 2.11.2(d). The second part (concerning the departures) now follows by time-reversal.

3.2. Phase approximation of a constant Let $\tau_1, \ldots, \tau_n$ be independent and exponentially distributed with rate n. Show that $\tau := \tau_1 + \cdots + \tau_n \approx 1$. *Hint:* Calculate $E\{\tau\}$ and $\mathrm{var}(\tau)$ (see Exercise 2.1(a)). One can then use the Chebyshev inequality (see Appendix A). This is known to be the best (least mean squares) approximation of the constant 1 with n phases. See Aldous-Schepp (1987).

3.3. Simple random variable Let τ be a random variable such that $P\{\tau = a_k\} = p_k$ for $1 \le k \le K < \infty$ for given values of a_k and p_k. Use the method of Remark 3.2.3 to approximate τ by the first hitting time of a continuous-time Markov chain. Show that the approximation is good. *Hint:* Let σ be the hitting time suggested in Remark 3.2.3. Estimate $P\{\sigma \approx a_k\}$.

3.4. Phase-type distributions Prove the formula (3.3). *Hint:* Let $g_i(t)$ be defined as in Definition 3.2.4. Using the first-step analysis, one finds that

$$P[\tau > t \mid \phi_0 = i] \approx (1 + S(i, i)\epsilon)\, P[\tau > t - \epsilon \mid \phi_0 = i] + \sum_{j \neq i} S(i, j)\epsilon\, P[\tau > t - \epsilon \mid \phi_0 = j] .$$

This leads to $\frac{d}{dt} g(t) = Sg(t)$ from which one concludes that $g(t) = \exp\{St\}$.

3.5. Phase-type distributions (cont.) Prove the formula (3.4). *Hint:* The first step equations for $a(i) := E[\tau \mid \phi_0 = i]$ are

$$a(i) = s(i)^{-1} + \sum_{j \neq i} \frac{S(i, j)}{s(i)} a(j)$$

where $s(i) := -S(i, i)$. These equations can be written as

$$-s(i)\, a(i) + \sum_{j \neq i} S(i, j)\, a(j) = -1 ,$$

or $Sa = -\mathbf{1}$. Alternatively, one can integrate (3.3).

3.6. Infinite mean queue length Let $x = \{x_t, t \geq 0\}$ be a Markov chain on $\mathbf{X} = \{0, 1, 2, \ldots\}$ and with rate matrix Q given by

$$q(i, i-1) = 1\{i > 0\},\ q(i, i+1) = a_i < 1,$$
$$q(i) = q(i, i-1) + q(i, i+1).$$

Show that for all $\epsilon > 0$ there is a choice of $\{a_i, i \geq 0\}$ for which

(a) $\mathbf{x}$ is ergodic with some invariant distribution π,

(b) $\lambda := \sum_{i=0}^{\infty} a_i \pi(i) \leq \epsilon$,

(c) $E\{N\} := \sum_{i=0}^{\infty} i\, \pi(i) = \infty$.

This shows that an arbitrarily small arrival rate can cause an infinite mean queue length.

Hint: Try $a_0 \ldots a_{i-1} = \frac{\beta}{i^2}$ for $i \geq 1$, with $\beta = \beta(\epsilon) > 0$.

3.7. Brumelle's result Prove the formula (3.8). *Hint:* Let $c(n, t)$ be the cost paid by customer n at time t (per unit of time). Assume that $c(n, t) = 0$ if the customer is not in the system at time t. Assume also that $c(n, t) \geq 0$. Then

$$C = \lim_{N \to \infty} \frac{1}{N} \sum_{n=1}^{N} \int_0^{\infty} c(n, t)\, dt \text{ and } S = \lim_{T \to \infty} \frac{1}{T} \int_0^{T} \sum_{n=1}^{\infty} c(n, t) dt \,.$$

The equation analogous to (3.21) is now

$$\sum_{i=1}^{D_t} \int_0^{\infty} c(n_i, u) du \le \int_0^{t} \sum_{n=1}^{\infty} c(n, u) du \le \sum_{n=1}^{A_t} \int_0^{\infty} c(n, u) du \,.$$

One then proceeds as in the proof of Little's result.

It must be assumed that (3.22) holds with θ_n replaced by $\int_0^{\infty} c(n, u) du$. This is immediate if $c(n, u)$ is bounded. If not, a separate justification is required.

3.8. Pollaczek-Khintchine formula Consider an M/GI/1 queue, that is, an FCFS queue with Poisson arrivals of rate λ and i.i.d. service times. Show that the average queueing delay (before the beginning of service) per customer $E\{D\}$ is given by

$$E\{D\} = \frac{\lambda E\{X^2\}}{2(1 - \rho)}$$

where $\rho = \lambda\mu^{-1}$, μ^{-1} is the mean service time and X is a random service time.

From this formula one finds that the average queue length is given by

$$E\{N\} = \rho + \frac{\rho^2 + \lambda^2\sigma^2}{2(1 - \rho)}$$

where σ^2 is the variance of the service times. This formula shows that the variability of the service times (as measured by the variance σ^2) increases the average queue length, and therefore the average delay $E\{\theta\} = \lambda^{-1}E\{N\}$.

Hint: Let $c(n, t)$ be equal to the remaining service time for a customer in the system. Then

$$S = E\{D\}$$

since $\sum_{n=1}^{\infty} c(n, t)$ is the total work-to-be-done in the queue at time t and since the average value of that quantity is the expected time needed to clear the customers ahead of an arriving customer (since Poisson arrivals see time averages). Also,

$$\int_0^\infty c(n,t)\,dt = D_n X_n + \frac{X_n^2}{2}$$

as Figure 3.15 shows.

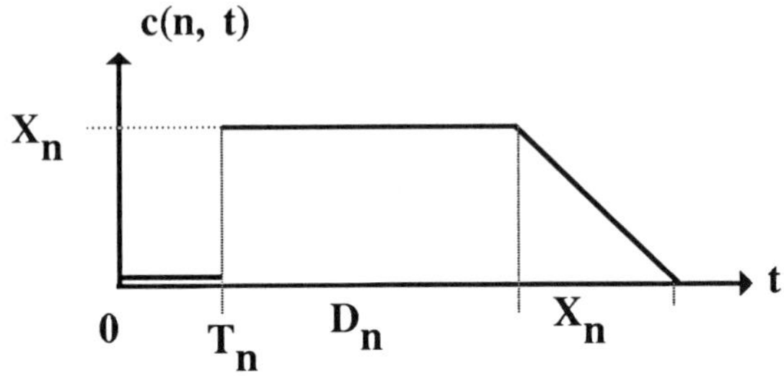

Figure 3.15. The remaining service time for customer n.

Hence,

$$C = E\{D_n X_n + \frac{X_n^2}{2}\} = \frac{E\{D\}}{\mu} + \frac{E\{X^2\}}{2}$$

where we used the independence of D_n and X_n. One concludes by using (3.8). See also Appendix B.

3.9. M/GI/1 queue with priorities Consider an M/GI/1 queue with customers of classes $j = 1, \ldots, J$. The classes correspond to nonpreemptive priorities. That is, whenever the server can start a new service, he chooses to serve a customer of the first represented class in the order $1, 2, \ldots, J$. Customers of different classes have different service time distributions and they arrive as independent Poisson processes with rate λ_j for class j. Let X_j denote a typical service time of a customer of class j. Show that the average queueing delay of a customer of class j is given by

$$E\{D_j\} = \frac{\sum_{i=1}^{J} \lambda_i E\{X_i^2\}}{2(1 - \sum_{i=1}^{J} \rho_i)(1 - \sum_{i=1}^{j-1} \rho_i)}$$

where $\rho_i = \lambda_i E\{X_i\}$ and where it is assumed that $\sum_{i=1}^{J} \rho_i < 1$ (stability condition). *Hint:* First argue that the expected delay of a customer of class j is equal to

$$\sum_{i=1}^{J} E\{N_i\}\,E\{X_i\} \;+\; \sum_{i=1}^{j-1} \lambda_i\,E\{D_j\}\,E\{V_i\}$$

$$+\;E\{\text{ residual service time of customer in service }\}\,.$$

To find the last term, say S, notice that the service time of the nth arrival is equal to X_i with probability $\lambda_i\lambda^{-1}$. Now, all the customers will be served so that the residual service time of the customer being served will be the same as the residual service time in an M/GI/1 FCFS queue with the above "mixed" service time distribution.

Let then $c(n, t)$ be the residual service time of the nth customer if in service in that queue at time t. Then $C = \sum_{i=1}^{J} \frac{\lambda_j}{\lambda} E\{X_i^2\}$ and $S = \lambda C$ is the value we are looking for. Hence,

$$S = \sum_{i=1}^{J} \lambda_i\,E\{X_i^2\}\,.$$

Combining the above expressions proves the result.

3.10. Phase-type Verify that $\mu_c = \nu_c s_{0c}$ (see Example 3.6.2).

3.11. Continuity Consider an M/GI/1 queue. Show that by approximating the service time distribution (e.g. by a phase-type) on also approximates the mean queue length. *Hint:* Call the original service times σ_n. Assume that one can find random variables θ_n such that

$$P\{\,|\,\theta_n - \sigma_n\,| > \epsilon\} \le \epsilon, \quad n \ge 1.$$

Estimate $P\{\,|\,x_n - y_n\,| \ge 1\}$ where x_n is the queue length at the nth arrival time in the original system while y_n is the same queue length in the approximating system.

3.12. Interfering queues Verify Fact 3.7.4. *Hint:* Use Kelly's Lemma 2.8.5.

3.13. Modified arrival rates Consider a network of J quasi-reversible queues, with exogenous Poisson arrival processes with rate γ_i^c for class $c \in \mathbf{C}$ into queue $i \in \{1, \ldots, J\}$. Assume that the customers do not change classes in the network.

When the state of the network is $x = (x^1, \ldots, x^J)$, denote by $n^c(x)$ the number of customers of class c in the network.

Modify the network by replacing the arrival rate of customers of class c into node i by

$$\gamma_i \frac{\Psi[n^d(x) + 1\{d = c\}, \quad d \in \mathbf{C}]}{\Psi[n^d(x), \quad d \in \mathbf{C}]}$$

for some nonnegative function Ψ.

Show that the modified network admits the invariant distribution

$$\tilde{\pi}(x) = \pi(x)\,\Psi[n^d(x), \quad d \in \mathbf{C}]\,.$$

Hint: Verify that the modified network reversed in time corresponds to the original network reversed in time but with the new arrival rates.

3.14. Mixed networks Show that a suitable choice of Ψ in **3.13** allows us to construct a mixed network from an open network.

3.15. Telephone exchange A telephone exchange consists of a groups of incoming lines that can be connected to d groups of outgoing lines. For $1 \le i \le a$, the incoming group i has a_i lines. For $1 \le j \le d$, the outgoing group j has d_j lines. An incoming line of group i can be connected to an outgoing line of group j if and only if $(i, j) \in \mathbf{C}$, a given subset of $\{1, \ldots, a\} \times \{1, \ldots, d\}$.

The interchange operates as follows. Calls arrive as a Poisson process with rate λ and are allocated randomly to an idle incoming line, if any. A call allocated to an incoming line of group i is randomly allocated to an idle line of a group j such that $(i, j) \in \mathbf{C}$, if any. Calls that are allocated have a mean duration equal to μ^{-1}; calls that cannot be allocated are lost.

Find the stationary probability of losing a call. *Hint:* Consider $\mathbf{C}$ as the set of classes of calls. The arrival rate of a call of class $c = (i, j)$ is then given by

$$\lambda \frac{a_i - \sum_k n^{ik}}{\sum_l (a_l - \sum_j n^{lj})} \times \frac{d_j - \sum_l n^{lj}}{\sum_k (d_k - \sum_l n^{lk})}$$

when n^c is the number of calls of class c for $c \in \mathbf{C}$.

This is of the form $\lambda \dfrac{\Psi[n^{11}, \ldots, n^{ij} + 1, \ldots, n^{ad}]}{\Psi[n^{11}, \ldots, n^{ij}, \ldots, n^{ad}]}$, with

$$\Psi[n^{11}, \ldots, n^{ij}, \ldots, n^{ad}]$$
$$= \frac{(\sum_i a_i - \sum_{ij} n^{ij})!}{\prod_i (a_i - \sum_j n^{ij})!} \times \frac{(\sum_j d_j - \sum_{ij} n^{ij})!}{\prod_j (d_j - \sum_i n^{ij})!} .$$

One can then apply the result of Exercise 3.13 to show that this system is a modified open network of M/GI/∞ queues.

3.16. Throughput of a system Let $x = \{x_t, t \geq 0\}$ be the state process of a quasi-reversible queueing system (queue or network) with arrival rates $\{\lambda_c, \ c \in \mathbf{C}\}$, invariant distribution π, and departure processes $\{D_t^c, \ t \geq 0, \ c \in \mathbf{C}\}$. For

$$n = (n^c, \ c \in \mathbf{C}) \in Z_+^{\mathbf{C}} := \{0, 1, 2, \ldots\}^{\mathbf{C}},$$

let

$$\pi(n) = \sum_x \pi(x) \, 1\{n^c(x) = n^c, \quad c \in \mathbf{C}\}$$

and

$$\sigma_d(n) = \lim_{\epsilon \downarrow 0} P_\pi \{D_{t+\epsilon}^d \geq D_t^d + 1 \mid n(x) = n\} ,$$

where $n^c(x)$ denotes the number of customers of class c in the system when its state is x and $n(x) := \{n^c(x), \ c \in \mathbf{C}\}$.

Show that $\{\pi(n), n \in Z_+^{\mathbf{C}}\}$ is uniquely determined by $\{\lambda_c, \ c \in \mathbf{C}\}$ and by $\sigma = \{\sigma_c(n), \ c \in \mathbf{C}, \ n \in Z_+^{\mathbf{C}}\}$. The collection σ will be called the **throughput** of the system. *Hint:* The quasi-reversibility is seen to imply that

$$\lambda_c^{-1} \, \pi(n) \, \sigma_c(n) = \pi(n - e_c) ,$$

where $n - e_c$ is the element of $Z_+^{\mathbf{C}}$ defined by $(n - e_c)^d = n^d - 1\{d = c\}$.

3.17. Norton's theorem Consider the network illustrated in Figure 3.16. It is built from two quasi-reversible systems Σ_1 and Σ_2.

For $i = 1, 2$ denote by $\pi_i(x^i)$ the invariant distribution of Σ_i when its arrival rates are λ_c for $c \in \mathbf{C}$. Fix $n \in Z_+^{\mathbf{C}}$ and assume throughout that $n(x_t^1) + n(x_t^2) = n$. Denote by $\mu(x^1, x^2)$ [resp. by $\mu(x^2)$] the invariant distribution of (x_t^1, x_t^2) [resp. of x_t^2] in the

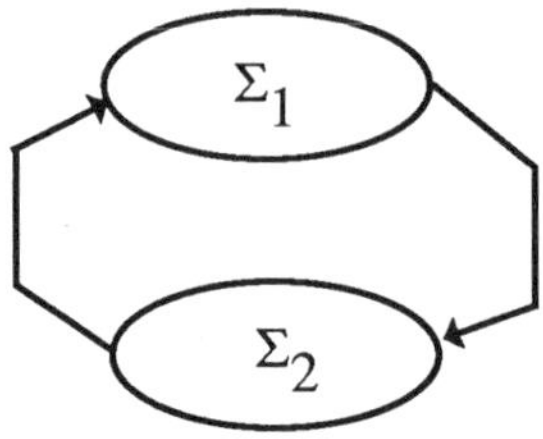

Figure 3.16. The network.

network.

Show that $\mu(x^2)$ depends on Σ_1 only through its throughput σ (defined in **3.16)**.

This result is referred to as **Norton's Theorem.** *Hint:* Clearly, $\mu(x^1, x^2) = \dfrac{\pi_1(x^1)\,\pi_2(x^2)}{\pi(n)}$. Also, $\mu(x^2) = \sum \mu(x^1, x^2)$, where the sum extends over $\{x^1 \mid n(x^1) + n(x^2) = n\}$. Hence,

$$\mu(x^2) \;=\; A\pi_1(n - n(x^2))\,\pi_2(x^2)\,,$$

where $\pi_1(n) := \pi_1\{x^1 \mid n(x^1) = n\}$ for $n \in Z_+^{\mathbf{C}}$, and where A is the normalizing constant.

3.18. Equilibrium forward recurrence time Verify formula (3.46). The distribution R_c is called the **equilibrium forward recurrence time distribution** associated with the distribution F_c. The interpretation is that if one uses light bulbs with i.i.d. lifetimes in a given socket, replacing a bulb by a new one as soon as it burns out, then, after a long time, the distribution of the residual lifetime of the bulb being used is the equilibrium forward recurrence time distribution. More about this in Chapter Seven. *Hint:* Recall the Definition 3.2.4 of the phase-type distribution. The invariant distribution of the phase of a customer of class c, given that it is in the M/PH/∞ queue is the same as the invariant distribution of the phase of the customer watched in the queue (in the sense of 2.9.9). Thus, one may look at the invariant distribution of the phase given that the customer re-enters the queue as soon as it leaves it. This invariant distribution is ν_c defined in (3.38). It corresponds to a residual service time that is at least equal to t with probability (we omit the subscript c for simplicity) νe^{St}. The invariant density of the residual service time is then

$$- \frac{d}{dt} \nu e^{St} 1 = -\nu S e^{St} 1 = \nu s_0 \alpha e^{St} 1 = \mu \alpha e^{St} 1$$

$$= \mu P\{\tau > t\} = \mu[1 - F(t)]$$

which then implies (3.46).

3.11. REFERENCES AND COMMENTS

Section 3.2 The concept of quasi-reversibility was understood algebraically by Muntz (1972) and probabilistically by Kelly (1979). See also Varaiya, et al. (1980b) and Melamed (1982a). For a detailed treatment of phase-type distributions, see Neuts (1981). See also Aldous-Schepp (1987) for the statement in Exercise 3.2.

Section 3.3 Theorem 3.3.2 is from Baskett, et al. (1975) and Kelly (1976a). The fact that the only quasi-reversible M/GI/I-FCFS queues are M/M/1 queues is proved in Disney, et al. (1973).

Section 3.4 The first proof of Little's result is due to Little (1961). The proof of section 3.4 is from Heyman and Stidham (1980). The extension (3.8) is due to Brumelle. See Heyman and Sobel (1982) for more details.

Section 3.5 The proof of Theorem 3.5.4 is adapted from Kelly (1982a). Proposition 3.5.5 is from Walrand (see the discussion of Kelly (1982a)). Fact 3.5.6 is proved in Walrand (1983).

Section 3.6 The model of the symmetric queue is due to Kelly (1979). For a justification of the phase-type approximations, see Barbour (1976).

Section 3.7 The examples of this section are borrowed from Kelly (1979), Burman, et al. (1984). See also Whittle (1986a) and Hordijk and van Dijk (1983).

Section 3.8 Norton's theorem was discovered by Chandy, et al. (1975). The heuristic derivation appears to be new.

Section 3.9 The first formal theory of insensitivity is due to Matthes (1962). See also Schassberger (1977-1978c), Burman (1981), Jansen and König (1980), and Whittle (1985) and (1986b). The connection with quasi-reversibility was presented by the author at Oberwolfach in May 1984. The idea of "age-dependent routing" in a GSMP was observed in Rumswick (1987).

CHAPTER FOUR

SOJOURN TIMES AND FLOWS

4.1. INTRODUCTION

This chapter discusses the **sojourn times** and the **flows** in networks. The results available so far are rather limited and point to the difficulty of the general problem.

In principle, the sojourn time of a customer can be viewed as a **first passage time** for the Markov chain that describes the evolution of the network together with that of a specific customer of interest. The distribution of the sojourn time can thus be obtained by solving the corresponding **backward equations.**

This method, discussed in section 4.9, is useful only for small closed networks. For a general network, there is little reason to suspect that sojourn times should have a simple distribution or characteristic function.

It turns out, however, that simple closed form results are available for **overtake-free** paths in open or in closed networks of quasi-reversible queues. In open networks, the sojourn times of a typical customer in the queues of an overtake-free path are **independent.** In closed networks, the **Laplace transform** of the joint distribution of these sojourn times is known and admits a simple form.

These results are described in section 4.2. They are proved in section 4.5. The proofs rely on the **Arrival Theorem** that is proved in section 4.4, and on **time reversal arguments.**

In paths with overtaking, the situation is substantially more difficult. Some **asymptotic results** (as the traffic intensities increase) are known. They are explained in section 4.8.

We will also discuss some results on the sojourn times in a closed network with one M/M/1 queue and one M/G/1 queue (see section 4.10). The **insensitivity** of sojourn times in some networks is discussed in section 4.7.

As was explained in Chapter One, most flows in networks are **not Poisson.** The ideas are reviewed in section 4.3 and the results are made precise in section 4.11. Section 4.12 deals with the properties of **interarrival times.**

4.2. EXAMPLES OF SOJOURN TIMES

The examples below describe the easy results about sojourn times. Example 4.2.1 shows that the sojourn time in an M/M/1 queue is exponentially distributed. Example 4.2.2 recalls that the sojourn times in M/M/1 queues in tandem are independent. The effect of overtaking is reviewed in Example 4.2.3. The case of closed networks is illustrated in Example 4.2.4. These examples were introduced in Chapter One.

Example 4.2.1 [M/M/1 queue] Consider the M/M/1 queue below, with arrival rate λ and service rate $\mu > \lambda$. Denote by S_n the nth positive arrival time in the queue.

The queue length $\{x_{S_n -}, n \geq 1\}$ observed just before the arrivals is an ergodic Markov chain. Indeed, it is easily seen to be irreducible and it is aperiodic, since it is possible for two successive values to be equal; it admits an invariant distribution which is known to be the invariant distribution π of $\{x_t, t \geq 0\}$. (Poisson arrivals see time averages; see Theorem 2.10.6.)

A customer who arrives and finds k customers in the queue has to wait for $k+1$ service times before leaving. Thus, if one denotes by θ_n the sojourn time of the customer who arrives at time S_n, one has, for $t \geq 0$ and $n \geq 1$,

$$P\{\theta_n \leq t\} = \sum_{k=0}^{\infty} P\{\sigma_1 + \cdots + \sigma_{k+1} \leq t\} P\{x_{S_n -} = k\} \tag{4.1}$$

where the σ_m are independent exponentially distributed random variables with rate μ (the service times). Since the distribution of $x_{S_n -}$ converges to π given by (2.16), one finds from (4.1) that

$$\lim_{n \to \infty} P\{\theta_n \leq t\} = \sum_{k=0}^{\infty} P\{\sigma_1 + \cdots + \sigma_{k+1} \leq t\} \rho^k (1 - \rho)$$

$$= 1 - e^{-(\mu - \lambda)t} . \tag{4.2}$$

The first equality is by dominated convergence (see Exercise 4.1) and the second results from a simple calculation outlined in Exercise 4.2. Also, if $x_{S_1 -}$ has distribution π, then (4.2) is the distribution of θ_n for all $n \geq 1$.

It is therefore legitimate to call **typical** a customer whose sojourn time is exponentially distributed with rate $\mu - \lambda$.

It should be pointed out that if x_0 has distribution π, then x_{S_1-} does not have distribution π. Indeed, if x_0 is not 0, then there is a positive probability that there will be departures between time $t = 0$ and time S_1. Therefore, $x_{S_1-} \le x_0$ and the inequality is strict with positive probability if $P\{x_0 > 0\} > 0$.

One can also show that the process $\{\theta_n, n \ge 1\}$ is ergodic, in addition to being asymptotically stationary. (See 2.6.4 and Chapter Seven for a discussion of ergodicity.) In particular, by Birkhoff's individual ergodic theorem,

$$\lim_{n \to \infty} \frac{1}{n} \sum_{m=1}^{n} 1\{\theta_m \le t\} = 1 - e^{-(\mu - \lambda)t}, \quad t \ge 0$$

and

$$\lim_{n \to \infty} \frac{1}{n} \sum_{m=1}^{n} \theta_m = \frac{1}{\mu - \lambda} .$$

Example 4.2.2 [M/M/1 queues in tandem] The two M/M/1 queues in tandem in Figure 4.1 have respective service rates μ_1 and μ_2. The arrival rate is $\lambda < \min\{\mu_1, \mu_2\}$.

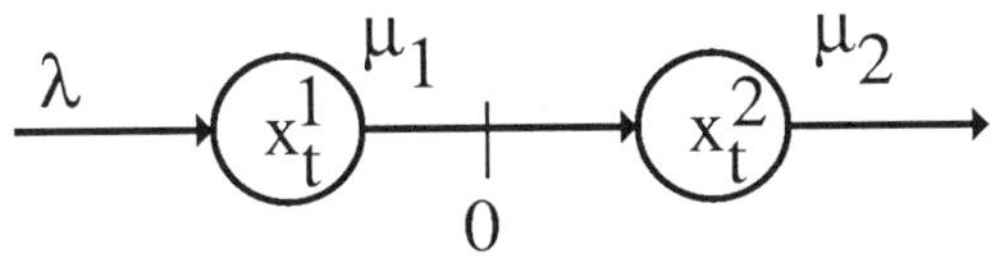

Figure 4.1. Two M/M/1 queues in tandem.

As was indicated in 1.11, the sojourn times of a typical customer (see below) in the two queues are independent and exponentially distributed with mean $(\mu_i - \lambda)^{-1}$ in queue i $(i = 1, 2)$.

A proof of this fact based on time reversal will be given below. The proof is important, since it will be repeated for the other results about overtake-free paths in this chapter.

Let us first review the concept of typical customer for this system. Recall that this Jackson network is such that the vector of queue lengths $x_t = (x_t^1, x_t^2)$ is ergodic and has the invariant distribution (see 2.5.6)

$$\pi(x^1, x^2) = \pi_1(x^1)\,\pi_2(x^2) \tag{4.3}$$

where

$$\pi_i(n) = \rho_i^n(1 - \rho_i), \quad \rho_i = \frac{\lambda}{\mu_i}, \quad \text{for } i = 1, 2 .$$

Let S_n be the nth time that a customer jumps from queue 1 to queue 2. The Markov chain $\{(x^1_{S_n}, x^2_{S_n-}), n \geq 1\}$ admits the invariant distribution (4.3), by the Arrival Theorem for Jackson networks (see 2.10.2). It can also be easily verified that this Markov chain is irreducible and aperiodic. It is therefore ergodic and asymptotically stationary. As in Example 4.2.1, this justifies the definition of a **typical** customer as being such that (x^1_S, x^2_{S-}) has distribution π given by (4.3), where S is the time when the customer jumps from queue 1 to queue 2.

Denote by θ^i_n the sojourn time in queue i of the nth customer who goes from queue 1 to queue 2 after $t = 0$.

It follows from the discussion above that (θ^1_n, θ^2_n) is asymptotically stationary, (see Exercise 4.3) and has a limiting distribution that corresponds to the distribution (4.3) for $(x^1_{S_n}, x^2_{S_n-})$. That distribution will be shown next to be that of two independent exponentially distributed random variables with rates $\mu_1 - \lambda$ and $\mu_2 - \lambda$, respectively.

This joint distribution also corresponds to the sojourn times of a customer who happens to **arrive** in the network and find the other customers with their invariant distribution. (For such a customer is also typical, and there can only be one typical distribution since it is the asymptotic distribution.)

It remains to verify the typical distribution. As in 1.11, say that the typical customer jumps from queue 1 to queue 2 at time 0. Thus, (x^1_0, x^2_{0-}) has distribution π given by (4.3).

Let us first explain why the argument of 1.11 requires some clarification. It was argued that the sojourn time θ_2 of the typical customer in queue 2 is a function of x^2_{0-} and of the service times in queue 2 after $t = 0$. Reversing the time, we find that the sojourn time θ_1 in queue 1 is a function of x^1_0 and of the service times in queue 1 before $t = 0$. By (4.3) one knows that x^1_0 and x^2_{0-} are independent. However, it is not clear that the service times in queue 1 before $t = 0$ and x^2_{0-} are independent. In fact, it could be argued that if queue 1 was faster than average just before time 0, then x^2_{0-} can be expected to be larger than average. Thus, a finer argument is needed. It is provided below.

Consider the queue length process $\{x_t, t \in \mathbb{R}\}$ and assume that

(x_0^1, x_{0-}^2) has distribution π. One way to construct this process is to glue back-to-back at time 0 a copy of $\{x_t, t \geq 0\}$ and an independent copy (given x_0) of the reversed Markov chain $\tilde{\mathbf{x}}$.

More precisely, let $\mathbf{y} = \{y_t, t \geq 0\}$ be a Markov chain with the same rate matrix as $\mathbf{x}$, and $\mathbf{z} = \{z_t, t \geq 0\}$ be a Markov chain with the same rate matrix as $\tilde{\mathbf{x}}$. Assume that given (y_0, z_0) the Markov chains $\mathbf{y}$ and $\mathbf{z}$ are independent. Assume also that $y_0 = x_0$ and $z_0 = x_{0-}$ where x_0 is as before. (See Figure 4.2.)

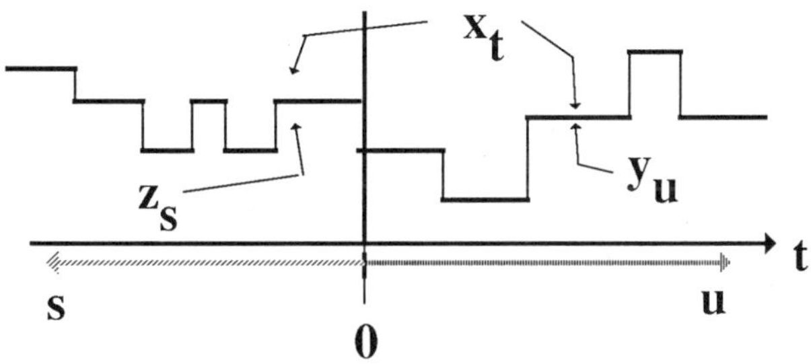

Figure 4.2. The equivalent construction of $\mathbf{x}$.

One can then consider that

$$x_t = \begin{cases} y_t \text{ for } t \geq 0 \\ z_{-t-} \text{ for } t < 0 . \end{cases}$$

Indeed, if one had chosen $y_0 = z_0$ with distribution π, then the resulting process $\mathbf{x}$ would be the stationary Markov chain with initial distribution π and rate matrix Q. Thus, the construction above can be viewed as the stationary process $\mathbf{x}$ conditioned on the fact that there is a jump from queue 1 to queue 2 at time 0.

Since, given x_0, the processes $\mathbf{y}$ and $\mathbf{z}$ are independent, it follows that the evolution of the queues in tandem corresponds to independent service times in queue 2 for $t \geq 0$ and in queue 1 for $t \leq 0$. One can therefore conclude as in 1.11.

Example 4.2.3 [Overtaking] It was indicated in 1.11 that if a customer can be **overtaken** along a given path in a network, either by a customer or by the effects of customers, then the sojourn times of the customer in the queues of that path will generally not be independent. (A precise definition of overtaking will be given in section 4.5.)

An intuitive justification of that fact was given in the case of a simple network in 1.11.

Example 4.2.4 [Closed network] There are N customers in the closed network of M/M/1 queues in Figure 4.3.

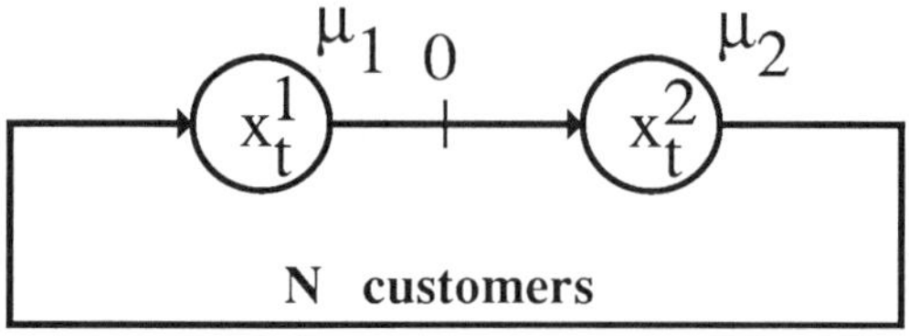

Figure 4.3. A closed network of M/M/1 queues.

An invariant distribution is $\pi(x^1, x^2) = A(\frac{1}{\mu_1})^{x^1}(\frac{1}{\mu_2})^{x^2}$. For $n \geq 0$, let $\mathbf{X}_n = \{x = (x^1, x^2) \mid x^1 \geq 0, x^2 \geq 0, x^1 + x^2 = n\}$. Thus, the vector of queue lengths x_t is an ergodic Markov chain on each X_n. As a consequence, by the Arrival Theorem for Jackson networks (see 2.10.2), the invariant probability distribution of the customers who do not jump at the jump times from queue 1 to queue 2 is given by

$$\pi_{N-1}(x) := \frac{\pi(x)}{\pi(X_{N-1})}, \quad x \in X_{N-1}.$$

Assume that a customer of interest jumps from queue 1 to queue 2 at time 0 in such a way that (x_0^1, x_{0-}^2) has distribution π_{N-1}. That customer is then typical. Denote by θ_2 the next sojourn time of that customer in queue 2, and by θ_1 the last sojourn time in queue 1. Given $\{(x_0^1, x_{0-}^2) = (m, n)\}$, θ_2 is the sum of $n + 1$ exponentially distributed random variables with mean μ_2^{-1} (the service times in queue 2); also, time reversal shows that θ_1 is distributed as the sum of $m + 1$ exponential random variables with mean μ_1^{-1}. Furthermore, given (x_0^1, x_{0-}^2), the random variables θ_1 and θ_2 are independent. This can be shown by a construction similar to that used in Example 4.2.2. Therefore,

$$E\{e^{-s_1\theta_1 - s_2\theta_2}\} = \sum_{x \in X_{N-1}} \pi_{N-1}(x)\left(\frac{\mu_1}{\mu_1 + s_1}\right)^{x^1+1}\left(\frac{\mu_2}{\mu_2 + s_2}\right)^{x^2+1} \tag{4.4}$$

for s_1 and s_2 positive.

To obtain (4.4) one used the fact that if σ is exponentially distributed with rate λ, then its Laplace transform is $E\{e^{-s\sigma}\} = \frac{\lambda}{\lambda + s}$, $s \geq 0$.

4.3. EXAMPLES OF FLOWS

As was indicated in section 1.9, the flows in most networks are **not Poisson processes,** even though their invariant distribution may be the same as if they were. The simple example below illustrates that point.

Example 4.3.1 [M/M/1 queue with feedback] The system in Figure 4.4 is an M/M/1 queue with an external arrival process with rate γ. Upon departure from the queue, each customer is independently fed back into the queue with probability $p \in (0, 1)$ and leaves otherwise, i.e., with probability $q = 1 - p$.

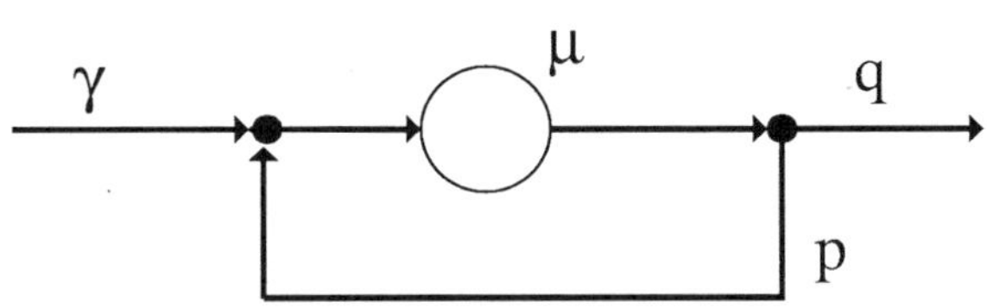

Figure 4.4. The M/M/1 queue with feedback.

Assume that $\lambda = \gamma q^{-1} < \mu$ and that the queue length process is stationary. The claim is that the process A_t that counts all the arrivals into the queue (external and feedbacks) is not a Poisson process.

To see this, notice that, for $t > \epsilon > 0$,

$$P\{A_{t+\epsilon} \geq A_t + 1\} = [\gamma + \mu\, p\, P\{x_t > 0\}]\, \epsilon + o(\epsilon)$$

and

$$P[A_{t+\epsilon} \geq A_t + 1 \mid A_t \geq A_{t-\epsilon} + 1] = [\gamma + \mu\, p]\, \epsilon + o(\epsilon)\,.$$

The second equation uses the fact that if an arrival occurred during $[t - \epsilon, t]$, then the queue is nonempty at time t with probability $1 - o(\epsilon)$.

These relations show that having observed an arrival just before time t increases the likelihood of another arrival just after time t. This implies that the process A_t does not have independent increments, and therefore cannot be Poisson. (This is the argument that was suggested in Exercise 1.5.)

A similar argument shows that the process A_t cannot be a **renewal process,** i.e., that it cannot have i.i.d. interarrival times. One can also show that the process that counts all the departures from the queue is not Poisson, and similarly for the process that counts the feedbacks.

Example 4.3.2 [M/M/1 queues in tandem with feedback] Consider the network of Figure 1.1, i.e., the two M/M/1 queues in tandem with feedback. In section 1.9 it was argued that the flows inside that network are not Poisson. Let us give some comments about the method that will be used in section 4.11 to prove that result.

Denote by A_t the total number of arrivals into queue 1 in $[0, t]$ (including the feedbacks). The idea is as follows: given that a customer entered queue 1 at time t the expected number of future jumps should be larger than its unconditional value. More precisely, one should expect that

$$E\{A_{\epsilon+t} - A_\epsilon \mid A_\epsilon = 1\} - E\{A_t\} \to \frac{p}{q}, \text{ as } t \to \infty \text{ and } \epsilon \to 0. \quad (4.5)$$

Indeed, for ϵ very small, the condition $A_\epsilon = 1$ says that a customer entered queue 1 just before time ϵ. That customer then has a positive probability of coming back into queue 1 a few times, thereby contributing to the future jumps of A_t. The mean number of times that a customer in queue 1 will come back to that queue is easily checked to be $\frac{p}{q}$. (Call this mean number m. Looking at the figure, one sees that with probability p the customer comes back once and then will come back an additional mean number of times equal to m. Hence $m = p(1 + m)$, so that $m = \frac{p}{q}$.)

It follows from (4.5) that A_t does not have independent increments, and is therefore not Poisson.

The technical difficulty in proving (4.5) is that one must show that the additional customer will come back fast enough, i.e, before the two terms that are subtracted "blow up." This will be shown in section 4.11 by a **coupling argument**.

4.4. ARRIVAL THEOREM

This section presents the generalization of the Arrival Theorem discussed in 2.10.2 for Jackson networks. It will be shown that the result applies to networks of quasi-reversible queues with Markov routing. This result will be used to analyze the sojourn times and the flows in networks.

Theorem 4.4.1 [Arrival Theorem] Consider the network of Definition 3.5.2 (see Figure 4.5). It admits the product-form invariant distribution π given by (3.24). Choose $c, d \in \mathbf{C}$ and $1 \leq i \neq j \leq J$ so that $\lambda_i^c r_{ij}^{cd} > 0$.

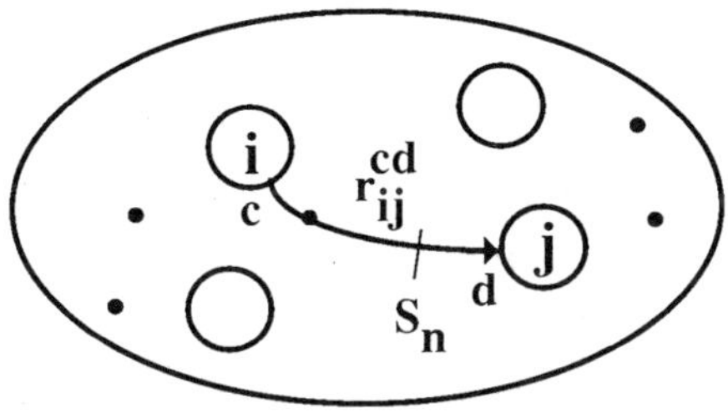

Figure 4.5. Network of quasi-reversible queues.

Denote by $S_1, S_2, \ldots$ the successive times when a customer of class c jumps from queue i to queue j where its class changes to d. Let

$$z_n = (x^1_{S_n}, \ldots, x^i_{S_n}, \ldots, x^j_{S_n-}, \ldots, x^J_{S_n}), \quad n \geq 1 . \quad (4.6)$$

Thus, z_n is the state of the customers who do not jump at time S_n. Then $\mathbf{z} = \{z_n, n \geq 1\}$ is a Markov chain with invariant distribution π.

In general, $\mathbf{z}$ is not irreducible on $\mathbf{X} = \mathbf{X}^1 \times \cdots \times \mathbf{X}^J$. One applies Fact 2.5.7 to find the invariant distribution on a closed irreducible subset. Before proving the theorem we give two examples of such normalizations.

Example 4.4.2 [Numbers may suffice] Queue 1 in the network in Figure 4.6 is M/PH/1-LCFS and queue 2 is M/PH/1-PS.

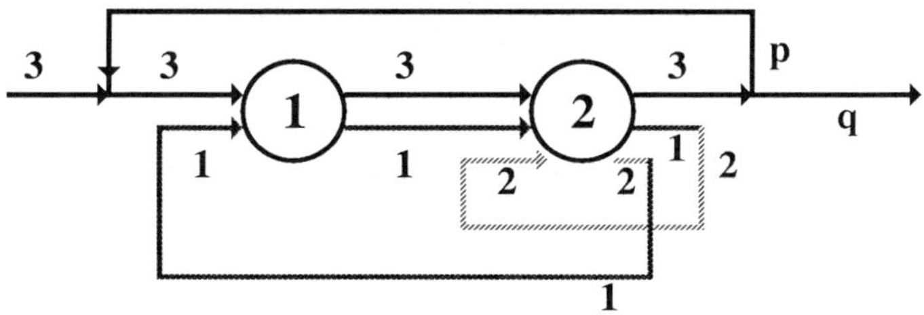

Figure 4.6. Example of a network of quasi-reversible queues.

Customers of class 3 enter the network via queue 1, then go to queue 2, and are fed back with probability p into queue 1. Customers of class 1 go to queue 1, then to queue 2, then are sent back to queue 2 as customers of class 2, then go to queue 1 as customers of class 1. This corresponds to the routing probabilities

$$r_{01}^{3} = 1, \quad r_{12}^{33} = 1, \quad r_{21}^{33} = p = 1 - r_{20}^{3},$$

$$r_{12}^{11} = 1, \quad r_{22}^{12} = 1, \quad r_{21}^{21} = 1 .$$

The **state** of the network specifies the **number** of customers of all classes in the queues, as well as their **positions** in the queues and their **phases of service.**

It can be seen that the state is irreducible on each set $\mathbf{X}_N$ that corresponds to having a total of N customers of class 1 and class 2 in queues 1 and 2, with 0 customer of class 2 in queue 1, and an arbitrary number of customers of class 3 in both queues.

Assume that $x_0 \in \mathbf{X}_N$ for some $N \geq 1$. Choosing the transitions $(i, c) \rightarrow (j, d)$ to be of the type $(1, 1) \rightarrow (2, 1)$, one finds that the chain $\mathbf{z}$ defined in Theorem 4.4.1 is irreducible on $\mathbf{X}_{N-1}$. But $\mathbf{X}_{N-1}$ is the set of states for the customers who do not jump at the times S_n.

One then concludes that the customer who jumps sees the others with their invariant distribution in the network.

This shows that the result of 2.9.1 extends to multi-class networks. This result holds because for this network the numbers of customers specify the closed irreducible subsets. The example below shows that this is not always the case.

Example 4.4.3 [Numbers may not suffice] Consider the following network of two M/M/1 queues (FCFS) with service rates μ_1 and μ_2, respectively. (See Figure 4.7.) There are four customers. Two customers are red and the other two are blue.

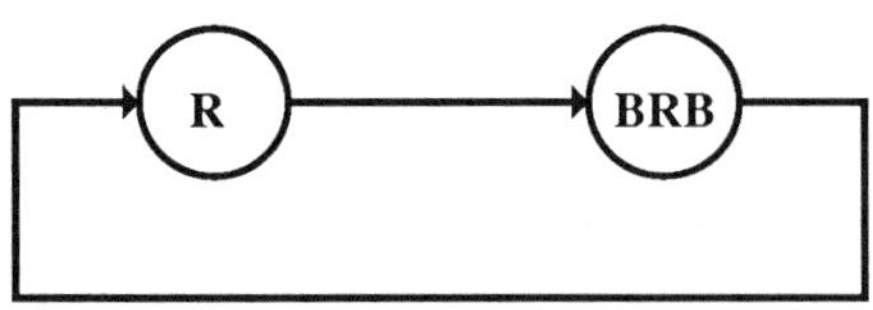

Figure 4.7. The network with two M/M/1 queues.

Since the multiclass M/M/1 queues are quasi-reversible (see section 3.6.1), the Arrival Theorem 4.4.1 applies to this network. Thus, an invariant distribution of $\mathbf{z}$ has a product form. Say that the S_ns are the arrival times of a red customer in the first queue. It can be observed that the configuration in Figure 4.8 is not possible for z_n, if the initial configuration of the customers is as in Figure 4.7.

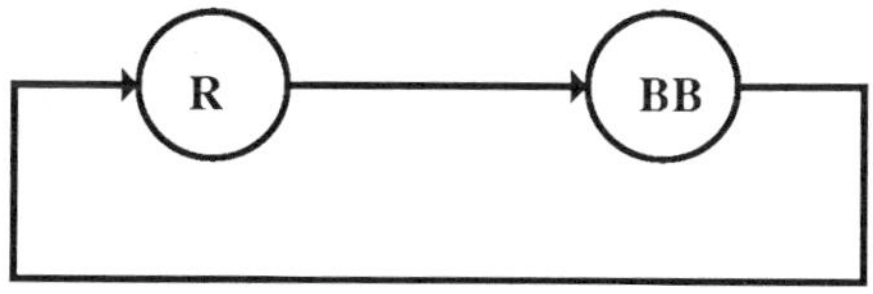

Figure 4.8. A configuration not possible for z_n.

However, this state would be possible for the network from which the red customer was removed. As a consequence, it is not true in general that a customer who jumps sees the others with their invariant distribution when that customer is removed from the network.

We now give two proofs of Theorem 4.4.1. (See also Remark 2.10.3.)

Proof 4.4.4 [Probabilistic] The network with delays illustrated in Figure 4.9 was introduced in 3.5.5 to prove the product-form results.

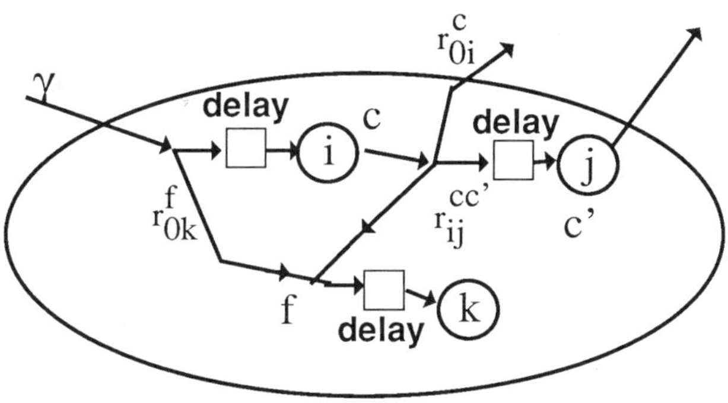

Figure 4.9. The network with delays.

Start the network at time 0 as in 3.5.5. That is, x_0 has distribution π and there are Poisson processes with rates λ_i^c in transit on the links. The jump times S_n were defined in 4.4.1. The fact that there is a jump on a given link during a small time interval $[t, t - \epsilon]$ where ϵ is smaller than the delay d, is independent of x_t. Indeed, the Poisson processes in transit at any time t are independent of x_t. Thus, for $\epsilon < d$,

$$P[x_t = x \mid t - \epsilon < S_n < t] = \pi(x), \quad x \in \mathbf{X}. \tag{4.7}$$

Letting $\epsilon \downarrow 0$ shows that in the network with delays $x_{S_n -}$ has distribution π.

Now, if d is small enough, then there is at most one customer in transit on the links at time t: the customer who jumped at time S_n. Thus, for d small, $x_t = z_n$. But the construction of 3.5.6 shows that, as $d \downarrow 0$, z_n converges to its value for $d = 0$. Therefore, letting $d \downarrow 0$ shows that the distribution of z_n in the network with $d = 0$ is given by π.

Proof 4.4.5 [Algebraic] The invariant probability $\nu(z)$ that $\{z_n = z\}$ is the invariant probability that

$$(x_{S_n -}, x_{S_n}) = (x, y)$$

for some pair (x, y) with

$$(x^i, y^i) \in \mathbf{D}_i^c, (x^j, y^j) \in \mathbf{A}_j^d, \text{ and}$$

$$x^j = z^j, y^i = z^i, x^l = y^l = z^l, \text{ for } i \neq l \neq j.$$

To see this, notice that both $x^i_{S_n -}$ and $x^i_{S_n}$ correspond to the customers who do not jump at time S_n. Denote by $A(z)$ this collection of pairs (x, y). From Theorem 2.10.6, it follows that

$$\nu(z) = \lambda^{-1} \sum_{(x,y) \in A(z)} \pi(x) \, q(x, y). \tag{4.8}$$

Recall that $\lambda := \lambda_i^c \, r_{ij}^{cd}$ is the rate of the S_ns.
Using the definition (3.5.2) of Q allows us to rewrite the right-hand side of (4.8) as

$$(\lambda_i^c \, r_{ij}^{cd})^{-1} \sum_{(x,y) \in A(z)} \pi(x) \frac{q_i(x^i, y^i) \, q_j(x^j, y^j)}{\lambda_j^d} \, r_{ij}^{cd}. \tag{4.9}$$

For $(x, y) \in A(z)$ one has $\pi(x) = \dfrac{\pi(z)}{\pi_i(z^i)} \pi_i(x^i)$ since $x^l = z^l$ for all $l \neq i$. One can then rewrite (4.9) as

$$\pi(z)\left[\frac{1}{\lambda_i^c\,\pi_i(z^i)}\sum_{x^i}\pi_i(x^i)\,q_i(x^i,z^i)\,1\{(x^i,z^i)\in D^{ic}\}\right]$$

$$\times\left[\sum_{y^j} q_j(z^j,y^j)\,1\{(z^j,y^j)\in A^{jd}\}\right]. \quad (4.10)$$

The quasi-reversibility of the queues i and j implies that the two terms between brackets are equal to 1, as seen in (3.1)–(3.2). Therefore,

$$\nu(z) = \pi(z),$$

as was to be shown.

4.5. OVERTAKE-FREE PATHS IN OPEN NETWORKS

As indicated in Example 4.2.3, the possibility of overtaking may create dependencies of sojourn times in different queues. Example 4.5.1 indicates that the overtaking may be one of influence and may not necessarily correspond to an actual overtaking by a customer. This motivates Definition 4.5.2.

Example 4.5.1 The customers of class 1 in the network in Figure 4.10 can be overtaken by the influence of other customers. This leads to suspect that the sojourn times in queues 1 and 3 of a typical customer of class 1 may not be independent. The queues are M/M/1.

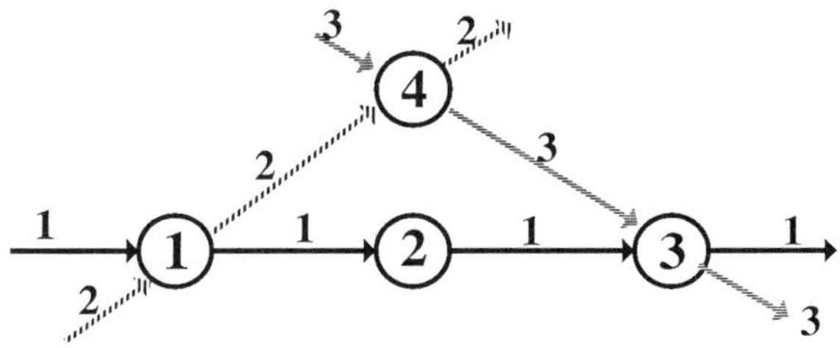

Figure 4.10. A network with overtaking.

One possible argument is that if the sojourn time of that customer in queue 1 is very large, then it is likely that the customer leaves behind many customers of class 2 in queue 1. This large number of customers is then more likely to congest queue 4 and to slow down the flow of customers of class 3 into queue 3. This will reduce the queue length seen by the customer of class 1 when reaching queue 3, and thereby reduce the sojourn time of the customer of class 1 in that queue. Thus, it appears that the sojourn times in queues 1 and 3 are negatively correlated.

Definition 4.5.2 [Overtake-free paths] Consider the network of J queues in Figure 4.11. The customers are of classes $c \in C$. The routing is Markov with probabilities $\{r_{ij}^{cd}\}$, as before, and the resulting network is assumed to be open.

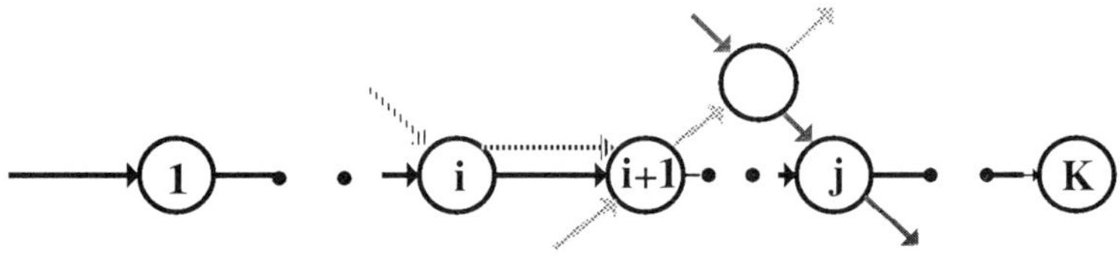

Figure 4.11. The network of quasi-reversible queues.

Fix a class c and let $(1, 2, \ldots, K)$ be a subset of $\{1, 2, \ldots, J\}$ such that it is possible for a customer of class c to go successively through the queues $1, 2, \ldots, K$. That is, $r_{ii+1}^{cc} > 0$ for $1 \le i \le K - 1$. Such a set of nodes is called a **path.**

That path is said to be **overtake-free** if it is not possible for a sequence of **influences** to overtake a customer during the passage of that customer through the path. To make this idea precise, one can define a **potential influence path** (PIP) as a sequence of queues $(i_1, i_2, \ldots, i_k)$ such that $\sum_{c_1, c_2 \in C} r_{i_l i_{l+1}}^{c_1 c_2} > 0$ for $1 \le l < k-1$.

The path $(1, 2, \ldots, K)$ is overtake-free if its queues are M/M/1-FCFS queues and if, whenever a PIP can go from queue i to queue j for some $1 \le i < j \le K$, it has to go from i to $i+1$.

This corresponds to the natural intuition: the influence cannot overtake the customer while in that path.

Theorem 4.5.3 [Overtake-free paths in open networks] Let $(1, 2, \ldots, K)$ be an overtake-free path for class c in a network of J quasi-reversible queues. Assume that this network is open and is ergodic. Then the sojourn times $\{\theta_1, \theta_2, \ldots, \theta_K\}$ of a typical customer of class c in the nodes of that path are **independent.** They are **exponentially distributed,** with mean $(\mu_i - \lambda_i)^{-1}$ in node i. (μ_i

is the service rate in queue i and $\lambda_i = \sum_c \lambda_i^c$ is the average rate of flow through that queue.)

Proof 4.5.4 Fix i in $\{1, \ldots, K\}$. The fact that the path is overtake-free implies that the following sets A, B, C of queues constitute a partition of $\{1, 2, \ldots, J\}$:

$$\begin{aligned} A &= \{j \mid \text{a PIP from } (1, \ldots, i-1) \text{ to } j \\ &\qquad \text{does not go through } i\} \cup \{1, \ldots, i\} \\ B &= \{j \mid \text{a PIP from } j \text{ to } (i+1, \ldots, K) \\ &\qquad \text{does not go through } i+1\} \cup \{i, \ldots, K\} \\ C &= \{1, \ldots, J\} - (A \cup B)\,. \end{aligned}$$

Those definitions are illustrated in Figure 4.12.

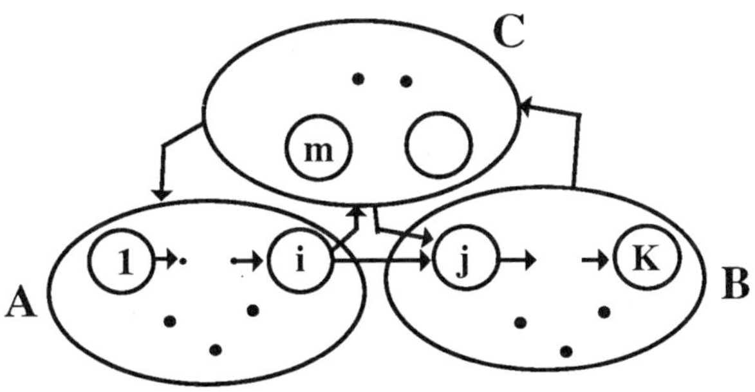

Figure 4.12. The subsets A, B and C of nodes.

be overtake-free.

Assume that the customer of interest leaves queue i at time 0. From the result on the distribution at the jumps it follows that x_0^A, the state of the queues in A at time 0, is independent of x_{0-}^B.

Now, the overtake-free assumption implies that the distribution of $(\theta_{i+1}, \ldots, \theta_K)$ is determined by x_{0-}^B, independently of x_0^A and x_0^C. Similarly, time reversal shows that the distribution of $(\theta_1, \ldots, \theta_i)$ is determined by x_0^A. (Time reversal reverses the routing and preserves the overtake-free property of the path.) More precisely, as indicated in Example 4.2.2, the two sets of sojourn times are functionals of two Markov chains that are independent given their value at time 0: one is distributed as $\{x_t^A, t < 0\}$ and the other one as $\{x_t^B, t \geq 0\}$.

This proves that $(\theta_1, \ldots, \theta_i)$ and $(\theta_{i+1}, \ldots, \theta_K)$ are independent for all $i \in \{1, \ldots, K - 1\}$. This implies the independence of the θ_is. The fact that each θ_i is exponentially distributed with rate $\mu_i - \lambda_i$ follows as for the M/M/1 queue.

Example 4.5.5 [Independent sojourn times and non-Poisson flows] In the network in Figure 4.13 there are two M/M/1 queues. Two classes of customers use this network, as indicated on the figure. The service rates in the queues do not depend on the customer class.

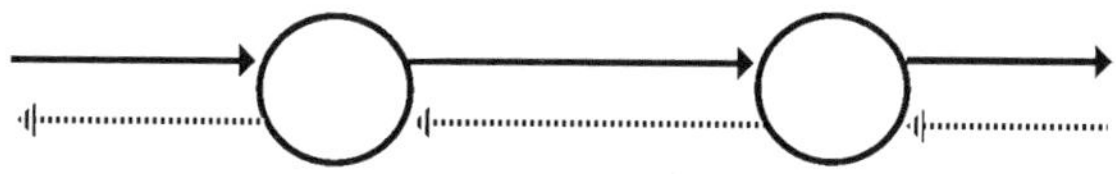

Figure 4.13. A network with non-Poisson flows.

It can be shown (see Example 4.11.6) that the internal flows are not Poisson. The path (1, 2) is overtake-free, so that the sojourn times of typical customers of class 1 in the two queues are independent. Similarly for class 2.

4.6. OVERTAKE-FREE PATHS IN CLOSED NETWORKS

The objective of this section is to show that the result of Example 4.2.4 extends to overtake-free paths in closed networks of quasi-reversible queues. The definition of overtake-free paths is slightly different from the open case; it reflects the fact that only the influence of the other customers can affect a given customer. This point is first illustrated on an example.

Example 4.6.1 [Overtake-free path] The following example illustrates the difference between the notions of overtake-free path used for open and for closed networks. (See Figure 4.14.)

There are three customers; two are of class 1 and one is of class 2. The routing of class 1 corresponds to the loop (1, 2, 3, 1, 2, 3, . . .); the customer of class 2 has a probability $\frac{1}{2}$ of being sent from 1 to 2 and a probability $\frac{1}{2}$ of going from 1 to 3; the other routing probabilities are the same as for class 1 (see figure). The path (1, 2, 3) is overtake-free for the customer of class

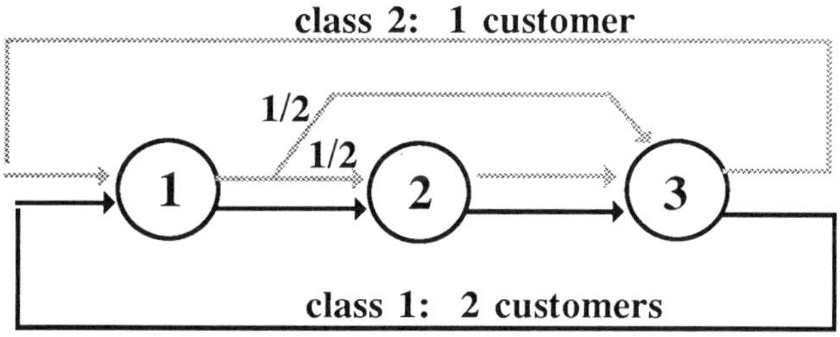

Figure 4.14. An overtake-free path in a closed network.

2, but it is not for the customer of class 1. Indeed, the customer of class 2 cannot be overtaken by influences due to the other customers, while the customers of class 1 can be.

Example 4.6.2 [Sojourn times] The point of this example is to illustrate the main idea of the general result.

The network of three M/M/1 queues in Figure 4.15 has N customers.

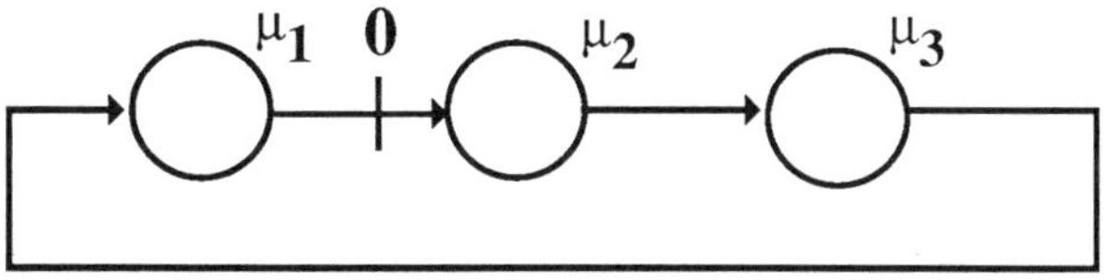

Figure 4.15. A network with three M/M/1 queues.

Consider a typical customer who leaves queue 1 at time 0. The successive sojourn times of that customer in 1 (just before time 0) and 2 and 3 are denoted respectively by θ_1, θ_2, and θ_3.

The key observation is as follows. Given that $x_0^1 = x^1$, θ_1 and $\{\theta_2, \theta_3\}$ are independent; moreover, $\{\theta_2, \theta_3\}$ is distributed as in a closed network with the two queues 2 and 3 with $N - x^1$ customers, and θ_1 is (by time reversal) distributed as the sum of $x^1 + 1$ exponentially distributed random variables with rate μ_1^{-1}. Indeed, once the customer of interest enters queue 2, one can close the loop from queue 3 to queue 2 without affecting the sojourn times in these two queues.

It is important to notice that the customer is again typical in the closed network with queues 2 and 3 and with $N - x^1$ customers. Indeed, the customer finds that closed network with its invariant probability distribution upon entering queue 2. This follows from

$$\pi_{N-1}(x^2, x^3 \mid x^1) = \frac{\pi_{N-1}(x^1, x^2, x^3)}{\pi_{N-1}(x^1)}$$

$$= \frac{\pi^{(2,3)}(x^2, x^3)}{\pi^{(2,3)}(x^2 + x^3 = N - x^1 - 1)}$$

$$= \pi^{(2,3)}_{N-x^1-1}(x^2, x^3).$$

The result of Example 4.2.4 can therefore be applied to the network (2, 3). Hence,

$$E\{e^{-s_1\theta_1 - s_2\theta_2 - s_3\theta_3}\}$$

$$= \sum_{x^1=0}^{N-1} \pi_{N-1}(x^1) \left(\frac{\mu_1}{\mu_1 + s_1}\right)^{x^1+1} \Phi(s_2, s_3; N - x^1) \tag{4.11}$$

where Φ is given by the result for two queues, i.e., $\Phi(s_2, s_3; N - x^1)$ is equal to

$$\sum_{(x^2,x^3) \in \mathbf{X}_{N-x^1-1}} \pi_{N-1}(x^2, x^3 \mid x^1) \left(\frac{\mu_2}{\mu_2 + s_2}\right)^{x^2+1} \left(\frac{\mu_3}{\mu_3 + s_3}\right)^{x^3+1}.$$

Therefore, (4.11) can be written as

$$\sum_{x \in \mathbf{X}_{N-1}} \pi_{N-1}(x) \left(\frac{\mu_1}{\mu_1 + s_1}\right)^{x^1+1} \left(\frac{\mu_2}{\mu_2 + s_2}\right)^{x^2+1} \left(\frac{\mu_3}{\mu_3 + s_3}\right)^{x^3+1}.$$

Theorem 4.6.3 [Sojourn times in overtake-free paths of closed networks]

Consider an ergodic closed network of J quasi-reversible queues with N customers. There is a customer (called M) of some class c who goes through a fixed path $(1, \ldots, K)$ with $K \le J$ whenever that customer enters queue 1.

That path is such that the influence of the other $N - 1$ customers cannot overtake M while in that path.

Denote by $\theta_1, \ldots, \theta_K$ the typical successive sojourn times of M in the queues of that path. Then

$$E\{e^{-\sum_{i=1}^{K} s_i\theta_i}\} = \sum_{x \in \mathbf{X}_{N-1}} \pi_{N-1}(x) \prod_{i=1}^{K} \left(\frac{\mu_i}{\mu_i + s_i}\right)^{x^i+1}. \tag{4.12}$$

Proof: The idea is similar to that of Example 4.5.2. For notational simplicity the proof is given in the case of a Jackson network. (See Figure 4.16.)

Fix i in $\{1, \ldots, K\}$ and define the sets A, B, C as in 4.5.4, with the understanding that the PIPs are now those corresponding to the $N - 1$ other customers.

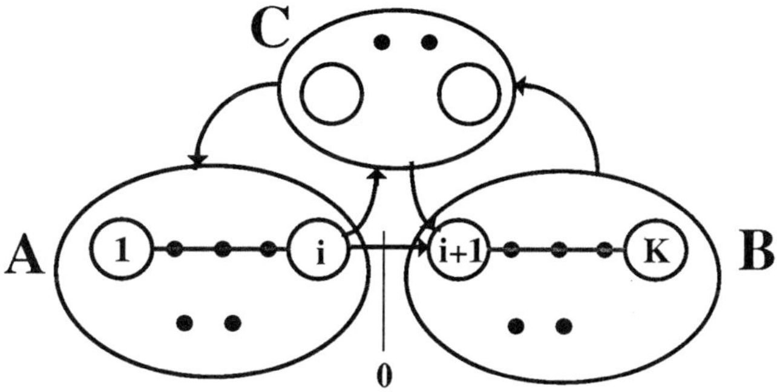

Figure 4.16. The network partition.

Assume that the customer M of interest jumps from queue i to queue $i + 1$ at time 0, and that the customer is typical. The key observation is again that as soon as customer M enters queue $i + 1$ one can close the subnetwork B without modifying the distribution of $(\theta_{i+1}, \ldots, \theta_K)$. Moreover, given the set of customers N_B in B, customer M is typical for the closed network B. This observation forms the basis of an induction on the size of the network, as was used in Example 4.6.2.

One has

$$E\{e^{-\sum_{j=1}^{K} s_j \theta_j}\} = E\{E[e^{-\sum_{j=1}^{K} s_j \theta_j} \mid N_A, N_B, N_C]\}$$

$$= E\{E[e^{-\sum_{j=1}^{i} s_j \theta_j} \mid N_A]\, E[e^{-\sum_{j=i+1}^{K} s_j \theta_j} \mid N_B]\} \quad (4.13)$$

where the last equality follows from the independence of X_0^A and X_{0-}^B given (N_A, N_B, N_C) and the fact that the distribution of $\{\theta_1, \ldots, \theta_i\}$ [resp. $\{\theta_{i+1}, \ldots, \theta_K\}$] depends only on X_0^A [resp. X_{0-}^B].

Assume now that Theorem 4.6.3 is known to hold for an arbitrary overtake-free path with fewer than K queues in a network with fewer than J queues. The closed network B with a given set N_B of customers is then a network for which the theorem holds.

Also, time reversal shows that the theorem holds for the closed network A with a set N_A of customers. That is,

$$E[e^{-\sum_{j=i+1}^{K} s_j \theta_j} \mid N_B] = \sum_{x^B \in \mathbf{X}^B_{N_B}} \pi^B_{N_B}(x^B) \prod_{j=i+1}^{K} \left(\frac{\mu_j}{\mu_j + s_j}\right)^{x^j + 1} \quad (4.14)$$

and

$$E[e^{-\sum_{j=1}^{i} s_j \theta_j} \mid N_A] = \sum_{x^A \in \mathbf{X}^A_{N_A}} \pi^A_{N_A}(x^A) \prod_{j=1}^{i} \left(\frac{\mu_j}{\mu_j + s_j}\right)^{x^j + 1} . \quad (4.15)$$

In these expressions, $\pi^A_{N_A}$ is the invariant distribution for the closed network A with a set N_A of customers, and $\mathbf{X}^A_{N_A}$ is the corresponding state space; the expressions for the closed network B are defined similarly.

It remains to combine these formulas. This is done by observing that

$$\pi^A_{N_A}(x^A)\, \pi^B_{N_B}(x^B) = \pi(x^A, x^B \mid N_A, N_B) . \quad (4.16)$$

The expressions (4.13)–(4.16) then show that the left-hand side of (4.12) is given by

$$E_{\pi_{N-1}}\{ \sum_{x^A,\, x^B} \pi(x^A, x^B \mid N_A, N_B) \prod_{i=1}^{K} \left(\frac{\mu_i}{\mu_i + s_i}\right)^{x^i + 1} \} ,$$

which is seen to be equal to the right-hand side of (4.12). One concludes this proof by induction upon observing that the result is trivially true when $K = 1$.

4.7. INSENSITIVITY OF SOJOURN TIMES

The objective of this section is to explain why the **mean sojourn time** of a customer *in a symmetric queue of an open network of quasi-reversible nodes depends on the service time distribution only through its mean.*

More will be shown. The conditional mean sojourn time given the service time will be shown to be proportional to that service time, in the above system. The emphasis is on the probabilistic understanding of this fact.

Proposition 4.7.1 [Insensitivity] Consider an **open** product-form network of quasi-reversible queues. (See Figure 4.17.) Assume that queue 1 is a symmetric queue and that class c is such that $\lambda_1^c > 0$ (see Definition 3.6.5 for the symmetric queues).

Denote respectively by θ and σ the **sojourn time** and the **service time** of a typical customer of class c in queue 1. Then

$$E[\theta \mid \sigma] = K\sigma \tag{4.17}$$

for a constant K that will not depend on the customer class.

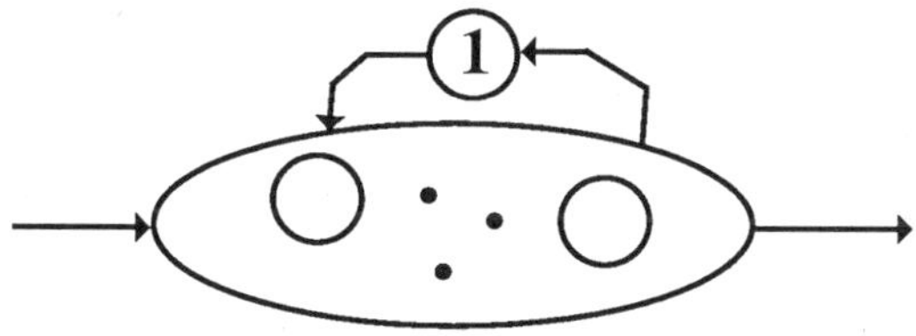

Figure 4.17.
Symmetric queue in a network of quasi-reversible queues.

In the case of a **closed** network, it can be shown that (4.17) continues to hold but with a constant K that may depend on the class.

Comments 4.7.2 [Outline of the argument] The argument essentially consists in considering two customers with respective sojourn and service times in queue 1 given by $\{\theta_1, \sigma_1\}$ and $\{\theta_2, \sigma_2\}$. It will be shown that if $\sigma_2 = n\sigma_1$, then $E\{\theta_2\} = nE\{\theta_1\}$.

This then implies the claim (4.17). Indeed, a small perturbation of the service time distributions can make sure that all the service times are multiples of a common ϵ.

Recall that deterministic service times can be approximated by the sum of a large number of i.i.d. exponentially distributed random variables. (See Remark 3.2.3.)

This then leads to the following idea. Assume that σ_1 and σ_2 correspond respectively to the sum of 1 and n i.i.d. exponentially distributed random variable(s) with mean μ^{-1}. It suffices to show that

$$E\{\theta_2\} = nE\{\theta_1\}. \tag{4.18}$$

Notation 4.7.3 Consider the second customer of Comments 4.7.2 with a service time equal to the sum of $n \geq 2$ i.i.d. exponentially distributed random variables with mean μ^{-1}. Those exponential random variables will be called the **stages** of service of the customer.

For $i = 1, \ldots, n$, denote by θ_2^i the sojourn time of that customer during the ith stage of service. Thus,

$$\theta_2 = \theta_2^1 + \cdots + \theta_2^n, \tag{4.19}$$

so that in order to show (4.18) it suffices to prove that

$$E\{\theta_2^i\} = E\{\theta_1\}, \quad 1 \le i \le n. \tag{4.20}$$

Definition 4.7.4 [A modification] To prove (4.20), the system will be slightly modified. The modification will affect the distribution of the sojourn time of the second customer, but will not affect its mean value.

For $1 \le i \le n-1$, assume that upon the completion of the ith stage of service, the second customer leaves the queue and is immediately fed back into the queue, as illustrated in Figure 4.18.

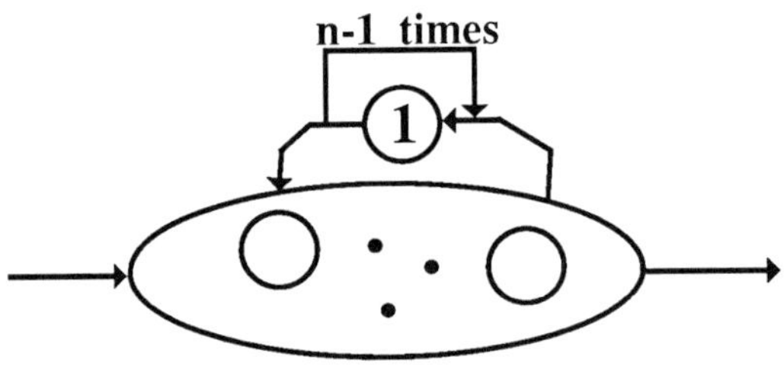

Figure 4.18. The modified network.

Denote by $\tilde{\theta}_2^i$ the new successive sojourn times of that customer, corresponding to the n stages of service ($1 \le i \le n$). The **claim** is that the $2n + 1$ random variables $\{\theta_1, \theta_2^i, \tilde{\theta}_2^i, 1 \le i \le n\}$ are identically distributed.

This will complete the proof of (4.20), and therefore of (4.17).

Proof of the claim Recall that a typical customer finds the others in equilibrium when entering a queue of a network of quasi-reversible nodes. (This is the Arrival Theorem 4.4.1.) As a consequence, θ_1 and $\theta_2^1 = \tilde{\theta}_2^1$ have the same distribution.

Now, assume that when customer 2 finishes the first stage of service there are m customers in the queue (including customer 2). It follows from the definition of the symmetric queue that the customer completes one stage of service in position l with probability $\gamma(m, l)$ for $1 \le l \le m$. The time θ_2^2 is then the sojourn time during the completion of one stage of service started in position l with probability $\gamma(m, l)$ when there are m customers in the queue. But, again by definition of the symmetric queue, if the customer were to leave and be instantaneously fed back, the probability that the customer would join the queue in position l would also be $\gamma(m, l)$. It follows that $\tilde{\theta}_2^2$ and θ_2^2 have the same distribution. Also, by the Arrival Theorem, the typical state of the queue seen at the beginning of $\tilde{\theta}_2^2$, upon the feedback, has the same distribution as that seen at the beginning of θ_1. It follows that $\tilde{\theta}_2^2$, θ_2^2, and θ_1 have the same distribution.

The argument can be repeated to complete the proof.

Comments 4.7.5

(a) The essential property of the symmetric queue used in the above argument is the following. The typical distribution at the completion of a stage of service is the same as the typical distribution at the beginning of a stage of service. Algebraically, this property is referred to as **station balance.** Thus, the above probabilistic argument explains why station balance implies (4.17), and therefore the fact that the mean sojourn time depends on the service time distribution only through its mean, i.e., the **insensitivity** of the mean service time.

(b) In the modification 4.7.4, the sojourn time is modified (indeed, $\tilde{\theta}_2^2 \neq \theta_2^2$). This is the case, except when the queue is either LCFS or PS. Indeed, in those two cases, one can assume that the customer comes back in the same position that he was in just before being fed back.

4.8. HEAVY TRAFFIC

This section deals with networks of M/M/1-FCFS queues in **heavy traffic,** i.e., when $\lambda_i \approx \mu_i$ for each queue. It is shown that a typical **sojourn time** in the network is almost distributed as a **random linear combination of independent exponentially distributed random variables.** The coefficients of the linear combination are the numbers of visits to the queues. The exponential random variables are typical sojourn times in the M/M/1 queues.

Example 4.8.1 [M/M/1 queue visited twice] Consider the system in Figure 4.19. There is a single M/M/1 queue that is visited exactly twice by customers who arrive as a Poisson process with rate λ. The service rate is equal to μ.

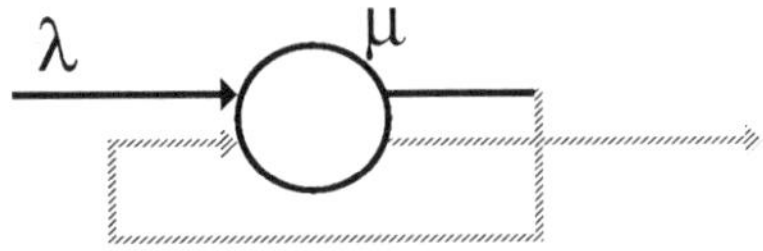

Figure 4.19. An M/M/1 queue visited twice.

If $\rho := \frac{2\lambda}{\mu} < 1$, then the invariant distribution of the queue length is given by

$$\pi(n) = \rho^n(1-\rho), \quad n \geq 0 .$$

Assume that a typical customer is fed back at time 0. It follows from the Arrival Theorem that

$$P\{x_0 = n\} = \pi(n-1), \quad n \geq 1 .$$

Denote by θ_1 and θ_2 the two successive sojourn times of this typical customer in the queue. By considering respectively the queue and the queue reversed in time (as in Example 4.7.2), one finds that given $\{x_0 = n\}$ the random variables θ_1 and θ_2 are i.i.d. and distributed as the sum of n i.i.d. service times.

Hence, the sojourn time $\theta = \theta_1 + \theta_2$ is such that for $s > 0$

$$\begin{aligned} E\{e^{-s\theta}\} &= \sum_{n=1}^{\infty} \pi(n-1) \left(\frac{\mu}{\mu + s}\right)^{2n} \\ &= \frac{\mu(1+\sqrt{\rho})}{s + \mu(1+\sqrt{\rho})} \times \frac{\mu(1-\sqrt{\rho})}{s+\mu(1-\sqrt{\rho})} . \end{aligned}$$

This calculation shows that the typical sojourn time is distributed as the sum of two independent exponentially distributed random variables with respective means $[\mu(1+\sqrt{\rho})]^{-1}$ and $[\mu(1-\sqrt{\rho})]^{-1}$.

Denote the last two random variables by τ_1 and τ_2, respectively. Notice that $E\{\theta\} = \frac{2}{\mu - 2\lambda}$, as implied by Little's result. Also, θ_1 and θ_2 are exponentially distributed with mean $\frac{1}{\mu(1-\rho)}$.

Observe that

$$\begin{aligned} \mathrm{var}(\theta) &= \mathrm{var}(\theta_1 + \theta_2) = 2\mathrm{var}(\theta_1) + 2\mathrm{cov}(\theta_1, \theta_2) \\ &= \mathrm{var}(\tau_1) + \mathrm{var}(\tau_2) = \frac{1}{\mu^2(1+\sqrt{\rho})^2} + \frac{1}{\mu^2(1-\sqrt{\rho})^2} \\ &= \frac{2(1+\rho)}{\mu^2(1-\rho)^2} . \end{aligned}$$

This shows that $\mathrm{cov}(\theta_1, \theta_2) = \dfrac{\rho}{\mu^2(1-\rho)^2}$. Simple algebra now yields

$$E\{ \mid \frac{\theta_1}{E\{\theta_1\}} - \frac{\theta_2}{E\{\theta_2\}} \mid^2 \} = 2(1-\rho) .$$

This calculation allows us to conclude that, as $\rho \to 1$, $\dfrac{\theta_1}{E\{\theta_1\}}$ converges in quadratic mean to $\dfrac{\theta_2}{E\{\theta_2\}}$. In particular, this implies that

$$\frac{\theta}{E\{\theta\}} \to \frac{2\theta_1}{E\{\theta\}}, \text{ in quadratic mean as } \rho \to 1 .$$

Comments 4.8.2

(a) This result says that, when properly normalized, θ_1 and θ_2 become very close to each other. As a consequence, θ becomes almost equal to $2\theta_1$, where θ_1 is exponentially distributed. Notice that θ does not become close to the sum of two i.i.d. exponential random variables; it becomes close to **twice** an exponential random variable (which is again exponentially distributed).

(b) The calculation extends easily to the case of n successive visits instead of two. To see this, denote by $\theta_1, \ldots, \theta_n$ the n successive sojourn times. (See Figure 4.20.)

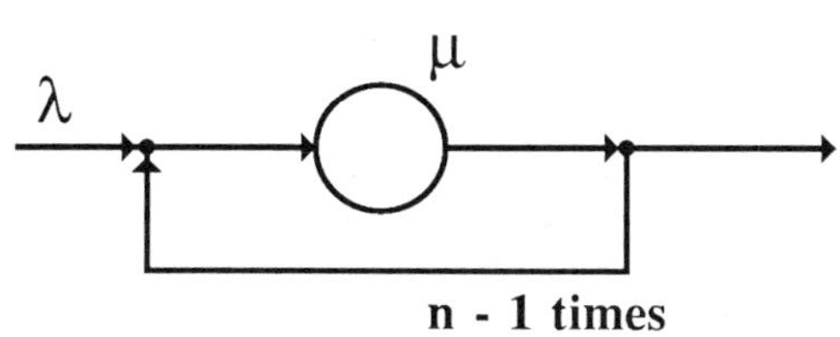

Figure 4.20. An M/M/1 queue visited n times.

Assume that the typical customer is fed back for the kth time at time $t = 0$. Arguing as above, one finds that

$$\frac{\theta_k}{E\{\theta_k\}} \to \frac{\theta_{k+1}}{E\{\theta_{k+1}\}}, \text{ in quadratic mean as } \rho := \frac{n\lambda}{\mu} \to 1 .$$

It follows easily that $\theta = \theta_1 + \cdots + \theta_n$ is such that

$$\frac{\theta}{E\{\theta\}} \to \frac{n\theta_1}{E\{\theta\}}, \text{ in quadratic mean as } \rho \to 1 .$$

In the above expression, θ_1 is exponentially distributed with mean $\frac{1}{\mu(1-\rho)}$. Thus, when properly normalized, θ becomes close to $n\theta_1$.

(c) Assume now that the routing is random, with a feedback probability p. (See Figure 4.21.)

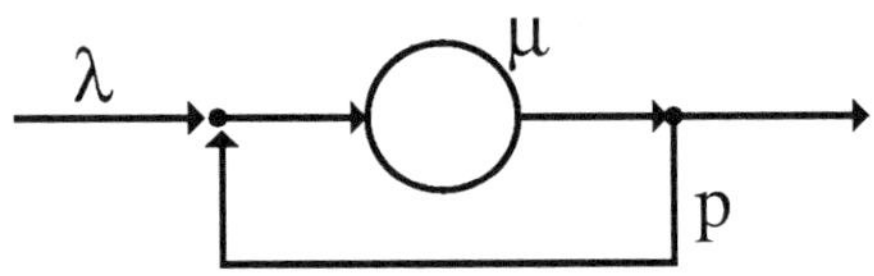

Figure 4.21. A queue with random feedback.

Let $\rho = \frac{\lambda}{\mu(1-p)}$. By conditioning on the number V of visits and by applying the above result, one finds that the sojourn time θ is such that $\frac{\theta}{E\{\theta_1\}}$ converges in quadratic mean to $V\frac{\theta_1}{E\{\theta_1\}}$ as $\rho \to 1$. It follows that (in quadratic mean, as $\rho \to 1$)

$$(1-\rho)\theta \to V \times (1-\rho)\theta_1 .$$

Justification 4.8.3 As can be anticipated, these results are a consequence of the Central Limit Theorem (see Appendix A).

For simplicity, consider Example 4.8.1. When ρ is almost one, x_0 is very likely to be very large. Now, when x_0 is large, the CLT implies that

$$\frac{\theta_i - \mu^{-1}x_0}{\mu^{-1}\sqrt{x_0}} \approx N(0, 1), \quad \text{for } i = 1, 2.$$

As a consequence, for large x_0,

$$\frac{|\theta_1 - \theta_2|}{\theta_1} \approx O(\frac{1}{\sqrt{x_0}}),$$

which is then very likely to be very small.

This argument applies, as before, to successive sojourn times in the case of multiple feedbacks. It can be extended to the case of networks. The essential idea is that, in heavy traffic, over an interval of duration t the queue length fluctuations are of order $O(\sqrt{t})$, by CLT. Thus, each time a customer comes back to one of the queues of the network, it finds its queue length to be the same as upon arriving into the network, to the order $O(\sqrt{\theta})$ with θ being the total sojourn time in the network. Upon the arrival into the network, the queue lengths are independent and are geometrically distributed. Thus, to the order $O(\sqrt{\theta})$, the random time θ is a random linear combination of independent exponentially distributed random variables.

4.9. SMALL NETWORKS – BACKWARD METHOD

For very small networks it may be possible to calculate passage time distributions by solving the corresponding backward equations. The point of this section is to illustrate that method.

Example 4.9.1 Consider the network of two M/M/1 queues in Figure 4.22. There are two customers in the network. Our objective is to determine the distribution of the *typical* time τ taken by one customer to go from point A to point B in the network.

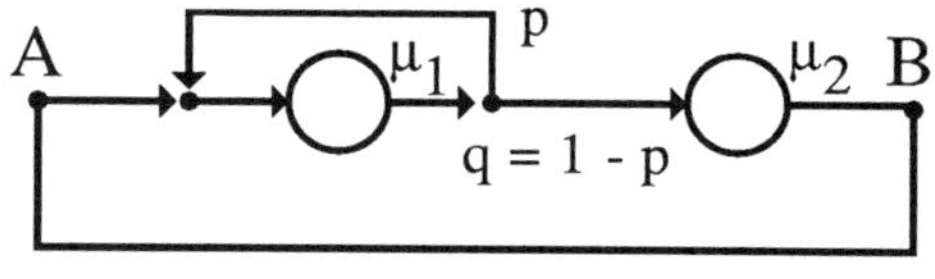

Figure 4.22. A simple closed network with two M/M/1 queues.

Definition 4.9.2 [State augmentation] The first step is to augment the state so as to keep track of the evolution of a given customer.

To do this, one of the customers will be marked. The customer of interest will be called 1 while the other will be called 0. By (01, .) one will indicate that the two customers are in the first queue, in the order 01. Similarly, (0, 1) indicates that customer 0 is in the first queue while customer 1 is in the second queue, etc. There are six possible values for the augmented state. The transition rates are readily verified to be as indicated in Figure 4.23.

Denote by Q this rate matrix on the state space $\{1, 2, \ldots, 6\}$ where the states are numbered as on Figure 4.23.

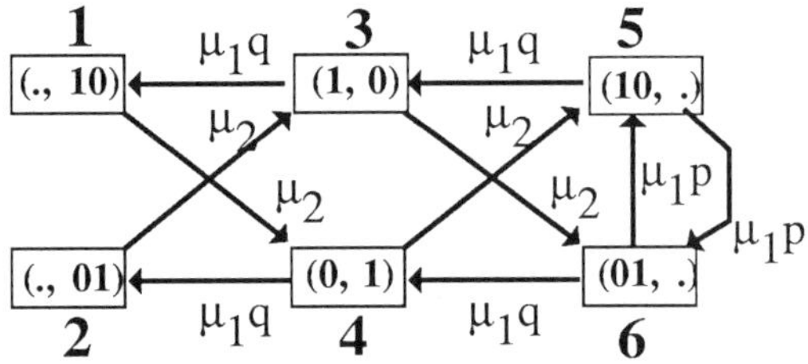

Figure 4.23. The state transition diagram of the network.

Problem 4.9.3 Observe that customer 1 crosses the points A and B when the state makes one of the transitions $2 \rightarrow 3$ or $4 \rightarrow 5$. Say that this happens first at time T_1. At time T_1 the state is either 3 or 5. The state then moves around in the state space until one of the above two transitions takes place, at time T_2. We are interested in the distribution of $\tau = T_2 - T_1$.

More specifically, we want to find the distribution of τ for a typical customer, i.e., for a customer who sees the other customer with its invariant distribution in the network when crossing point A. (By the Arrival Theorem 4.4.1.)

When 0 is alone in the network, its motion is described by a Markov chain on $\{1, 2\}$ with the transition matrix $q(1, 2) = \mu_1 q$ and $q(2, 1) = \mu_2$. Thus, the invariant probability that customer 1 will find customer 0 in the second queue is $\mu_1 q(\mu_1 q + \mu_2)^{-1}$.

Therefore one has to find the distribution of

$$\tau = \min\{t \geq 0 \mid (x_{t-}, x_t) \in \{(2, 3), (4, 5)\}\}$$

given that

$$P\{x_0 = 3\} = \frac{\mu_1 q}{\mu_1 q + \mu_2} = 1 - P\{x_0 = 5\} ,$$

and that x_t has the rate matrix indicated on the state transition diagram above.

Solution 4.9.4 Denote by y_t the Markov chain x_t killed whenever it makes one of the transitions $2 \rightarrow 3$ or $4 \rightarrow 5$. This Markov chain y_t has the same rate matrix as x_t except that the arrows $2 \rightarrow 3$ and $4 \rightarrow 5$ are replaced by transitions to a "coffin" state δ. (See Figure 4.24.)

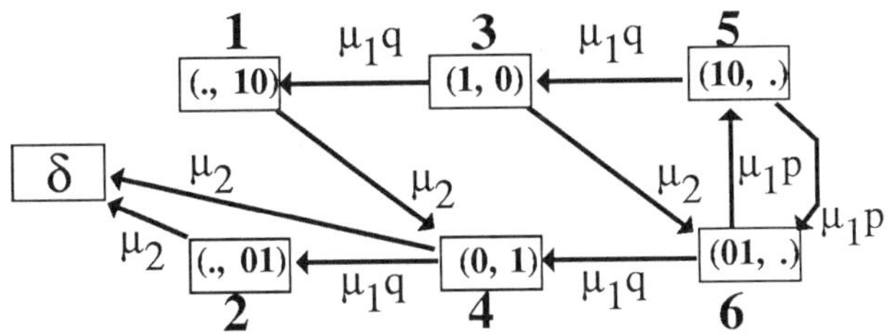

Figure 4.24. The transition diagram of the killed process.

Let the transition probabilities of y_t on $\{1, 2, \ldots, 6\}$ be denoted by $\tilde{P}_t$ for $t \geq 0$. Then

$$\frac{d}{dt}\tilde{P}_t = \tilde{Q}\,\tilde{P}_t ,$$

by the Kolmogorov equation.

Now, $\tau > t$ if and only if y_t is still in $\{1, \ldots, 6\}$. As a consequence,

$$P_\nu\{\tau > t\} = \sum_{i=1}^{6} P_\nu\{y_t = i\} = \nu\,\tilde{P}_t\mathbf{1} ,$$

where $\mathbf{1}$ is a column vector of ones and $P_\nu\{.\}$ is the probability that corresponds to the initial distribution ν for y_0.
The solution is then

$$F(t) = \mathbf{1} - \nu e^{\tilde{Q}t}1, \quad t \geq 0, \quad \nu = (0, 0, \alpha, 0, 1 - \alpha, 0) ,$$

$$\alpha = \frac{\mu_1 q}{\mu_1 q + \mu_2} .$$

A slight simplification is possible. Define $g_t = \tilde{P}_t\mathbf{1}$. Then g_t can be found by solving $\frac{d}{dt}g_t = \tilde{Q}g_t$, $g_0 = \mathbf{1}$. This last system is of order 6, while the Kolmogorov equations are of order 6 × 6.

4.10. OTHER RESULTS

This section presents two other results on sojourn times in networks. The first result shows that the joint distribution of the sojourn times in a cycle of one M/M/1 and one M/D/1 queue **depends on the order** of the queues. The second result concerns the joint distribution of the sojourn times in a path that starts and ends with an **M/M/s queue.**

Example 4.10.1 [M/D/1-M/M/1 cycle] The system is as indicated in Figure 4.25. There are two customers in the cycle, called C_1 and C_2, respectively. Queue 1 is an M/D/1 and queue 2 is an M/M/1. C_1 arrives in queue 1 at time 0. Denote by θ_1 and θ_2 the sojourn times of C_1 in queues 1 and 2, respectively. The claim is that θ_1 and θ_2 are independent.

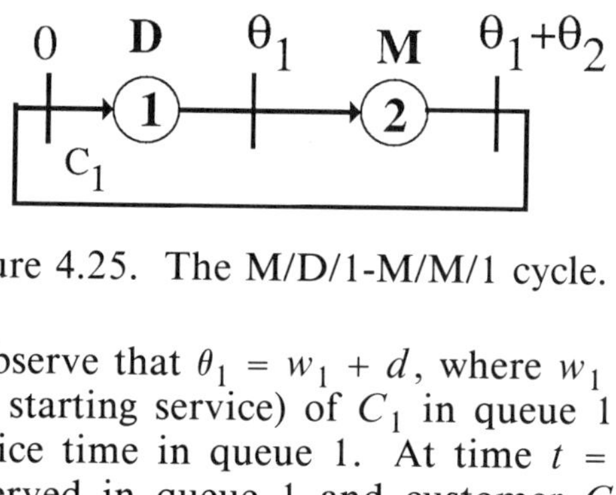

Figure 4.25. The M/D/1-M/M/1 cycle.

To see this, observe that $\theta_1 = w_1 + d$, where w_1 is the waiting time (time before starting service) of C_1 in queue 1 and d is the deterministic service time in queue 1. At time $t = w_1$, customer C_1 starts being served in queue 1 and customer C_2 starts being served in queue 2. The evolution of the system after w_1 is then independent of w_1 since the state is specified at that time. In particular, θ_2 is independent of w_1. Therefore, θ_2 is independent of $w_1 + d = \theta_1$, as claimed.

Next, interchange the order of the two queues. That is, queue 1 is now the M/M/1 and queue 2 is the M/D/1. (See Figure 4.26.) Define θ_1 and θ_2 as before. Thus, $\theta_1 = w_1 + \sigma_1$ and $\theta_2 = w_2 + d$. The claim is that θ_1 and θ_2 are negatively correlated, so that they cannot be independent.

To see this, observe once again that the state is specified at time w_1, so that the evolution after w_1 is independent of w_1. Now,

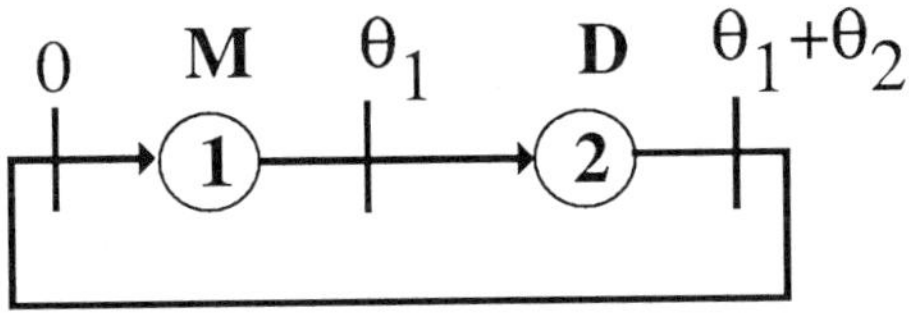

Figure 4.26. The M/M/1-M/D/1 cycle.

w_2 and σ_1 are negatively correlated. Indeed, $w_2 = \max\{d - \sigma_1, 0\}$. Hence, $\theta_1 = w_1 + \sigma_1$ and $\theta_2 = w_2 + d$ are negatively correlated, as claimed.

Example 4.10.2 [M/M/s-M/M/1-...-M/M/1-M/M/s cycle]
The cycle in Figure 4.27 consists of one M/M/1 queue between two M/M/s queues. The result extends to an arbitrary number of M/M/1 queues. There are N customers in the system and queues 1 and 3 have a and b servers, respectively.

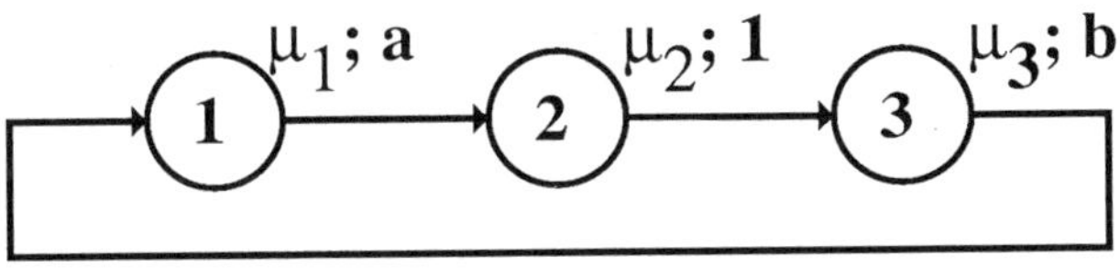

Figure 4.27. A cycle with multiserver queues.

Denote by $\theta_1, \theta_2, \theta_3$ the successive sojourn times of a typical customer in the three queues.
The claim is that

$$E\{e^{-s_1\theta_1 - s_2\theta_2 - s_3\theta_3}\} = \sum_{x \in \mathbf{X}_{N-1}} \pi_{N-1}(x)\, F_1(s_1, x^1)\, F_2(s_2, x^2)\, F_3(s_3, x^3)$$

where π_{N-1} is the invariant distribution of the network on the set $\mathbf{X}_{N-1}$ that corresponds to $N-1$ customers and

$$F_1(s, n) = \prod_{m=1}^{n+1} \frac{\mu_1 \min\{m, a\}}{\mu_1 \min\{m, a\} + s}$$

$$F_2(s, n) = \left(\frac{\mu_2}{\mu_2 + s_2}\right)^{n+1}$$

$$F_3(s, n) = \prod_{m=1}^{n+1} \frac{\mu_3 \min\{m, b\}}{\mu_3 \min\{m, b\} + s} .$$

The proof is similar to the case of M/M/1 queues. The essential observation is that the overtakings in the first and last queues do not play a role. To see this, observe that for the last queue this follows from the fact that future arrivals do not affect the sojourn time of the customers already present in the queue (even though they may be overtaken). For the first queue, one argues that reversing the time in the network is the same as reversing the arrows. This is the case since the M/M/s queues are time-reversible.

It remains only to derive F_1 (F_3 is similar), and the proof will follow by the same induction argument as in Example 4.6.2 for three queues, and as in Theorem 4.6.3 for the extension to an arbitrary number of M/M/1 queues.

To find F_1, consider a customer who enters the first queue and finds n customers upon arrival. Denote by $\theta(n)$ the sojourn time of that customer. Then $\theta(n)$ can be decomposed as the sum of the time until the first service completion and an independent random variables distributed as $\theta(n-1)$. Indeed, the evolution starts afresh after the first service completion. The first service completion is the minimum of $\min\{n+1, a\}$ independent and exponentially distributed service times with rate μ_1. It is therefore exponentially distributed with rate $\min\{n+1, a\}\mu_1$. Thus,

$$\theta(n) = \theta(n-1) + \tau(\min\{n+1, a\}\mu_1), \quad n \geq 2$$

where $\tau(\beta)$ denotes an exponentially distributed random variable with rate β that is independent of $\theta(n-1)$. Also, $\theta(0) = \tau(\mu_1)$. Hence,

$$F_1(s, n) = F_1(s, n-1)\frac{\mu_1 \min\{n+1, a\}}{\mu_1 \min\{n+1, a\} + s},$$

$$n \geq 0 \text{ with } F_1(s, -1) := 1 .$$

This gives the formula for F_1.

4.11. FLOWS

The objective of this section is prove the "loop criterion" for Poisson flows in networks of quasi-reversible queues. The basic idea was introduced in Example 4.3.2.

Let us first state the result.

Theorem 4.11.1 [Loop Criterion] Consider a stationary open network of quasi-reversible queues with Markov routing. A flow on the link (i, j) of the network is not Poisson if the link can be visited more than once by the same customer.

Example 4.11.2 In the case of a single-class network of quasi-reversible queues (see Figure 4.28) one can conclude that a flow on a link is Poisson if and only if the link is **not part of a loop.**

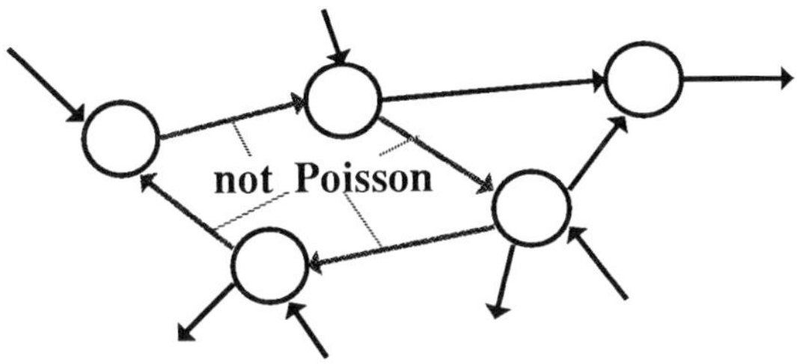

Figure 4.28. A single class network.

Indeed, either a link can be visited more than once by the same customer or it can be viewed as the output of a subnetwork. Also, we know that networks of quasi-reversible queues are quasi-reversible. (See Theorem 3.5.4.)

Example 4.11.3 The link (1, 2) in the multiclass network in Figure 4.29 cannot be visited more than once by the same customer. However, Theorem 4.11.1 cannot decide whether the flow on that link is Poisson or not.

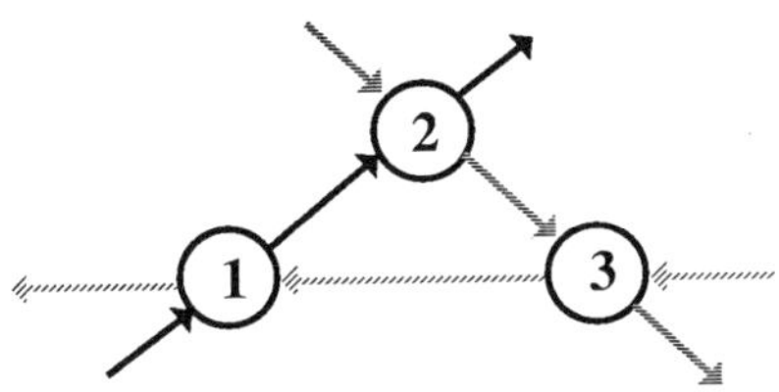

Figure 4.29. A link in a multiclass network.

See Example 4.11.6 for a related situation.

Definition 4.11.4 [Coupled networks] Denote by x_t the state at time $t \geq 0$ of the stationary network (i.e., with its invariant product-form initial distribution π). Let ν be the distribution just after a jump of a customer of class c on the link (i, j). By the Arrival Theorem 4.4.1, the distribution ν corresponds to π to which one adds one customer of class c in queue j.

Define the process $\{(x_t, y_t), t \geq 0\}$ so that y_0 has one more customer of class c in queue j than x_0 and the same number of customers in the other queues. In addition, the evolutions of y_t and x_t are independent for $t > 0$. Since the Markov chain (x_t, y_t) admits the invariant distribution $\pi \times \pi$ it follows that it will visit the state $(\mathbf{0}, \mathbf{0})$ after some random time τ that has a finite mean. (By $\mathbf{0}$ we denote the state of the empty network.) Modify the definition of y_t so that x_t and y_t agree after time τ, and call the new process z_t. Clearly, $\{z_t, t \geq 0\}$ and $\{y_t, t \geq 0\}$ have the same law.

Fact 4.11.5 Let A_t [resp. B_t] be the number of customers of class c that cross the link (i, j) in $[0, t]$ for the network x_t [resp. z_t]. Then

$$E\{B_\tau - A_\tau\} = \alpha > 0 \tag{4.21}$$

where α is the average number of visits to link (i, j) made before leaving the network by a customer of class c initially in queue j.

Proof: Let $a(k, d)$ be the average number of visits to link (i, j) as a customer of class c made by a customer of class d started in queue k before leaving the network. These numbers can be calculated by solving the first-step equations

$$a(k, d) = \sum_{(l,e)} r_{kl}^{de} \left[1\{(k, l) = (i, j), e = c\} + a(l, e)\right] .$$

The only fact that is needed is that these numbers are finite for an open network.

Observe that

$$E\{A_\tau\} = \sum_{(k,d)} a(k, d)\, E\{n_k^d(x_0) + \tau\, \gamma_k^d\} \tag{4.22}$$

where γ_k^d is the rate of exogenous arrivals of class d into queue k and $n_k^d(x)$ is the number of customers of class c in queue k when the network is in state x.

The interpretation of this formula is that the customers that cross the link (i, j) as customers of class c are those that were in the network at time 0 and those who arrived in $[0, \tau]$. A formal proof of (4.22) is left as an exercise.

A similar formula can be written for $E\{B_\tau\}$. Subtracting gives (4.21) with $\alpha = a(j, c)$.

Proof of Theorem 4.11.1 We are now ready to prove the Theorem. Since τ has finite mean, one finds that (see Exercise 4.10)

$$E\{B_t - A_t\} \rightarrow E\{B_\tau - A_\tau\} \text{ as } t \rightarrow \infty . \tag{4.23}$$

Therefore, there is a finite t such that $E\{B_t\} > E\{A_t\}$. As discussed in Example 4.3.2, this implies that having observed a jump time of the counting process on link (i, j) before time ϵ affects the future jump times. Therefore, the flow on link (i, j) is not a Poisson process.

Example 4.11.6 Consider the network in Figure 4.30 introduced in Example 4.5.5.

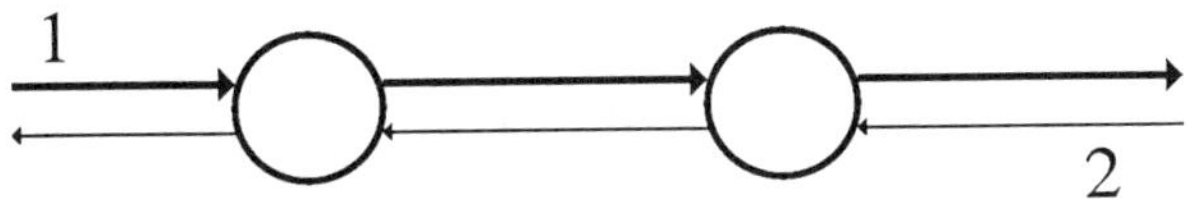

Figure 4.30. A two class network.

There are two classes of customers. The queues are M/M/1, and the two classes are served at the same rate in the queues. The claim is that the internal flows are not Poisson.

The argument is as follows. The Arrival Theorem shows that the distribution just after one jump from queue 1 to queue 2 is the invariant distribution plus one customer of class 1 in queue 2.

One can extend this fact and show that if n jumps from queue 1 to queue 2 have been observed in $(0, \epsilon)$ and if the network has its invariant distribution at time 0, then the conditional distribution at time ϵ is the invariant distribution plus the n customers of class 1 in queue 2. These customers will slow down the flow of customers of class 2 from queue 2 to queue 1. In turn, this will make queue 1 more available to serve customers of queue 1. Hence, given that many jumps were observed from queue 1 to queue 2 in $(0, \epsilon)$, the number of future jumps from queue 1 to queue 2 is likely to be larger. Thus, that process does not have independent increments and it cannot be Poisson.

4.12. INTERARRIVAL TIMES

This section discusses a few simple results on the interarrival times in networks of queues. We start with an example to illustrate the method of calculation of interarrival time distributions. Then we show that the interarrival times and interdeparture times have the same distribution at an M/M/1 queue in a network of quasi-reversible queues.

Example 4.12.1 Consider the M/M/1 queue with feedback illustrated in Figure 4.31. Assume that the system is stationary and denote the invariant queue length distribution by π. Thus,

$$\pi(n) = \rho^n(1-\rho),\ \rho := \frac{\gamma}{q\mu},\quad n \geq 0. \tag{4.24}$$

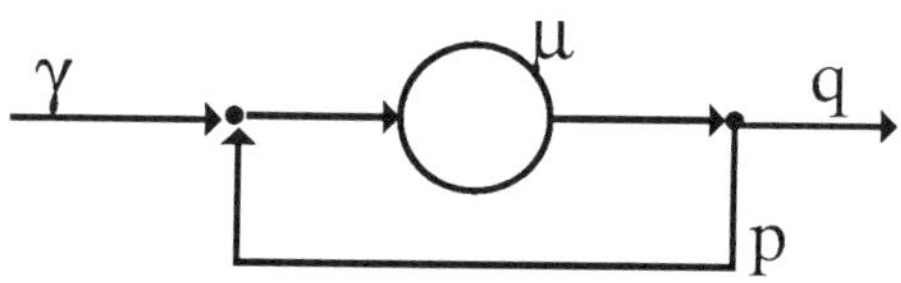

Figure 4.31. An M/M/1 queue with feedback.

We will show the following result.

Fact 4.12.2 Denote by τ a typical interdeparture time from the M/M/1 queue. Then

$$E\{e^{-s\tau}\} = \frac{\mu p}{\mu-\gamma}\frac{\mu}{\mu+s} + \frac{\mu q-\gamma}{\mu-\gamma}\frac{\gamma}{\gamma+s}. \tag{4.25}$$

Moreover, the distribution of a typical interarrival time is the same as that of the typical interdeparture time τ.

Proof: A typical interdeparture time is the time between a typical departure time and the next departure time. Let ν be the distribution of the queue length just after that typical departure time.

Now, by the "ratio of rates" formula (2.50), one has (with $\lambda := \gamma q^{-1}$)

$$\begin{aligned}\nu(n) &= \lambda^{-1}\left[\pi(n+1)\mu q + \pi(n)\mu p\, 1\{n>0\}\right]\\ &= \pi(n)q + \pi(n-1)p,\quad n \geq 0\end{aligned}$$

where $\pi(-1) := 0$. In particular,

$$\nu(0) = \pi(0)q = (1 - \rho)q = q - \frac{\gamma}{\mu}. \tag{4.26}$$

The first departure time τ from the queue started at time 0 with distribution ν can be determined as follows. If $x_0 > 0$, then τ is a service time, i.e., is exponentially distributed with rate μ. If $x_0 = 0$, then τ is the sum of an arrival time and a service time. These random variables are independent and exponentially distributed with rates γ and μ, respectively. Thus, the Laplace transform of τ is equal to

$$E\{e^{-s\tau}\} = \nu(0)\frac{\gamma}{\gamma + s}\frac{\mu}{\mu + s} + (1 - \nu(0))\frac{\mu}{\mu + s}$$

where we used the fact that the Laplace transform of an exponential distribution with rate α is equal to $\alpha(\alpha + s)^{-1}$.

Using (4.26), one finds, after some algebraic simplifications, that this expression agrees with (4.25).

This shows that τ can be viewed as a random variable that is exponentially distributed with rate μ with probability $\mu p(\mu - \gamma)^{-1}$ and is exponentially distributed with rate γ otherwise. In particular, we see that τ is exponentially distributed if and only if $p = 0$. This confirms that the departure process from the queue is not Poisson if $p > 0$.

The statement about the interarrival times is immediate by time reversibility. Indeed, the departure process from the queue has the same distribution as the arrival process into the queue reversed in time. Also, reversing the time of a point process does not modify the distribution between successive jump times. Thus, the interarrival times and the interdeparture times must have the same stationary distribution.

Example 4.12.3 Consider the situation illustrated in Figure 4.32.

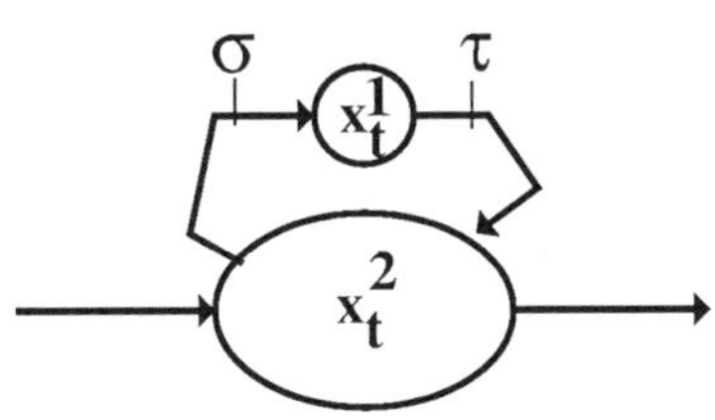

Figure 4.32. An M/M/1 queue in a quasi-reversible network.

Queue 1 is an M/M/1 queue with service rate μ. The rest of the system is a quasi-reversible network modeled by a Markov chain $\mathbf{x}^2$. Let $x_t = (x_t^1, x_t^2)$. The invariant distribution is $\pi(x^1, x^2) = \pi_1(x^1)\,\pi_2(x^2)$. We will show the following result.

Theorem 4.12.4 The typical interarrival time σ and the typical interdeparture time τ at the M/M/1 queue have the same distribution.

Proof:

(i) First notice that the arrival process into queue 1 reversed in time has the same law as the departure process from the same queue in the network reversed in time. Indeed, by reversing the time, the departures after time t become the arrivals before time t. Also, reversing the time does not affect the stationary distribution of the time between successive jump times. As a consequence, it suffices to show that the typical interdeparture distribution from queue 1 has the same distribution when the network is reversed in time.

(ii) Consider the typical interdeparture distribution τ (for the network in forward time). Say that $\tau = T_{n+1} - T_n$, where the T_ms are the departure times. To make sure that τ is typical, one has to guarantee that the sequence $\{x_{T_m}, m \in \mathbb{Z}\}$ is stationary, i.e., that T_n is a typical departure time. The corresponding distribution of x_{T_n} is given by the Arrival Theorem. It is such that the customers who do not jump have distribution π. Now, if $x_{T_n}^1 > 0$, then τ is a service time in queue 1, i.e., τ is exponentially distributed with rate μ. Also, if $x_{T_n}^1 = 0$, then τ is the sum of an idle time for queue 1 and of an independent exponentially distributed service time with rate μ.

(iii) The same discussion applies to the network reversed in time.

(iv) To conclude the proof it remains to show two facts. First, the probability that $x_{T_n}^1 = 0$ is the same for the network and the network reversed in time. Second, the distribution of an idle period for queue 1 is the same for the network and for the network reversed in time. This is done in the two paragraphs below.

(v) The invariant distribution π is obviously the same for the network and for the network reversed in time. Applying the Arrival Theorem then shows that the distribution of the state of queue 1 just after a typical departure is the same in both networks.

(vi) A typical idle period for queue 1 in the network (in forward time) is the duration of a time interval where $x_t^1 = 0$. Reversing the time does not change those durations.

The proof is complete.

4.13. EXERCISES

4.1. Asymptotic delay Justify the limit in (4.2).
Hint: Let

$$P(i, j) := P[x_{S_{n+1}-} = j \mid x_{S_n -} = i]$$
$$\phi_j(t) := P\{\sigma_1 + \cdots + \sigma_{j+1} \leq t\}.$$

First assume that the distribution ν_1 of $x_{S_1 -}$ is concentrated on a given value i, i.e., $\nu_1(j) = 1\{j = i\}$ for $j \geq 0$. Then

$$\nu_1 \leq K\pi$$

with $K = [\pi(i)]^{-1}$. As a consequence, if ν_n denotes the distribution of $x_{S_n -}$, then

$$\nu_n = \nu_1 P^{n-1} \leq K\pi P^{n-1} = K\pi, \quad n \geq 2.$$

Moreover, $\nu_n \rightarrow \pi$ as $n \rightarrow \infty$. One concludes from the Dominated Convergence Theorem (see Appendix A) that

$$\psi_i(n, t) := P[\theta_n \leq t \mid x_{S_1 -} = i]$$
$$= \sum_j \nu_n(j)\, \phi_j(t) \rightarrow \sum_j \pi(j)\, \phi_j(t) =: F(t).$$

Next consider the general case for ν_1. One has

$$P\{\theta_n \leq t\} = \sum_i \nu_1(i)\, \psi_i(n, t).$$

Also, $\psi_i(n, t) \leq 1$ and $\sum_i \nu(i)\, 1 < \infty$. Therefore, again by dominated convergence,

$$\sum_i \nu_1(i)\, \psi_i(n, t) \rightarrow \sum_i \nu_1(i)\, F(t) = F(t).$$

This is the first equality in (4.2).

4.2. Delay in M/M/1 queue Prove that the typical delay in an M/M/1 queue is exponentially distributed with rate $\mu - \lambda$. *Hint:* The standard proof goes as follows. Let τ be the typical delay (with distribution $F(t)$ defined in Exercise 4.1). Since τ is the sum of $j + 1$ i.i.d. exponentially distributed random variables with rate μ with probability $\rho^j(1 - \rho)$, one has

$$E\{e^{-s\tau}\} = \sum_{j=0}^{\infty} \rho^j (1-\rho) E\{e^{-s(\sigma_1 + \cdots + \sigma_{j+1})}\}$$

$$= \sum_{j=0}^{\infty} \rho^j (1-\rho) [\frac{\mu}{\mu+s}]^{j+1}$$

$$= (1-\rho)\frac{\mu}{\mu+s} [1-\rho\frac{\mu}{\mu+s}]^{-1} = \frac{\mu-\lambda}{\mu-\lambda+s}.$$

In this calculation we used three basic facts. The first is that σ is exponentially distributed with rate α if and only if $E\{e^{-s\sigma}\} = \alpha(\alpha+s)^{-1}$ for complex numbers s with a real part larger than $-\alpha$. The second is that the mean value of a product of independent random variables is the product of their mean values. The third fact is that if $|a| < 1$, then $\sum_{n=0}^{\infty} a^n = (1-a)^{-1}$.

4.3. Limiting delays in queues in tandem Prove the result corresponding to that of Exercise 4.1 for two M/M/1 queues in tandem. That is, prove that the vector of sojourn times (θ_n^1, θ_n^2) of the nth customer converges in distribution to two independent exponentially distributed random variables with rates $\mu_1 - \lambda$ and $\mu_2 - \lambda$, respectively. *Hint:* First show that

$$\phi_x(n, t_1, t_2) := P[\theta_n^1 \leq t_1,\ \theta_n^2 \leq t_2 \mid x_{S_1-} = x] \rightarrow F(t_1, t_2)$$

by dominated convergence, as in Exercise 4.1.

Next, conclude that

$$\sum_x \nu_1(x)\, \phi_x(n, t_1, t_2) \rightarrow F(t_1, t_2),$$

again by dominated convergence.

4.4. Limiting mean queue length Consider an M/M/1 queue with $\lambda < \mu$. Show that

$$E[x_t \mid x_0 = i] \rightarrow \frac{\lambda}{\mu-\lambda}$$

for all $i \geq 0$. *Hint:* Use dominated convergence as in Exercise 4.1.

4.5. Correlation of delays Show that in an M/M/1 queue the random variables θ_n and θ_{n+1} are positively correlated.

4.6. Delay in an M/M/s queue Analyze the distribution of a typical delay in an M/M/s queue. *Hint:* See Example 4.10.2.

4.7. Cycle time Assume that $N = 2$ in (4.4). Find the variance of $\theta_1 + \theta_2$.

4.8. Small networks Assume that the queues in Figure 4.22 are LCFS and that the two customers have different service rates in each of the two queues. Calculate the distribution of the cycle time.

4.9. Loop criterion Prove (4.23). *Hint:* One has $E \mid B_t - A_t \mid \leq E\{B_\tau\} + E\{A_\tau\} < \infty$ for $t \geq 0$. Conclude by using the Dominated Convergence Theorem.

4.14. REFERENCES AND COMMENTS

Section 4.2 The result on sojourn times in tandem queues is due to Reich (1957). Example 4.2.3 is discussed in Simon and Foley (1979). The results on the queues in a cycle are due to Schassberger and Daduna (1983). The time-reversal proofs are due to Kelly and Pollett (1983).

Section 4.3 The fact that flows are generally not Poisson in an open network was discovered by Burke (1976).

Section 4.4 The Arrival Theorem can be found as Theorem 3.12 in Kelly (1979). See also Cohen (1957), Descloux (1967), Melamed (1982a), Sevcik and Mitrani (1981), and Lavenberg and Reiser (1980).

Section 4.5 The results of this section are from Varaiya, et al. (1980a). See Melamed (1982b) for a different approach.

Section 4.6 The results for closed networks are from Daduna (1982). The proofs given in this section are from Kelly and Pollett (1983).

Section 4.7 This section presents results discovered in Barbour and Schassberger (1981) and Jansen (1984). The proof given here is an extension of the argument of Tsoucas, et al. (1984a).

Section 4.8 This section is inspired from Reiman (1982).

Section 4.10 The results of 4.10.1 are due to Boxma (1983). That of 4.10.2 is from Daduna (1986).

Section 4.11 See Melamed (1979) for the loop criterion in networks of M/M/1 queues. Example 11.6 is from Varaiya, et al. (1981), where the details of the analysis can be found. Theorem 4.11.1 is given here with weaker assumptions than previously.

Section 4.12 The calculation of Fact 4.12.2 is from Burke (1976). The extension is a generalization of the result for networks of M/M/1 queues in Labetoulle, et al. (1981). The proof is due to Kiessler (1987). For recent surveys of results on flows in networks, see Disney and König (1985) and Disney and Kiessler (1987). A few additional results on flows will be given in Chapter Ten as applications of the nonlinear filtering approach.

CHAPTER FIVE

NUMERICAL METHODS

5.1. INTRODUCTION

This chapter discusses some of the **numerical methods** that have been developed to calculate invariant distributions of queueing systems. Many useful methods are not covered here.

Section 5.2 presents the **matrix-geometric** solutions that arise in a number of systems.

In section 5.3 we review the **mean value analysis** for multiclass networks. The basic idea of that method was introduced in section 1.8.

The method of **asymptotic expansions** is explained in an example in section 5.4.

Section 5.5 describes the method of **fixed-population-mean** that approximates a closed network by an open one.

5.2. MATRIX-GEOMETRIC SOLUTIONS

Comments 5.2.1 Roughly the idea is that if a rate matrix Q or a transition matrix P on the nonnegative integers has a geometric invariant distribution $\{\pi(n) = a\rho^n, n \geq 0\}$, then replacing the scalars in Q or P by $N \times N$ matrices leads to an invariant distribution of the form

$$\nu(n, i) = \nu_n(i) = \nu_0 R^n(i), \quad 1 \leq i \leq N, \quad n \geq 0,$$

where the ν_n are $1 \times N$ vectors and R is an $N \times N$ matrix.

We first look at a simple example.

Example 5.2.2 [Two queues with a finite waiting room] Consider the network in Figure 5.1. The first queue is M/M/1 with service rate μ_1. The second queue is M/M/1/N with service rate μ_2. Thus, the second queue has a waiting room that can accommodate at most N customers.

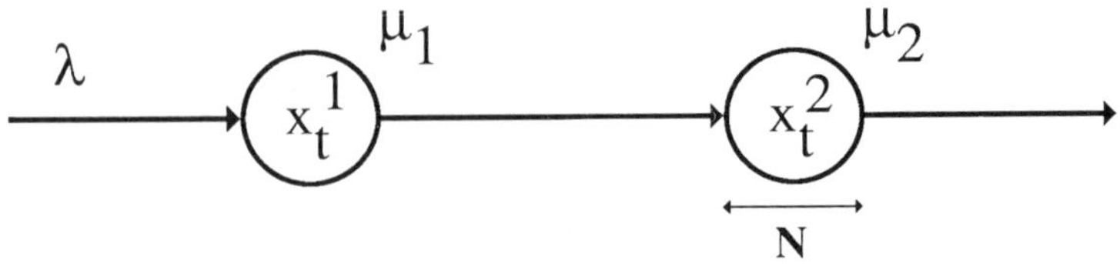

Figure 5.1. The system M/M/1-M/M/1/N.

The **blocking mechanism** is such that server 1 stops working whenever the second queue is full. This is called **communication blocking.**

Denote the states by

$$x = (x^1, x^2) \in \mathbf{X} = \{0, 1, 2, \ldots\} \times \{0, 1, \ldots, N\}.$$

In this notation, x^i is the number of customers in queue i ($i = 1, 2$). Order the states as

$$(0, 0), (0, 1), \ldots, (0, N), (1, 0), (1, 1), \ldots, (1, N), (2, 0), \ldots,$$

i.e., in the **lexicographic** order with x^1 as the most significant symbol.

One finds that the rate matrix Q of the state process is

$$Q = \begin{bmatrix} B_0 & A_0 & 0 & 0 & 0 & . \\ A_2 & A_1 & A_0 & 0 & 0 & . \\ 0 & A_2 & A_1 & A_0 & 0 & . \\ 0 & 0 & A_2 & A_1 & A_0 & . \\ 0 & 0 & 0 & A_2 & A_1 & . \\ . & . & . & . & . & . \end{bmatrix} \tag{5.1}$$

where the $(N + 1) \times (N + 1)$ matrices B_0 and A_i are as follows:

$$A_2 = \begin{bmatrix} 0 & \mu_1 & 0 & . & 0 \\ 0 & 0 & \mu_1 & . & 0 \\ . & . & . & . & . \\ 0 & 0 & 0 & . & \mu_1 \\ 0 & 0 & 0 & . & 0 \end{bmatrix}, \quad A_1 = \begin{bmatrix} 0 & 0 & . & 0 & 0 & 0 \\ \mu_2 & 0 & . & 0 & 0 & . \\ . & . & . & . & . & 0 \\ 0 & . & . & \mu_2 & 0 & 0 \\ 0 & . & . & 0 & \mu_2 & 0 \end{bmatrix} - \Delta \tag{5.2}$$

and

$$\Delta = \text{diag}\{\mu_1 + \lambda, \mu_1 + \mu_2 + \lambda, \ldots, \mu_1 + \mu_2 + \lambda, \mu_2 + \lambda\}, \tag{5.3}$$

$$A_0 = \lambda I$$

where I denotes the $(N + 1) \times (N + 1)$ identity matrix, and

$$B_0 = \begin{bmatrix} -\lambda & 0 & 0 & . & 0 & 0 \\ \mu_2 & -\lambda - \mu_2 & 0 & . & 0 & 0 \\ 0 & \mu_2 & -\lambda - \mu_2 & . & 0 & 0 \\ . & . & . & . & . & . \\ 0 & 0 & 0 & . & \mu_2 & -\lambda - \mu_2 \end{bmatrix}. \tag{5.4}$$

By partitioning the vector π on $\mathbf{X}$ as $\pi = [\pi_0, \pi_1, \pi_2, \ldots]$ where $\pi_n = [\pi(n, 0), \pi(n, 1), \ldots, \pi(n, N)]$ for $n \geq 0$, one finds that the balance equations $\pi Q = 0$ are solved by

$$\pi_n = \pi_0 R^n, \quad n \geq 0, \tag{5.5}$$

provided that the vector π_0 and the $(N + 1) \times (N + 1)$ matrix R solves the equations

$$\begin{aligned} A_0 + RA_1 + R^2 A_2 &= 0 \\ \pi_0 B_0 + \pi_0 RA_2 &= 0. \end{aligned} \tag{5.6}$$

Therefore, the procedure for finding an invariant distribution is to solve the first equation (5.6) for R, then to find a positive vector π_0 such that the second equation (5.6) holds. Finally, it remains to normalize π by a suitable choice of a multiplicative factor in π_0. This will be possible if λ is small enough.

A consequence of the form of the invariant distribution (5.6) is that the average number in the system L can be calculated as

$$L = \sum_{n=0}^{\infty} \sum_{i=0}^{N} (n+i)\, \pi(n,i) = \sum_{n=0}^{\infty} \sum_{i=0}^{N} (n+i)\, \pi_0\, R_n(i)$$

$$= \sum_{n=0}^{\infty} \pi_0\, R^n [v + n\mathbf{1}] ,$$

where $v = [0, 1, \ldots, N]^T$ and $\mathbf{1} = [1, 1, \ldots, 1]^T$. This gives

$$L = \pi_0 (I - R)^{-1} v + \pi_0 (I - R)^{-2} R\, \mathbf{1} .$$

Using $L = \lambda W$ allows us to calculate the average delay W through the system.

Example 5.2.3 [Markov modulated arrivals] Consider the system illustrated in Figure 5.2. One is given a Markov chain $\{y_t, t \geq 0\}$ with rate matrix S on $\{1, 2, \ldots, N\}$. This is used to define an arrival process $\{A_t, t \geq 0\}$ that has rate $\lambda(y_t)$. These arrivals enter a FCFS queue with independent service times that are exponentially distributed with rate μ.

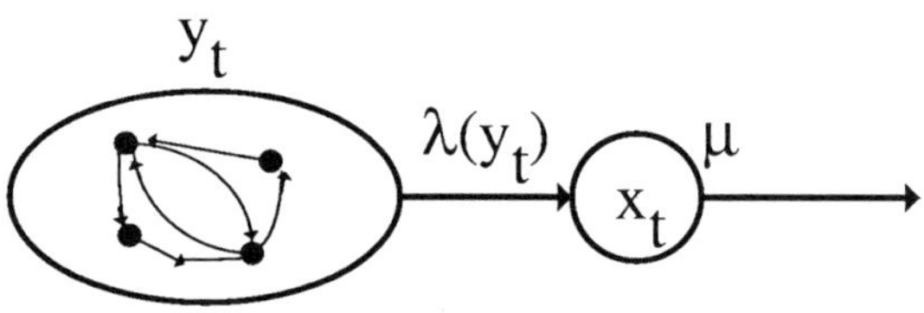

Figure 5.2. An M/M/1 queue with Markov modulated arrivals.

Denote by x_t the number of customers in the queue at time t, for $t \geq 0$. A more precise definition is that the process (x_t, y_t) is a Markov chain with rate matrix Q given below. Order the states in the lexicographic order with x being the most significant symbol. Then Q is as in (5.1) with

$$\begin{aligned} A_0 &= \text{diag}\{\lambda(1), \ldots, \lambda(N)\} \\ A_1 &= \text{diag}\{-\lambda(1) - \mu, \ldots, -\lambda(N) - \mu\} + S \\ A_2 &= \text{diag}\{\mu, \ldots, \mu\} \\ B_0 &= \text{diag}\{-\lambda(1), \ldots, -\lambda(N)\} + S \end{aligned} \tag{5.7}$$

where $\text{diag}\{a(1), \ldots, a(N)\}$ denotes an $N \times N$ diagonal matrix with diagonal elements $\{a(n)\}$.

The formulas (5.5)-(5.6) apply to this case. One can show that the Markov chain (x_t, y_t) is ergodic if y_t is ergodic and if the **average arrival rate** is less than μ. The average arrival rate is equal to

$\sum_{y=1}^{N} \nu(y)\,\lambda(y)$ where ν is the invariant distribution of $\mathbf{y}$.

The **average queue length** is now calculated as

$$L = \sum_{x=0}^{\infty} \sum_{y=1}^{N} x\,\pi(x, y) = \sum_{x=0}^{\infty} x\,\pi_0 R^n \mathbf{1}$$

$$= \pi_0 (I - R)^{-2} R\,\mathbf{1}\,.$$

This allows us to calculate the **average sojourn time** as

$$W = [\sum_{y=1}^{N} \nu(y)\,\lambda(y)]^{-1} L\,,$$

by Little's result.

Comments 5.2.4 Similar results hold when the capacity of the first queue in Example 5.2.2 is finite and when that of the queue in Example 5.2.3 is finite.

5.3. MEAN VALUE ANALYSIS

Mean value analysis is one of a number of algorithms for **closed** or **mixed product-form networks.** The basic ideas were illustrated in section 1.8.

For notational simplicity we will restrict attention to multiclass networks of M/M/1-FCFS (with equal service rates for all classes), M/PH/1-LCFS, M/PH/1-PS, and M/PH/∞ queues. The method extends to general symmetric queues.

Basic Formulas 5.3.1 As seen in sections 3.6.1–3.6.4, the invariant probability for the single and infinite server queues considered here is given by

$$\pi(c_1, \phi_1; \ldots; c_n, \phi_n) = \begin{cases} \dfrac{\rho^n}{n!} e^{-\rho} \prod_{m=1}^{n} p(c_m, \phi_m) \\ \qquad \text{for M/PH/}\infty \text{ queues} \\ (1 - \rho)\rho^n \prod_{m=1}^{n} p(c_m, \phi_m) \\ \qquad \text{for single server queues}\,. \end{cases} \tag{5.8}$$

In these expressions, $\rho := \sum \lambda_c\,\mu_c^{-1}$ is the traffic intensity and

$p(c, \phi)$ is the probability that a given customer in the queue is of class c and is in phase ϕ of service.

Definitions 5.3.2 [Customer chains] Let $R = [r_{ij}^{cd}]$ be the routing probabilities of the network. It will be assumed that the network is **mixed.** That is, the Markov chain with transition matrix R on $\{0, 1, \ldots, J\} \times \mathbf{C}$ admits closed irreducible subsets $\mathbf{C}_1, \ldots, \mathbf{C}_K$ of $\{1, 2, \ldots, J\} \times \mathbf{C}$. (By $(0, c)$ we indicate a customer outside the network.) The sets $\mathbf{C}_k$ are called **closed customer chains.**

The networks that we will consider admits irreducible sets of states that are characterized by the **number of customers in the closed chains.** For that to be the case, one has to avoid the possibility of closed loops of M/M/1-FCFS queues that would preserve the order of customers of different classes. (See Example 4.4.3.)

This is a natural restriction since the service rates in such a loop could not depend on the customers' classes, so that one may assume that the loops of M/M/1-FCFS queues are used by customers of the same class.

Examples 5.3.3 The networks in Figure 5.3 will illustrate the definition of closed customer chains.

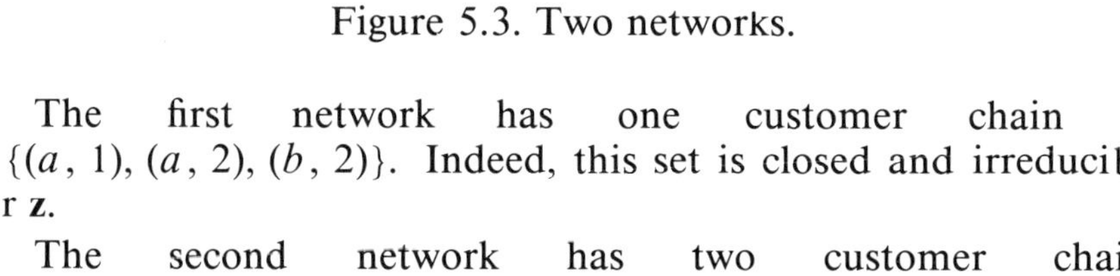

Figure 5.3. Two networks.

The first network has one customer chain $\mathbf{C}_1 = \{(a, 1), (a, 2), (b, 2)\}$. Indeed, this set is closed and irreducible for $\mathbf{z}$.

The second network has two customer chains $\mathbf{C}_1 = \{(a, 1), (a, 2), (a, 3)\}$ and $\mathbf{C}_2 = \{(b, 1), (b, 2)\}$.

Also, if one eliminates the artificial case of M/M/1-FCFS loops, then the first network has closed irreducible subsets of states that are determined by the number N_1 of customers of chain $\mathbf{C}_1$. Similarly, the closed irreducible subsets of the second network are specified by the numbers N_1 and N_2 of customers of chain 1 and 2, respectively.

Notation 5.3.4 Consider a network with K chains $\mathbf{C}_1, \ldots, \mathbf{C}_K$. Let $\mathbf{N} = (N_1, \ldots, N_K)$ be the numbers of customers of the various chains. Let also e_k denote the unit vector in direction k in $\{0, 1, 2, \ldots\}^K$. That is, $\mathbf{N} - e_k$ corresponds to N_l customers of chain l for $l \neq k$ and to $N_k - 1$ customers of class k.

Denote by $W_{ki}(\mathbf{N})$, $L_{ki}(\mathbf{N})$ and $\lambda_{ki}(\mathbf{N})$ the **average time** spent by a customer of chain k in node i, the **average number** of customers of chain k in node i, and the **average rate** of flow of chain k through node i, under the invariant distribution $\pi_{\mathbf{N}}$. Let also $L_i(\mathbf{N})$ be the average number of customers in node i and μ_{ki}^{-1} be the average service time for a customer of chain k in queue i.

Finally, denote by $\{\lambda_c, c \in \mathbf{C}\}$ a solution of the flow conservation equations, and let a_{ki} be the sum of λ_c over all the classes c that are of chain k and that go through node i. Thus, $\dfrac{a_{kj}}{a_{ki}}$ is the average number of visits to node j made per visit to node i by a customer of chain k. In particular, $\lambda_{kj}(\mathbf{N}) = \dfrac{a_{kj}}{a_{ki}} \lambda_{ki}(\mathbf{N})$.

Fact 5.3.5 [Mean value analysis equations] One has

$$
\begin{aligned}
W_{ki}(\mathbf{N}) &= \mu_{ki}^{-1}[1 + L_i(\mathbf{N} - e_k)\ 1\,\{i \text{ is a single server queue}\}] \\
L_{ki}(\mathbf{N} - e_k) &= \lambda_{ki}(\mathbf{N} - e_k)\ W_{ki}(N - e_k) \qquad (5.9) \\
\lambda_{ki}(\mathbf{N} - e_k)\,[\sum_{j=1}^{J} a_{kj}\ W_{kj}(\mathbf{N} - e_k)] &= a_{ki}\, N_k\ .
\end{aligned}
$$

These formulas can be used to calculate recursively (as $\mathbf{N}$ increases) the average queue lengths, delays, and throughputs of customers of the various chains at all the nodes.

Proof: The first equation is obvious if the queue is an M/PH/∞ queue. If it is an M/M/1 queue with the same service rate for all the customers, then the formula follows as in section 1.8.

If it is an M/PH/1-LCFS of -PS queue, then one can argue as follows. First, the formulas (5.8) show that the average number of customers and therefore also the average delay of a typical customer is the same as for an M/M/1-FCFS queue. Second, we know from Proposition 4.7.1 that, given the service time, the average delay of a customer in an M/PH/1-LCFS or -PS queue is proportional to the service time.

As a consequence, the average delay of a customer of class k in an M/PH/1-LCFS or -PS queue must be equal to

$$\frac{\mu_{ki}^{-1}}{\mu_i^{-1}} \times [\mu_i^{-1}\{1 + L_i(\mathbf{N} - e_k)\}] .$$

Indeed, the term between brackets is equal to the average delay of a typical customer in an M/PH/1-LCFS or -PS or an M/M/1-FCFS, by the Arrival Theorem, while the multiplicative factor takes into account the proportionality with the average service time of a customer of class k in an M/PH/1-LCFS or -PS queue.

The last two equations in (5.9) follow from Little's result applied to customers of chain k in queue i and in the network, respectively.

Comments 5.3.6 A host of approximations have been proposed for dealing with the equations (5.9). The interested reader should consult the literature on performance evaluation.

5.4. ASYMPTOTIC EXPANSIONS

The complexity of the algorithm of section 5.3 and of related ones makes them exceedingly slow when the product $\prod_{k=1}^{K} (1 + N_k)$ becomes large. The asymptotic expansion method explained in this section obtains estimates of the normalization constants $\pi(\mathbf{X}(\mathbf{N}))$ when the populations N_k are large.

For simplicity of notation, the method will be described only in the case of single-class closed networks.

Model 5.4.1 Consider a single-class closed network with J queues. Assume also that the queues $1, \ldots, I := J-1$ are M/M/1-FCFS or M/PH/1-LCFS or -PS, and that queue J is an M/PH/∞ queue.

The formulas (5.8) apply to these queues. Let N be the total number of customers in the closed network. It is assumed that the network state is irreducible on the set $\mathbf{X}(N)$ which corresponds to a total of N customers.

Denote by ρ_i the traffic intensity in queue i, for $1 \le i \le J$. That is, $\rho_i = \dfrac{\lambda_i}{\mu_i}$ where the λ_i solve the usual flow conservation equations. Let

$$\pi(\mathbf{X}(N)) = e^{-\rho_J} \prod_{i=1}^{I} (1 - \rho_i)\, G(N)\,.$$

Then the equations (5.8) show that

$$G(N) = \sum_{\mathbf{n} \in \mathbf{A}(N)} \frac{\rho_J^{n_J}}{n_J!} \prod_{i=1}^{J-1} \rho_i^{n_i}\,, \tag{5.10}$$

where

$$\mathbf{A}(N) = \{\mathbf{n} = (n_1, \ldots, n_J) \mid n_i \geq 0 \text{ and } n_1 + \cdots + n_J = N\}\,.$$

The problem is then to evaluate (5.10) when N is large.

Assumptions 5.4.2 [Normal usage] It will be assumed that

$$\begin{aligned} \rho_J &= bN \\ 0 < \rho_i &< b, \quad 1 \leq i \leq I := J - 1\,. \end{aligned} \tag{5.11}$$

One possible interpretation of the first equation is that the average service time μ_J^{-1} is proportional to N. Thus, when N is very large, the customers each require a long service time in queue J, and this makes the throughput of that queue converge as $N \to \infty$. (Recall that queue J is M/PH/∞.)

Preliminaries 5.4.3 Recall Euler's integral

$$n! = \int_0^\infty e^{-s} s^n \, ds, \quad n \geq 0\,,$$

which is readily verified by integration by parts. It follows that

$$a^n = \int_0^\infty e^{-s} \frac{(a\,s)^n}{n!} \, ds\,.$$

One can therefore rewrite (5.10) as

$$G(N) = \int_0^\infty \cdots \int_0^\infty e^{-(s_1 + \cdots + s_I)}$$

$$\times \Big[\sum_{\mathbf{n} \in \mathbf{A}(N)} \frac{(bN)^{n_J}}{n_J!} \prod_{i=1}^{I} \frac{(\rho_i\, s_i)^{n_i}}{n_i!} \Big] ds_1 \ldots ds_I\,.$$

The advantage of this rewriting is that the term between brackets is equal to

$$\frac{(bN + \rho_1 s_1 + \cdots + \rho_I s_I)^N}{N!}.$$

Thus,

$$G(N) = \int_0^\infty \cdots \int_0^\infty e^{-(s_1 + \cdots + s_I)}$$

$$\times \frac{(bN + \rho_1 s_1 + \cdots + \rho_I s_I)^N}{N!} ds_1 \ldots ds_I .$$

This formula can be rewritten as

$$G(N) = \frac{1}{N!} \int_0^\infty \cdots \int_0^\infty e^{-(s_1 + \cdots + s_I)} \tag{5.12}$$

$$\times (bN)^N (1 + \frac{\alpha}{N})^N ds_1 \ldots ds_I$$

where $\alpha = b^{-1}(\rho_1 s_1 + \cdots + \rho_I s_I)$.

Asymptotic Expansion 5.4.4 A Taylor expansion of $(1 + \frac{\alpha}{N})^N$ in $\frac{1}{N}$ shows that

$$(1 + \frac{\alpha}{N})^N = e^\alpha [1 - \frac{1}{N}\frac{\alpha^2}{2} + \frac{1}{N^2}(\frac{1}{3}\alpha^3 + \frac{1}{8}\alpha^4)$$

$$- \frac{1}{N^3}(\frac{1}{8}\alpha^4 + \frac{1}{6}\alpha^5 + \frac{1}{48}\alpha^6) + \cdots].$$

As a consequence, (5.12) is of the form

$$G(N) = \frac{(bN)^N}{N!}[1 - \frac{1}{N}\{\frac{1}{2}D(2)\} + \frac{1}{N^2}\{\frac{1}{3}D(3) + \frac{1}{8}D(4)\} \tag{5.13}$$

$$- \frac{1}{N^3}\{\frac{1}{8}D(4) + \frac{1}{6}D(5) + \frac{1}{48}D(6)\} - \ldots]$$

where

$$D(m) = \int_0^\infty \cdots \int_0^\infty e^{\alpha - (s_1 + \cdots + s_I)} \alpha^m ds_1 \ldots ds_I . \tag{5.14}$$

It remains to explain how the terms $D(m)$ can be calculated.

Pseudonetworks 5.4.5 From (5.14) one can write

$$D(m) = B\int_0^\infty \cdots \int_0^\infty e^{-(u_1 + \cdots + u_I)}$$

$$\times\ (b_1u_1 + \cdots + b_Iu_I)^m du_1 \ldots du_I \quad (5.15)$$

where

$$u_i = (1 - b^{-1}\rho_i)s_i$$

$$B = \prod_{i=1}^{I} (1 - b^{-1}\rho_i)^{-1}$$

$$b_i = b^{-1}\rho_i(1 - b^{-1}\rho_i)^{-1},$$

for $1 \le i \le I$. Notice that the change of variables corresponds to $u_i \ge 0$ because of the normal usage assumption $0 < \rho_i < b$.

The **key observation** is that (5.15) corresponds to the normalization constant of a network of I single server nodes with m customers, called a **pseudonetwork.** Indeed, reversing the calculation that led to (5.12) allows us to write successively

$$\frac{D(m)}{Bm!} = \int_0^\infty \cdots \int_0^\infty \sum_{\mathbf{n}\in E(m)} \prod_{i=1}^{I} \frac{(b_iu_i)^{n_i}}{n_i!} du_1 \ldots du_I$$

$$= \sum_{\mathbf{n}\in E(m)} b_1^{n_1} \ldots b_I^{n_I} \quad (5.16)$$

where

$$E(m) := \{\mathbf{n} = (n_1, \ldots, n_I) \mid n_i \ge 0 \text{ and } n_1 + \cdots + n_I = m\}.$$

This last expression (5.16) can be evaluated efficiently by the convolution algorithm when m is not too large. Notice also that (5.13) shows that when N is very large, it is likely that only a few terms in the expansion are needed and these terms involve only very small values of m. For instance, in order to obtain the coefficients of N^{-1}, N^{-2}, and N^{-3}, one needs only to evaluate the expressions (5.16) for $m = 2, 3, 4, 5, 6$.

5.5. FIXED-POPULATION-MEAN

This section explains the *fixed-population-mean* approximation method for closed product-form networks. The idea is to approximate a closed network by an open network having a mean population equal to the population of the closed network. This approximation is justified by limit theorems.

For simplicity, we again explain the ideas in the case of a single-class closed network.

Model 5.5.1 [Jackson Network] Consider the closed Jackson network of Figure 5.4.

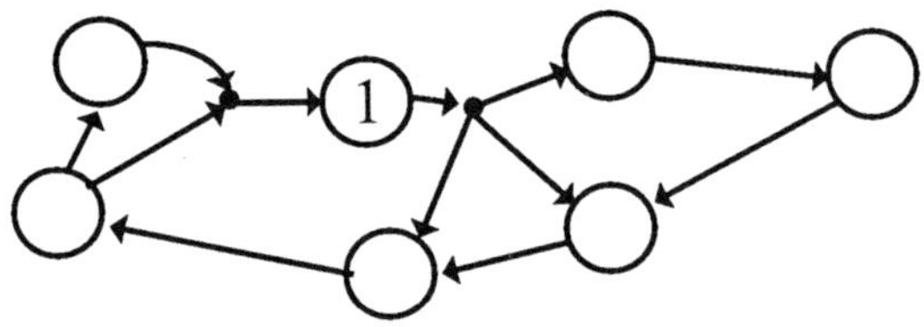

Figure 5.4. Closed Jackson network.

The network is assumed to be irreducible on each set $\mathbf{X}_N$ that corresponds to having N customers in the network. That is, the solutions of the flow conservation equations are equal up to a multiplicative constant.

Assume, for simplicity, that the J queues have the same traffic intensity, i.e., that the routing matrix is such that

$$\frac{\alpha_i}{\mu_i} = \beta, \quad 1 \le i \le J$$

where the α_i solve the flow conservation equations and are uniquely determined by the condition that $\alpha_1 = 1$. That is,

$$\beta = \frac{1}{\mu_1} .$$

Definition 5.5.2 [Associated open network] One can associate an open network to the network above. This is done by considering one queue, say queue 1, and by replacing the flow into that queue by a Poisson process with an arbitrary rate λ and by forcing the customers who would enter that queue from other queues to leave the network. This associated network is illustrated in Figure 5.5.

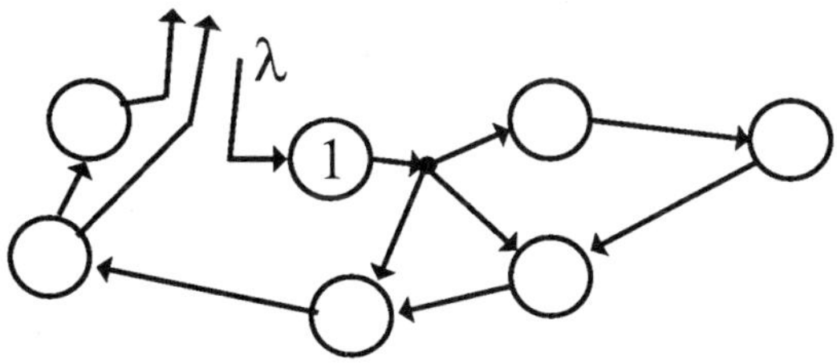

Figure 5.5. The associated open network.

Since the rate in queue 1 of the closed network is equal to λ, it follows from the definition of β that the traffic intensities in all the queues of the open network are now equal to

$$\rho = \lambda\beta = \frac{\lambda}{\mu_1}. \tag{5.17}$$

Average population 5.5.3 The average number of customers in the open network is equal to

$$L(\lambda) = J\frac{\rho}{1-\rho} = J\frac{\lambda\mu_1^{-1}}{1-\lambda\mu_1^{-1}} \tag{5.18}$$

where we used the fact that the J queues have the same traffic intensity ρ.

Fixed-population-mean 5.5.4 In order to analyze the closed network of Figure 5.4, one will approximate it by the open network of Figure 5.5, where λ is chosen so that $L(\lambda)$ is equal to the population in the closed network.

This method is called the **fixed-population-mean method.** It can be applied to more complicated examples, but the analysis of the quality of the approximation becomes more difficult.

Analysis 5.5.5 In order to determine how well this approximation is doing, we will calculate the error that it makes on the estimation of the throughput of queue 1.

Say that there are N customers in the closed network. The value of λ chosen by the fixed-population-mean method is such that $L(\lambda) = N$. Using (5.18), we find

$$\lambda = \mu_1\frac{N}{J+N}. \tag{5.19}$$

Thus, the above value of λ is the throughput of queue 1 in the closed network, as **estimated** by the fixed population mean method.

The actual throughput of queue 1 in the closed network can be calculated for this simple network. It is μ_1 times the probability that queue 1 is nonempty in the closed network. This probability is equal to

$$\sum_{x \in \mathbf{Y}} \pi_{X_N}(x) \tag{5.20}$$

where

$$\pi_{X_N}(x) = \frac{\pi(x)}{\pi(\mathbf{X}_N)} 1\{x \in \mathbf{X}_N\}$$

and $\mathbf{Y}$ is the set of states in $\mathbf{X}_N$ such that queue 1 is nonempty. Now,

$$\pi(x) = (1 - \rho)^J \prod_{i=1}^{J} \rho^{x^i},$$

so that

$$\pi(x) = (1 - \rho)^J \rho^N, \; x \in \mathbf{X}_N .$$

That is, all the states of $\mathbf{X}_N$ have the same probability.

It follows that the probability (5.20) that queue 1 is nonempty is given by the ratio of the number of states in $\mathbf{Y}$ over the number of states in $\mathbf{X}_N$. The number of states in an irreducible closed Jackson network with J queues and N customers is seen to be equal to $\binom{N+J-1}{J-1}$. This is, therefore, the number of states in $\mathbf{X}_N$. The number of such states where queue 1 is empty is the number of states in a network with $J-1$ queues and N customers. Therefore, the probability that queue 1 is nonempty is equal to

$$1 - \frac{\binom{N+J-2}{J-2}}{\binom{N+J-1}{J-1}} = \frac{N}{N + J - 1} .$$

Therefore, the actual throughput of queue 1 in the closed network is equal to

$$\mu_1 \frac{N}{N + J - 1} . \tag{5.21}$$

Comparing (5.19) and (5.21) shows that the approximation is very good when the number of queues is large.

Comments 5.5.6 The number of customers in the queues of a stationary open Jackson network are independent random variables. Assume that the traffic intensities are equal in all the J queues, as in the example above. Denote the random queue lengths by $\{X^i, i \leq i \leq J\}$.

Assume that the mean population is equal to N in the open network, then

$$E\{X^i\} = \frac{N}{J} \quad \text{and} \quad \text{var}(X^i) = (\frac{N}{J})^2 ,$$

where the verification of the second identity is left as a simple exercise.

It follows that the variance of the total population is equal to

$$\text{var}(X^1 + \cdots + X^J) = \frac{N^2}{J} ,$$

so that the coefficient of variation of the total population (the standard deviation divided by the mean) is equal to $\frac{1}{\sqrt{J}}$.

This shows that the total population in the open network is likely to be close to N if J is not small.

5.6. EXERCISES

5.1. Matrix-geometric solutions Consider the following naive model of an **integrated communication system** where voice and data communications share two transmission channels.

One is given an M/M/2 queue that serves customers of class 2 with rate μ_2. These customers arrive at rate λ_2. Customers of class 1 arrive as a Poisson process with rate λ_1. They require independent exponentially service times with rate μ_1.

The queue accepts at most one customer of class 1 at a time. It then instantaneously (preemptively) allocates one of the servers to the customer of class 1. The other arriving customers of class 1 are rejected if there is already one such customer in the queue.

The services of the customers of class 2 that were interrupted by customers of class 1 start afresh when those customers leave.

(a) Draw the transition diagram of this Markov chain on $\{0, 1\} \times \{0, 1, 2, \ldots\}$.

(b) Solve for the invariant distribution by using the methods of section 5.2.

(c) Explain why this is a naive model of a system where customers of class 1 correspond to voice calls and customers of class 2 are data packets.

5.2. Mixed networks Consider one of the queues of section 5.3.1. Let $\mathbf{C}_0, \ldots, C_K$ be disjoint subsets of $\mathbf{C}$. Show that the probability that the queue contains n_k customers with a class in $\mathbf{C}_k$ for $1 \le k \le K$ and possibly some customers of a class in $\mathbf{C}_0$ is seen from (5.8) to be given by

$$\begin{cases} e^{-[\rho - \rho(\mathbf{C}_0)]} \prod_{k=1}^{K} \frac{\rho(\mathbf{C}_k)^{n_k}}{n_k!} \\ (1 - \frac{\rho}{1 - \rho(\mathbf{C}_0)}) [(\sum_{k=1}^{K} n_k)!] \prod_{k=1}^{K} \frac{\{\rho(\mathbf{C}_k)[1 - \rho(\mathbf{C}_0)]^{-1}\}^{n_k}}{n_k!} \end{cases} .$$

5.7. REFERENCES AND COMMENTS

Section 5.2 For a detailed discussion and many applications of the matrix-geometric ideas, see Neuts (1981). The application in Exercise 5.2 is from Fischer (1977). See Chapter 12 in Schwartz (1987) for other related models.

Section 5.3 A presentation of numerical methods for product-form networks can be found in Lavenberg (1983). A survey is in Heidelberger and Lavenberg (1984). See also Reiser (1977)-(1981), Reiser and Lavenberg (1980), Salza and Lavenberg (1981), Schweitzer (1979), Tijms (1986), and Courtois (1977).

Section 5.4 The asymptotic expansions are explained in McKenna, et al. (1981), McKenna and Mitra (1982), and Morrison and Mitra (1985).

Section 5.5 The fixed-population-mean is discussed in Whitt (1984).

CHAPTER SIX

NON-PRODUCT-FORM NETWORKS

6.1. INTRODUCTION

Generically, networks are **not** product-form. Most networks are not tractable and, in fact, relatively little can be said about them. There are, of course, many particular examples that have been investigated in the literature. However, it is difficult to extract general principles from these examples. As a consequence, this section is not very satisfactory and will, we hope, be rapidly outdated. The topics discussed here are the **heavy traffic** and **light traffic** approximations for networks, some simple **bounds** based on Little's result, a few **approximation** ideas, **comparison results,** and queues in tandem with **dependent service times.**

To a large extend, this chapter discusses some work in progress. We hope that this will make the readers aware of some interesting research directions, even if most of the sections do not conclude with readily usable formulas.

To motivate the ideas, consider the case of two M/GI/1 queues in tandem. The invariant distribution of this system is not known. When the **traffic intensity** in both queues approaches one, the queues are nonempty most of the time. In that case, the increments of the queue lengths can be approximated by linear combinations of **independent renewal processes.** In turn, these renewal processes, when properly normalized, can be approximated by **Brownian motions with drift.** This leads to a diffusion approximation of the network in heavy traffic. Such results permit us to approximate the distribution of the network, and also to obtain some qualitative results about the behavior of the queue length process. See section 6.2.

When the traffic intensity is very small for each queue, one can estimate the probability that a typical customer will find n other customers when entering a given queue, for n small. These estimates can then be used to approximate the sojourn time of a customer in the network. This method is explained in section 6.3.

The departure process from a GI/GI/1 queue is not a renewal process (unless it is an M/M/1 queue). By assuming that it is, one can get estimates of the queue length in the second queue. This is the idea of the **renewal approximation** discussed in section 6.4. Other methods are also explained.

One of the few generally applicable results in networks is Little's result (and its variations). Not surprisingly, that result can be used to obtain simple, albeit often crude, **bounds** on network quantities. An example as well as related ideas are discussed in section 6.5.

In a number of applications, the service times of a given customer in the two queues are **not independent.** For instance, in communication networks, the customers represent messages and the service time is proportional to the length of the message, which does not change when the message moves through the network. Thus, even if it could be argued that the message lengths are independent and exponentially distributed, the network would not be a network of M/M/1 queues. Some results are given in section 6.6.

In some cases it is possible to **compare** networks that cannot be analyzed. This general idea is of some potential usefulness in design applications. A few simple comparison results are discussed in section 6.7.

6.2. HEAVY TRAFFIC APPROXIMATIONS

One considers a single class network of J single-server and FCFS queues. (See Figure 6.1.) The arrival process from outside into queue i is a **renewal process** with an interarrival time distribution that has mean γ_i^{-1} and variance u_i^2. (See Lemma 6.2.3 for a definition of renewal processes.)

The exogenous **arrival processes** and the **service times** are independent. The service times in queue i have mean μ_i^{-1} and variance v_i^2. The routing is **Markov,** with probabilities r_{ij}. It will be assumed that $r_{ii} \equiv 0$. This assumption is not essential.

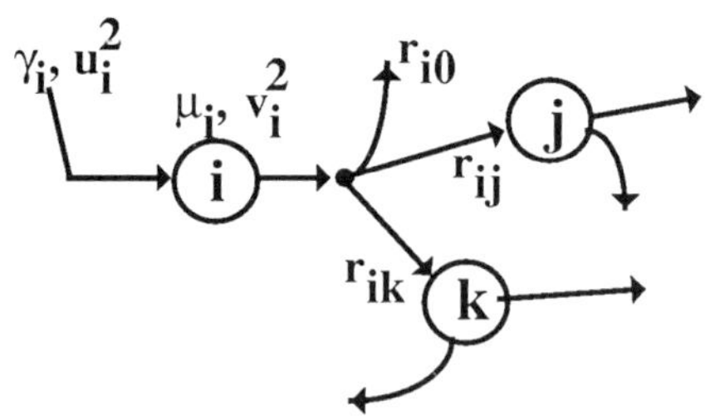

Figure 6.1. The Network.

It is assumed that the network is in **heavy traffic.** By this, one means that the rate at which work enters the network for queue i is close to 1, for all i. The rate at which work enters the network for queue i is seen to be equal to $\rho_i := \lambda_i \mu_i^{-1}$ where λ_i solves the usual flow conservation equations. This network will be denoted by (R, γ, u, μ, v).

The objective is to show that under those conditions, the queue length process, when properly normalized, can be approximated by a reflected **Brownian motion with drift.** To make this idea precise, one shows that as the parameters of the network are such that each queue approaches saturation, the normalized queue length process converges to a reflected diffusion. Thus, one must consider a sequence of networks and show that the corresponding sequence of queue length processes converges. The result will first be stated and a heuristic justification will follow.

Theorem 6.2.1 [Diffusion Approximations for Single-Class Networks]
For $n \geq 1$ one is given a network $(R, \gamma_{(n)}, u, \mu_{(n)}, v)$ such that

$$(\mu_{(n)} - \lambda_{(n)}) \sqrt{n} \to c$$

$$\gamma_{(n)} \to \gamma, \ \mu_{(n)} \to \mu$$

where $\lambda_{(n)}$ solves the flow conservation equations for the nth network.

Assume that the moments of order $2 + \epsilon$ of the interarrival times and service times of all the networks are bounded, for some $\epsilon > 0$.

Then, the vector $L_t^{(n)}$ of queue lengths of the nth network is such that (by $(.)^T$ one denotes transposition)

$$\frac{1}{\sqrt{n}} L_t^{(n)} \Rightarrow Z_t := mt + AB_t + (I - R^T) Y_t$$

where

$m = -(I - R^T)c$

B is a vector of J independent standard Brownian motions

Y is a vector of minimum nondecreasing functions such that $Z \geq 0$ and

A is a matrix such that $AA^T = \Sigma = [\sigma_{ij}]$ with

$$\sigma_{ii} = \sigma_i^2 = \gamma_i^3 u_i^2 + \mu_i^3 v_i^2 + \sum_k \mu_k r_{ki}(1 - r_{ki} + r_{ki} \mu_k^2 v_k^2)$$

$$\sigma_{ij} = -\mu_i^3 v_i^2 r_{ij} - \mu_j^3 v_j^2 r_{ji} + \sum_k \mu_k r_{ki} r_{kj}(\mu_k^2 v_k^2 - 1) .$$

Comments 6.2.2

(a) Recall that a process $\{W_t, t \geq 0\}$ is a **standard Brownian motion** if its trajectories are continuous and if its increments $W_{t_{i+1}} - W_{t_i}$ for $0 \leq t_1 < t_2 < \cdots$ are independent and Gaussian with mean zero and variance $t_{i+1} - t_i$, respectively.

(b) In Theorem 6.2.1, the symbol $\Rightarrow$ indicates **weak convergence** of processes on finite time intervals. Intuitively, this means that the processes on the left-hand side of $\Rightarrow$ become more and more similar to that on the right-hand side. Specifically, weak convergence means the convergence of the distribution of functionals of the processes that are continuous at continuous functions (with respect to the topology of uniform convergence). For instance, the maximum value of $\frac{1}{\sqrt{n}} L_t^{(n)}$ for $0 \leq t \leq T$ has a distribution that converges to that of the corresponding quantity for Z_t.

This weak convergence is **stronger** than convergence of the finite dimensional distributions. For instance, if X is a random variable that is uniformly distributed on [0, 1], then $X_t^{(n)} := 1\{X - n^{-1} < t < X + n^{-1}\}$ defines a process on $[-1, 2]$ for $n \geq 1$. The finite dimensional distributions of that process are seen to converge to that of a process that is identically zero. However, the maximum value is not converging to that of the latter in distribution, so that the process cannot be converging weakly to zero.

It is known that weak convergence to a continuous process is equivalent to the convergence of the finite dimensional distributions and the uniform convergence in probability to zero of the **modulus of continuity** of the process. Intuitively, this means that the variations of the process on small intervals should go to zero (uniformly, in probability). For the processes such as $\frac{1}{\sqrt{n}} L_t^{(n)}$, this latter condition can be shown to be satisfied. Thus, weak convergence will follow from the convergence of the finite dimensional distributions.

(c) The process Z_t is a vector of correlated Brownian motions with drift m that is **reflected** so that it does not become negative. The reflections are **oblique** with respect to the boundaries of $\mathbb{R}_+^J$. The angles of reflections are given by the routing matrix R. (See Figure 6.2.)

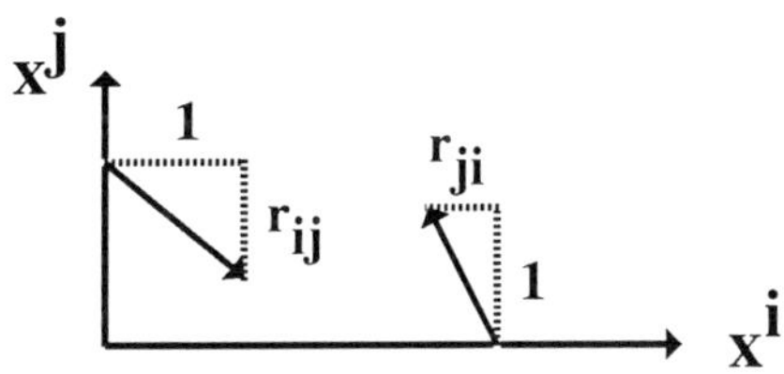

Figure 6.2. The reflections.

Before giving an intuitive discussion of Theorem 6.2.1, one has to recall two basic results.

Lemma 6.2.3 [Central Limit Theorem for Renewal Processes] Let $\{\tau_n, n \geq 1\}$ be a sequence of i.i.d. random variables with mean $\lambda^{-1} \in (0, \infty)$ and with variance $\sigma^2 < \infty$. Assume also that $E\{\tau^\alpha\} < \infty$ for some $\alpha > 2$.

The **renewal process** with interarrival times $\{\tau_n\}$ is defined by

$$X_t = \max\{n \mid \sum_{m=1}^{n} \tau_m \leq t\}, \quad t \geq 0,$$

with the convention that $X_t = 0$ if $t < \tau_1$.

Then

$$Z_t^{(n)} := \frac{X_{nt} - \lambda nt}{\sqrt{n}} \Rightarrow \sqrt{\lambda^3 \sigma^2}\, B_t \text{ as } n \to \infty$$

where $\{B_t, t \geq 0\}$ is a standard Brownian motion.

Proof (outline): It can be shown that the modulus of continuity converges to zero uniformly in probability. Let us explain why the finite dimensional distributions converge to that of the indicated limit. This is done by showing that the increments of $\{Z_t^{(n)}\}$ converge to independent Gaussian random variables with the desired variances. The independence comes from the fact that the increments are determined asymptotically by distinct interarrival times.

It remains to show why the increments converge to Gaussian random variables. To do this, fix $t > 0$ and $x \in (-\infty, +\infty)$. As

$n \to \infty$ choose $m \to \infty$ in such a way that $\dfrac{m - \lambda nt}{\sqrt{n\lambda^3\sigma^2 t}} \to -x$. In particular, $\dfrac{\lambda nt}{m} \to 1$. With $T_n = \tau_1 + \cdots + \tau_n$ one has

$$P\{\frac{X_{nt} - \lambda nt}{\sqrt{n\lambda^3\sigma^2 t}} \geq -x\} \approx P\{T_m \leq nt\}$$

$$= P\{\frac{T_m - \lambda^{-1}m}{\sigma\sqrt{m}} \leq \frac{nt - \lambda^{-1}m}{\sigma\sqrt{m}} \approx x\}$$

$$\to P\{N(0, 1) \leq x\}$$

where $N(0, 1)$ is a Gaussian random variable with mean zero and variance 1. (The last assertion follows from the Central Limit Theorem. See Appendix A.)

Lemma 6.2.4 [Central Limit Theorem for Sampled Renewal Processes]

Let N_t be a renewal process. Denote the mean and the variance of the interarrival times by λ^{-1} and u^2, respectively. Assume that the interarrival times have a finite moment of order $2 + \epsilon$, for some $\epsilon > 0$.

Define the process $\{X_t = (X_t^1, X_t^2)^T, t \geq 0\}$ by **sampling** the process $\{N_t\}$ with the probabilities α and β ($\alpha + \beta \leq 1$), as indicated in Figure 6.3.

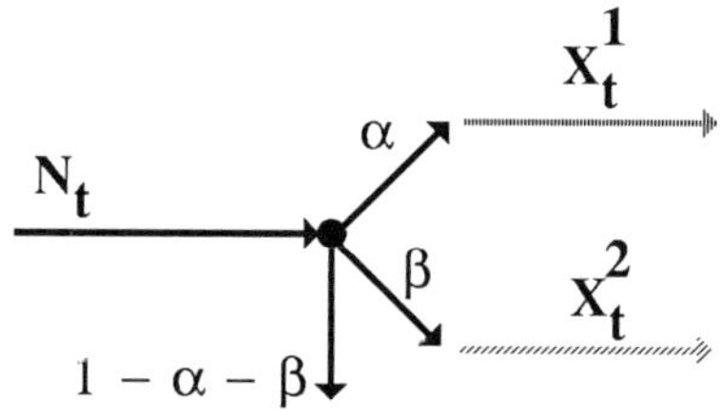

Figure 6.3. Sampling a renewal process.

Then

$$Z_t^{(n)} := \frac{X_{nt} - \lambda nt}{\sqrt{n}} \Rightarrow \tilde{X}_t := MB_t, \quad \text{as } n \to \infty$$

where B_t is a vector of two independent standard Brownian motions and the matrix M is such that

$$\Sigma := MM^T = \begin{bmatrix} \lambda\alpha(1 - \alpha + \alpha\lambda^2 u^2) & \lambda\alpha\beta(\lambda^2 u^2 - 1) \\ \lambda\alpha\beta(\lambda^2 u^2 - 1) & \lambda\beta(1 - \beta + \beta\lambda^2 u^2) \end{bmatrix}$$

This will be summarized by writing that $\text{var}(Z) = \Sigma$.

The above result is a simple consequence of the Central Limit Theorem for renewal processes (see Lemma 6.2.3). Indeed, first observe that if the T_n are the successive interarrival times of N_t, then the first interarrival time S_1 of X_t^1 is equal to $T_1 + \cdots + T_n$ with probability $\alpha(1 - \alpha)^{n-1}$ for $n \geq 1$. This implies that $E\{S_1\} = (\lambda\alpha)^{-1}$ and that $\text{var}(S_1) = \alpha^{-1}u^2 + \alpha^{-2}\lambda^{-2} - \alpha^{-1}\lambda^{-2}$, as a simple calculation shows. Applying Lemma 6.2.3 with (λ, u^2) replaced by $([E\{S_1\}]^{-1}, \text{var}(S_1))$ gives the first diagonal entry of Σ. The second diagonal entry is obtained by symmetry. To find the covariance, one uses the fact that $\text{var}\,(\tilde{X}_t^1 + \tilde{X}_t^2) = \text{var}\,(\tilde{X}_t^1) + \text{var}\,(\tilde{X}_t^2) + 2\text{cov}\,(\tilde{X}_t^1, \tilde{X}_t^2)$. Also, the variance of $\tilde{X}_t^1 + \tilde{X}_t^2$ is obtained by replacing α by $\alpha + \beta$ in the calculation of the variance of $\tilde{X}_t^1$. This gives the remaining entries of Σ.

Justification of Theorem 6.2.1 Denote by F^{ij} the renewal process obtained by sampling with probability r_{ij} the process with interarrival times equal to independent service times in queue i of the nth network. The time subscript and the superscript n are omitted for simplicity. (See Figure 6.4.)

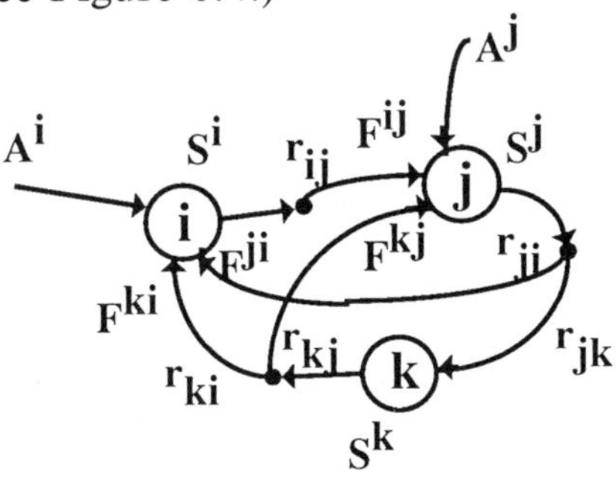

Figure 6.4. The network.

When the queues are **positive,** the increments of the queue lengths evolve as

$$\begin{aligned} X^i &= A^i + \sum_{k \neq j} F^{ki} + F^{ji} - S^i \\ X^j &= A^i + \sum_{k \neq i} F^{kj} + F^{ij} - S^j . \end{aligned} \tag{6.1}$$

(One neglects the fact that some residual service or interarrival times may not be complete by deciding that the jumps occur at the times of the renewal processes.)

One then has

$$X^i_{nt} = (A^i_{nt} - \gamma_i nt) + \sum_k (F^{ki}_{nt} - \mu_k r_{ki} nt) - (S^i_{nt} - \mu_i nt) .$$

Denote by $\mathrm{var}(V)$ the asymptotic variance per unit of time of the increments of a renewal process V normalized as in Lemma 6.2.4. One finds

$$\begin{aligned} \sigma_i^2 &= \mathrm{var}(A^i) + \mathrm{var}(S^i) + \sum_k \mathrm{var}(F^{ki}) \\ &= \gamma_i^3 u_i^2 + \mu_i^3 v_i^2 + \sum_{k \neq i} \mu_k r_{ki}(1 - r_{ki} + r_{ki} \mu_k^2 v_k^2) \end{aligned}$$

where the result of Lemma 6.2.4 was applied to the each process F^{ki} obtained by sampling S^k with probability r_{ki}.

The covariance $\mathrm{cov}(X^i, X^j)$ of the increments of X^i and X^j is, after normalizing,

$$\sigma_{ij} = \sum_k \mathrm{cov}(F^{ki}, F^{kj}) - \mathrm{cov}(F^{ij}, S^i) - \mathrm{cov}(F^{ij}, S^j) .$$

Now, according to Lemma 6.2.4,

$$\mathrm{cov}(F^{ki}, F^{kj}) = \mu_k r_{ki} r_{kj}(\mu_k^2 v_k^2 - 1)$$

and

$$\begin{aligned} \mathrm{cov}(F^{ij}, S^i) &= \mathrm{cov}(F^{ij}, F^{ij} + \sum_{k \neq j} F^{ik}) \\ &= \mathrm{var}(F^{ij}) + \sum_{k \neq i} \mathrm{cov}(F^{ij}, F^{ik}) . \end{aligned}$$

Combining the above observations shows that

$$\sigma_{ij} = -\mu_i^3 v_i^2 r_{ij} - \mu_j^3 v_j^2 r_{ji} + \sum_k \mu_k r_{ki} r_{kj}(\mu_k^2 v_k^2 - 1) .$$

The times when the **queues become empty** are handled as follows. When queue i is empty, the equations (6.1) keep subtracting customers from that queue and adding them to the other queues. A fraction r_{ij} of the customers mistakenly subtracted from queue i are sent to queue j, for $j \neq i$. To correct that error, one should add an increasing function to the process constructed above so that it does not become negative, while subtracting a fraction r_{ij} from queue j of whatever is added to queue i.

Thus, the queue length vector will be approximated by the process Z_t defined in Theorem 6.2.1.

Remarks 6.2.5 Theorem 6.2.1 is for the queue lengths in single-class open networks that are "uniformly" in heavy traffic. The result has been extended in a number of directions:

(a) Diffusion approximations are available for the **work-to-be-done** in the queues and for the **sojourn times.**

(b) Consider a queueing network in heavy traffic. As Theorem 6.2.1 shows, the nth network in the converging sequence is such that the variation of the queue length in nt time units is of the order $\sqrt{nt}$. (Recall that the standard deviation of a Brownian motion is $\sqrt{t}$ in t time units.) Thus, if a customer spends T time units in the nth network, then the variations of the queue lengths during the sojourn time of that customer will be of the order $\sqrt{Tn^{-1}}$. For $n \gg 1$, this says that **the queue lengths remain essentially constant during the visit of a given customer.** Combining this with the strong law of large number (using the fact that the queue lengths are very large), one finds that the vector of sojourn times in the queues divided by $\sqrt{n}$ for a customer will **converge to** $\sum_i n_i \, \mu_i^{-1} \, Z_\tau^i$ where n_i is the number of visits to queue i and τ is the arrival time of the customer in the network.

(c) If some queues have a traffic intensity that is negligible compared to that of the others, their queue lengths converge to zero in the above normalization. This leads to a diffusion of a **dimensionality** that is less than the original number of queues.

(d) The network may be **controlled** in a way that makes certain queue lengths be **asymptotically equal.** For instance, if customers join the shortest of J parallel GI/GI/1 queues, then the limiting diffusion is one-dimensional (as long as the heavy traffic assumptions are satisfied). To see this, consider the case of two queues that are initially equal. For the first queue to exceed the other by $\epsilon\sqrt{n}$ over a time interval of duration nt, it is necessary that the number of arrivals plus service times of the second queue lag behind the departures of the first queue. The probability of that event is negligible if the sum of the arrival and service rate in the second queue exceeds the service rate in the first queue, as can be checked by observing that the diffusion approximation of the corresponding renewal processes has a drift that is proportional to $\sqrt{n}$.

(e) The dimensional reduction observed above also occurs in **multiclass networks.** For instance, if there are two classes of customers, and if one has a preemptive-resume priority over the other at all the queues, then the high-priority queue lengths are asymptotically negligible. However, the high-priority customers reduce the effective service rates for the low-priority customers. As another example, if the various customer classes share the service capacities (e.g., by processor-sharing), then their queue lengths are asymptotically proportional at each node.

(f) The limiting diffusion $\{Z_t\}$ of Theorem 6.2.1 can be shown to admit a **product-form invariant density,** which is then given by

$$\pi(z) = \prod_{i=1}^{K} \alpha_i e^{-\alpha_i z^i}, \; z = (z^1, \ldots, z^J) \geq 0$$

where $\alpha_i = -2c_i\sigma_{ii}^{-1}$, if and only if $\alpha_i > 0$ and

$$\sigma_{ij} = -\frac{1}{2}[\sigma_{ii}r_{ij} + \sigma_{jj}r_{ji}]$$

for $1 \leq i \neq j \leq J$.

It can be verified that these conditions hold for limits of Jackson networks (which clearly must admit a product-form invariant distribution). It is worth noticing that these conditions also hold for some limits of networks that are not product-form.

Pause for Thought 6.2.6 To extract useful information from Theorem 6.2.1, one must be able to characterize the limiting diffusion process Z, but the theory of such processes is only partially developed at this point. For instance, the density of the diffusion at time t is known to solve second order partial differential equations (the Fokker-Planck equations) with boundary conditions expressed as first order partial differential equations. A similar result also holds, of course, for the stationary density when it exists.

Unfortunately, these equations cannot be solved explicitly, in general. One can then justify the approach by the possibility of solving numerically partial differential equations. This justification would require an investigation of the **numerical complexity** of the procedure, since it is also possible to approximate the network by a Markov chain (using PH distributions), and to solve numerically the balance equations for the invariant distribution, or the Kolmogorov equations for the transient distribution. The diffusion approach may be more easily justified in the cases of dimensional reductions of Remark 6.2.5.

Another potential use of the diffusion approximation in the case of dimensional reduction is in **simulation experiments.** The reduced diffusion may be easier to simulate than the original network.

Even in the few cases where the partial differential equations can be solved explicitly, it is a nontrivial task to use the solution to infer properties of a heavily loaded network. As can be expected, knowing the limit of a sequence may not give much information about a given term in the sequence. In fact, a number of examples have shown that the diffusion approximations of the queue length process, the sojourn time process, and the work-to-be-done, lead to three different approximate formulas for the queue length in a heavily loaded queue. These approximate formulas may not even be consistent with known bounds in particular cases (such as the M/GI/1 queue).

A successful approach has been to combine heavy traffic diffusion approximations with light traffic approximations. The idea is to obtain the performance measures of a given queueing network by suitably **interpolating** between the heavy traffic and the light traffic limits of that network.

6.3. LIGHT TRAFFIC APPROXIMATIONS

As discussed in section 6.2.1, it is possible to obtain heavy traffic limits of normalized network queue lengths or waiting times by analyzing the limiting diffusions. This section explains how to derive light traffic approximations.

The idea is to consider a measure of performance as a function of the arrival rate, say λ. It will be shown that the nth derivative with respect to λ, at $\lambda = 0$, of such a quantity can be found by analyzing the network with n arrivals during $(-\infty, +\infty)$. In principle, this permits us to write a **Taylor expansion** at 0 of the measure of performance. The resulting approximation can then be combined with the heavy traffic limit to obtain estimates for moderate traffic conditions.

Formulation 6.3.1 Let **x** be a Markov chain with rate matrix

$$Q_\lambda = Q + \lambda(A - I)$$

on some state space **X**. In this expression, λ is an arbitrary positive number and both Q and $A - I$ are rate matrices. It is assumed that A is nonnegative and that its diagonal elements are equal to zero. In addition, it is assumed that $Q(i, j) = 0$ whenever $A(i, j) > 0$.

The **interpretation** is that a transition (i, j) with $A(i, j) > 0$ must correspond to an arrival, and that Q is the rate matrix when $\lambda = 0$.

Observe that the probability that **x** jumps from i to j in $\epsilon \ll 1$ time unit and that there is an arrival is equal to $\lambda A(i, j)\epsilon$, while the probability of an arrival in a time interval of duration ϵ is equal to $\lambda\epsilon$. As a consequence, the probability that $x_{t+\epsilon} = j$ given that $x_t = i$ and that there is an arrival in $(t, t + \epsilon)$ is equal to $A(i, j)$.

Definition 6.3.2 [Performance measure] We will consider performance measures of the form

$$W_S(\lambda) = E_\lambda [F \mid x_{-S} = \theta, 0] \quad \text{with } F := \int_0^T f(z_t)\,dt \qquad (6.2)$$

where $S, T > 0$ and $\theta \in \mathbf{X}$ are given, and so is the function $f : \mathbf{X} \to [0, 1]$. In this expression, $E_\lambda[. \mid x_{-S} = \theta, 0]$ is the conditional expectation given that $x_{-S} = \theta$ and that there is an arrival at time 0.

The interpretation is that x_t for $t > 0$ is the **state** of the system given that a **tagged customer entered at time** 0.

Taylor Expansion 6.3.3 One can write

$$W_S(\lambda) = E_\lambda\{E[F \mid x_{-S} = \theta, 0; \tau_1, \tau_2, \ldots]\}$$

where $-S < \tau_1 < \tau_2 < \cdots T$ are the arrival times in $[-S, T]$ (other than the arrival at time 0). Notice that the conditional expectation given the arrival times does not depend on λ. Since the arrivals are Poisson with rate λ, one finds that

$$W_S(\lambda) = \sum_{n=0}^{\infty} \frac{[\lambda(T+S)]^n}{n!} e^{-\lambda(T+S)} E[F \mid x_{-S} = \theta, n, S]$$

where $E[. \mid n, S]$ is the conditional expectation given that there are exactly n arrivals, independent and uniformly distributed in $[-S, T]$ (in addition to the arrival at time 0).

Taking the first derivative with respect to λ, at $\lambda = 0$, of this uniformly convergent series gives

$$\frac{d}{d\lambda} W_S(0) = (S + T)\, E[F \mid x_{-S} = \theta, 1, S] - (S + T)\, W_S(0) \quad (6.3)$$

$$= \int_{-S}^{T} \{E[F \mid x_{-S} = \theta, t] - W_S(0)\}\, dt$$

where $E[. \mid t]$ is the conditional expectation given that there is only one arrival and that it occurs at time t (in addition to the arrival at time 0).

Higher-order derivatives can be calculated similarly.

Variation 6.3.4 In most applications, one is interested in a performance measure that corresponds to

$$V(\lambda) := E_\lambda\{F\} \quad (6.4)$$

where E_λ is the expectation with respect to the system started at $-\infty$ with the arrival rate λ, given that there is an arrival at time 0.

If the Markov chain $\mathbf{x}$ is **asymptotically stationary** for λ, then this mean value is the limit when $S \to \infty$ of (6.2), as can be seen by dominated convergence. Interchanging the limit and the differentiation then shows that

$$\frac{d}{d\lambda} V(0) = \int_{-\infty}^{T} \{E[F \mid t] - V(0)\}\, dt \quad (6.5)$$

where $E[. \mid t]$ is the expectation with respect to the law of $\mathbf{x}$ started from $-\infty$, given that there was an arrival at time t.

It is necessary to assume that $\mathbf{x}$ is asymptotically stationary for $0 \le \lambda < \lambda_0$ for some $\lambda_0 > 0$.

Calculations 6.3.5 For simplicity, we will assume that θ is absorbing for $\mathbf{x}$ when $\lambda = 0$ and that $f(\theta) = 0$ when θ denotes the empty state. One can then write (6.5) as

$$\frac{d}{d\lambda}V(0) = \int_{-\infty}^{0}\{\int_{0}^{T} F[s \mid t, 0]\,ds\}\,dt + \int_{0}^{T}\{\int_{0}^{t} F[s \mid 0]\,ds\}\,dt + \int_{0}^{T}\{\int_{t}^{T} F[s \mid 0, t]\,ds\}\,dt$$

where $F[s \mid t, 0]$ denotes the mean value of $f(x_s)$ given that the system was empty at time $t < 0$, that an arrival occurred at time t, and another at time 0, and similarly for $F[s \mid 0]$ and $F[s \mid 0,t]$.

Example 6.3.6 Consider the Jackson network with J queues in Figure 6.5.

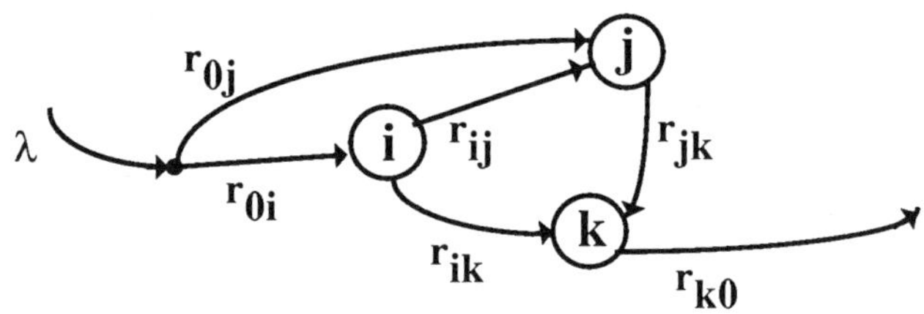

Figure 6.5. Jackson network.

Let $V(\lambda)$ be the expected delay of a tagged customer who enters the system at time 0.

To calculate the first derivative, we will have to analyze the evolution of the network with the tagged customer C and with another customer A. A possible state will be (i, j), where i is the position of C and j is the position of A. If $i = j$, then we have to distinguish the state ai where C is ahead of A in i, and the state bi where C is behind A in i. There are therefore $N(N+1) + 2N$ states, including the state (0, 0) corresponding to the empty network. The rate matrix Q on this state space **X** is easily written down.

The function $f(x)$ takes the value one if C is in the network when its state is x, and the value zero otherwise. The network is started with the customer A entering at time t and the customer C entering at time 0. One can absorb the state in (0, 0). It is then possible to do the calculations discussed in section 6.3.3. Unfortunately, the details are complicated.

6.4. AD HOC APPROXIMATIONS

A number of ad hoc approximations, without precise analytical justifications, have been suggested for networks. This section will review three such ideas.

LCFS Approximation 6.4.1. A simple approximation method for open networks of M/GI/s_i queues with Poisson exogenous arrivals is based on the observation that the network would be product-form if the queues were LCFS instead of being FCFS, as seen in section 6.3.

The difficulty is then to relate the performance measures of the network of LCFS queues to those of the network of FCFS queues. One way to do this is to assume that the same relationship will hold between the LCFS and the FCFS quantities for each queue as if it were alone. For instance, the following approximation for the average waiting time W in a M/GI/s queue is known (see Exercise 6.3):

$$W \approx \frac{1}{2}(1 + \beta^2)W_e$$

where β is the coefficient of variation of the service times, and W_e is the average waiting time in a M/M/s queue with the same arrival rate and mean service time. Also, one knows that the average waiting time in the M/GI/s-LCFS queue is equal to W_e. This leads to the idea of calculating the average waiting time W_e^i in queue i of the product-form network of M/GI/s_i- LCFS queues, and of approximating the waiting time in queue i of the network of M/GI/s_i queues with the same exogenous arrival rates and routing probabilities by $\frac{1}{2}(1 + \beta_i^2)\ W_e^i$ where β_i is the coefficient of variation of the service time in queue i.

Renewal Approximation 6.4.2 Some results are available for the GI/GI/1 queue. One idea is then to approximate the flows in a network of G/GI/1 queues by **renewal processes.** A suitable choice of the interarrival time distributions should give better results than assuming that the flows are Poisson (for this is a particular case).

Consider then a network with renewal exogenous arrivals and Markov routing. The **sum of independent renewal processes** with rates $\{\lambda_i, i = 1, \ldots, N\}$ and coefficients of variations $\{\beta_i, i = 1, \ldots, N\}$ is approximated by a renewal process with rate $\lambda = \sum \lambda_i$ and squared coefficient of variation

$1 - w + w\sum \lambda_i \lambda^{-1}\beta_i^2$ where $w^{-1} = 1 + 4(1 - \rho)^2(\sum \lambda_i^2 \lambda^{-2} - 1)$ with ρ being the traffic intensity of the queues into which the sum of the processes enters. This specific expression has been fine-tuned on the basis of a number of examples.

The **departure process** of a GI/GI/1 queue with traffic intensity ρ and interarrival coefficient of variation β is approximated by a renewal process with squared coefficient of variation $\rho^2\alpha^2 + (1 - \rho^2)\beta^2$, where α is the coefficient of variation of the service times.

Finally, a process obtained by **sampling** with probability p a renewal process with coefficient of variation β is approximated by a renewal process with the obvious rate and with squared coefficient of variation $p\beta^2 + 1 - p$.

This procedure permits us to calculate the mean values and the variance of the interarrival times at each queue (that is, for the renewal approximations). This is done by solving consistency equations for the network. For instance, the consistency equations for the rates are the usual flow conservation equations.

In order to calculate performance measures of the queues, one has to associate a **distribution** with the first two moments. This can be done by approximating the distribution by a suitable PH distribution. Alternatively, one can use a **two-moment approximation** for the performance measure of interest (e.g., the mean queue length or the mean delay).

Perhaps the most surprising fact about these methods is that they result in remarkably good approximations, when compared with simulation results. This is surprising, even though the number of degrees of freedom of the approach allows to fit a large number of examples, in the sense that the actual flows are probably far from being renewal.

Let us note that some asymptotic results have been proposed to justify the approximations. For instance, it is known that the sum of many independent renewal processes with small intensities can be approximated by a Poisson process. Also, the heavy traffic approximations can be used to make plausible some renewal approximations. However, more work is certainly needed to determine the accuracy of these approximations and their domain of validity.

The method has been extended to multiclass networks and variations are still being developed.

It is not difficult to imagine models of processes that introduce a dependence between interarrival times. For instance, by considering that the jump times correspond to a subset of possible transitions of a continuous-time Markov chains, one obtains a set of point processes that is dense in the class of stationary point processes. However, trying to fit the superposition, sampling, or departure processes in this larger class is computationally difficult.

Blocking Approximations 6.4.3 Except in a few exceptional cases such as the restriction of a reversible product-form network, or doubly stochastic routing, networks with **blocking** do not admit known invariant distributions. A number of approximation heuristics have been proposed for networks with blocking.

Consider a network of G/GI/1/N queues, i.e., single-server queues with i.i.d. service times and a finite waiting room. Assume that the routing is Markov. There are two commonly used models for what happens when a customer who completes service is being sent to a queue that is full.

The first model, called **manufacturing blocking** considers that the customer is served and waits until the destination is no longer full, at which time the server becomes available to the next customer. The second model, called **communication blocking,** assumes that the service cannot start until the destination is ready to accept the customer. The terminology is motivated by the blocking behavior in the corresponding physical systems.

A simple heuristic is to assume that all the flows are independent Poisson processes. To determine the rates, one proceeds as follows. Let $B_i(\lambda_i)$ be the blocking probability of queue i when its arrival process is Poisson with rate λ_i. The actual rate through queue i is then given by $\lambda_i(1 - B(\lambda_i))$. The flow conservation equations are then

$$\lambda_i[1 - B_i(\lambda_i)] = \sum_j \lambda_j[1 - B_j(\lambda_j)]\, r_{ji} \, .$$

No bounds are available for the errors resulting from this and other similar approximation ideas. One can attempt to justify a proposed approximation method by comparing the resulting values with simulation experiments.

6.5. SIMPLE BOUNDS

There are relatively few results known to hold for general nonproduct-form networks, other than Little's result and its variations. There are also a few results that are intuitively obvious and that can be proved without too much effort. This section will discuss some examples.

Example 6.5.1 [Acyclic Network of G/GI/s/N Queues] Consider an acyclic network of G/GI/s/N queues with Markov routing and deterministic arrivals. The blocking model is either manufacturing or communication blocking. It is natural to expect that increasing the waiting room or the number of servers in one queue will speed up the network, i.e., will increase the number of departures from the network in any $[0, T]$, for any $T > 0$, in a sense to be made precise.

What can be shown is that one can construct the stochastic processes corresponding to the original network and to the network with one queue being modified in such a way that the modified network outputs more customers in $[0, T]$ with probability one, for all $T > 0$.

Example 6.5.2 [Bounds by Little's Result] Little's result can be used to derive simple bounds on average network quantities. The example below will illustrate the procedure. Consider a closed network with N customers and two queues. (See Figure 6.6.)

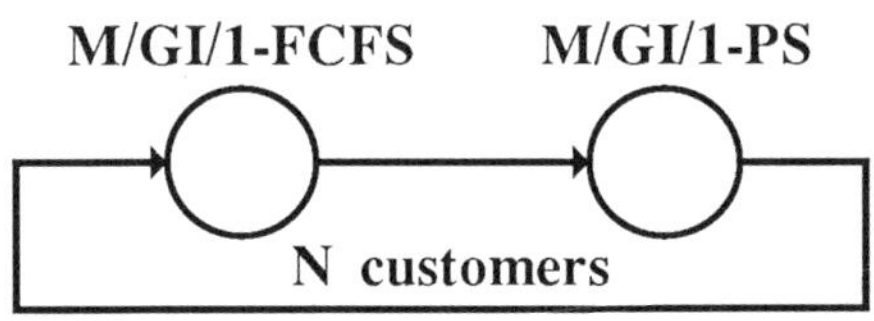

Figure 6.6. A closed network of two queues.

The average service time for queue i is μ_i^{-1} for $i = 1, 2$. Queue 1 is FCFS and queue 2 is Processor-Sharing. Say that the service times have PH distributions such that the closed network is ergodic. The network is largely untractable unless the service times are exponentially distributed in the first queue.

Denote by θ_i the average sojourn time of a customer in queue i, and by μ_i^{-1} the average service time in that queue. The claim is that the throughput λ is such that

$$(\mu_1^{-1} + \mu_2^{-1})^{-1} \leq \lambda \leq N(\mu_1^{-1} + \mu_2^{-1})^{-1} .$$

Indeed, $\mu_i^{-1} \leq \theta_i \leq N\mu_i^{-1}$ for $i = 1, 2$. Also, $N = \lambda(\theta_1 + \theta_2)$, by Little's result.

Examples 6.5.3 [Other bounds] Other bounds can be obtained by comparing service disciplines. For instance, if the server in a single-server queue can observe the service times of the arriving customers, then the discipline that always serves the customer with the smallest remaining processing time (SRPT) is easily seen to minimize the departure times. Combining this with the fact that speeding up arrivals speeds up departures, one finds that a network of single-server queues with i.i.d. processing times at all the queues is the fastest when the SRPT discipline is used at all the queues.

One can also compare the processor-sharing and the FCFS disciplines, under suitable assumptions on the service time distributions. As an example, assume that the service times are i.i.d. and such that the residual processing time decreases stochastically with the service time already received. One then finds that it is preferable to keep working at maximum rate on the customer who has received the largest amount of service. As a consequence, the FCFS discipline is preferable to the processor-sharing discipline.

6.6. DEPENDENT SERVICE TIMES

Multiclass networks of quasi-reversible queues permit us to analyze systems where the service times of a customer in successive queues are not independent. Unfortunately, this is not possible for FCFS queues. This section will describe some results on FCFS queues with dependent service times and with blocking.

Say that there are n nodes, each with a waiting room **capacity** of B customers. The service times of a customer are identical in all the queues, and the service times of the different customers are i.i.d. with distribution $F(.)$ with unit mean.

The question of interest is at what rate customers can go through the system. Thus, it will be assumed that the first node is always full. Denote by $\lambda(n, B)$ [resp. by $\mu(n, B)$] the average departure rate (or **throughput**) from the system when the communication [resp. manufacturing] blocking model is used. The main results are as follows:

$$\lambda(n, B) \le \mu(n, B) \le \lambda(n, B+1) \le (B+1)\,\mu(n, 1)$$

$$\lambda(n, B) = O([\int_0^\infty (1 - F(t)^n)\,dt]^{-1}), \quad \text{as } n \to \infty$$

$$\mu(n, B) = O([\int_0^\infty (1 - F(t)^n)\,dt]^{-1}), \quad \text{as } n \to \infty$$

$$[1 + q + (n-1) \int_{(B-1)q}^{\infty} t dF(t)]^{-1} \le \lambda(n, B)$$

$$\le \frac{B}{q} + [n \int_q^\infty t dF(t)]^{-1}, \quad 0 < q < \infty .$$

These estimates can then be used to determine how fast the buffer size B should grow with n for the throughput λ to be bounded away from zero. As particular cases, one has the following facts:

(a) Assume that $1 - F(t) = O(t^{-\rho})$ as $t \to \infty$ with $\rho > 1$. Then

$$\lim_{n \to \infty} \lambda(n, B(n)) = 1 \text{ [resp. 0] if } \lim_{n \to \infty} \frac{B(n)}{n^{\frac{1}{\rho-1}}} = \infty \text{ [resp. 0]} .$$

(b) If $-\log[1 - F(t)] = O(t^\rho)$ as $t \to \infty$, then

$$\lim_{n \to \infty} \lambda(n, B(n)) = 1 \text{ [resp. 0] if } \lim_{n \to \infty} \frac{B(n)}{(\log n)^{\frac{1}{\rho}}} = \infty \text{ [resp. 0]} .$$

Denote by $\nu(n, B)$ the throughput of the system with the communications blocking when the service times at the different nodes are i.i.d. It is known that if $F(t) \equiv 1 - e^{-t}$, then

$$\lim_{n \to \infty} \nu(n, 1) = \frac{1}{4} .$$

Moreover, for each $n \ge 1$, the infimum value of $\nu(n, B)$ over all the distributions $F(.)$ with unit mean is equal to n^{-1}.

Also, if $\int_0^\infty t^\rho\, dF(t) = \infty$ for some $\rho > 2$, then $\lim_{n \to \infty} \nu(n, B) = 0$.

6.7. COMPARISON RESULTS

This section discusses a few **comparison results** for networks. The results were selected because they illustrate typical arguments that are used for such approximations. We will explain the **stochastic ordering** results that were introduced in section 1.18.

Definitions 6.7.1 [Stochastic ordering] Let $X = (X^1, \ldots, X^d)$ and $Y = (Y^1, \ldots, Y^d)$ be two random vectors that take values in $\mathbb{R}^d$ for some $d \geq 1$. One says that X is **stochastically smaller** than Y and one writes $X \underset{stoch}{\leq} Y$ if

$$P\{X^i \geq a^i, \quad 1 \leq i \leq d\} \leq P\{Y^i \geq a^i, \quad 1 \leq i \leq d\} \tag{6.6}$$

for all $a^i \geq 0$, $1 \leq i \leq d$.

One writes $X \underset{stoch}{=} Y$ to indicate that X and Y have the same distribution function.

Notice that these definitions do not involve the joint distribution of X and Y.

Fact 6.7.2 [Pathwise ordering] If $d = 1$, then $X \underset{stoch}{\leq} Y$ if and only if one can find two random variables $\tilde{X}$ and $\tilde{Y}$ defined on the same probability space such that

$$\tilde{X} \underset{stoch}{=} X, \quad \tilde{Y} \underset{stoch}{=} Y, \quad P\{\tilde{X} \leq \tilde{Y}\} = 1\,. \tag{6.7}$$

Proof: Let U be a random variable that is uniformly distributed on $[0, 1]$. One can choose

$$\tilde{X} = f(U) \quad \text{and} \quad \tilde{Y} = g(U)\,,$$

with

$$f(u) = \min\{x \in \mathbb{R} \mid P\{X \leq x\} \geq u\}$$

$$\text{and} \quad g(u) = \min\{x \in \mathbb{R} \mid P\{Y \leq x\} \geq u\}\,.$$

The verification that these choices satisfy (6.7) is left to the reader.

Remark 6.7.3 This property **does not carry over** to the multidimensional case.

To see this, let $d = 2$ and let X and Y be such that

$$P\{X = (0,0)\} = P\{X = (1,1)\} = \frac{1}{6},$$

$$P\{X = (1,0)\} = P\{X = (0,1)\} = \frac{1}{3},$$

$$P\{Y = (0,0)\} = P\{Y = (1,1)\} = \frac{1}{3},$$

$$P\{Y = (1,0)\} = P\{Y = (0,1)\} = \frac{1}{6}.$$

One can verify that $X \underset{stoch}{\leq} Y$. If one could find $\tilde{X}$ and $\tilde{Y}$ such that (6.7) holds, then this would imply that $P\{X = (0,0)\} \geq P\{Y = (0,0)\}$, which is not the case.

This observation shows that stochastic ordering on $\mathbb{R}^d$ for $d \geq 2$ will not be proved by simple pathwise arguments.

Example 6.7.4 [Networks of two M/M/1 queues] Consider the two networks in Figure 6.7.

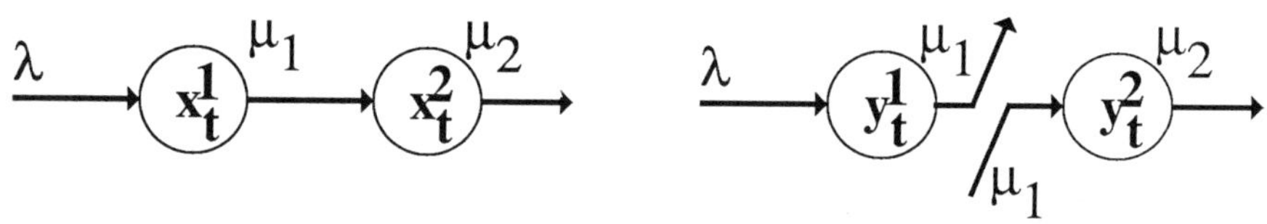

Figure 6.7. Two networks.

The queues are M/M/1 and the arrivals are Poisson, with the indicated service and arrival rates.

Assume that the two networks are started with the same vector of queue lengths. It is intuitively plausible that the vector of queue lengths y_t in the second network will increase faster than the corresponding vector x_t in the first network. Indeed, the arrival rate in the second queue is always at least as large in the second network as in the first one.

However, one cannot expect a pathwise ordering $x_t \leq y_t$. Indeed, say that $x_0 = y_0 = (1, 0)$. If the first transition to happen (at time τ, say) is a service completion in the first queue, then the states after the first transition are $x_\tau = (0, 1)$ and $y_\tau = (0, 0)$.

The point is that if one conditions on the fact that the first transition is either a service completion in the first queue or an exogenous arrival in the second queue (each with probability one half), then the distributions of the states just after the first transition are given by

$$P\{x_\tau = (1, 0)\} = P\{x_\tau = (0, 1)\} = \frac{1}{2}$$

$$P\{y_\tau = (0, 0)\} = P\{y_\tau = (1, 1)\} = \frac{1}{2}.$$

In this case, one sees that $x_\tau \underset{stoch}{\leq} y_\tau$.

This simple observation is the key to the stochastic ordering property of networks.

Example 6.7.5 Consider the two networks illustrated in Figure 6.8.

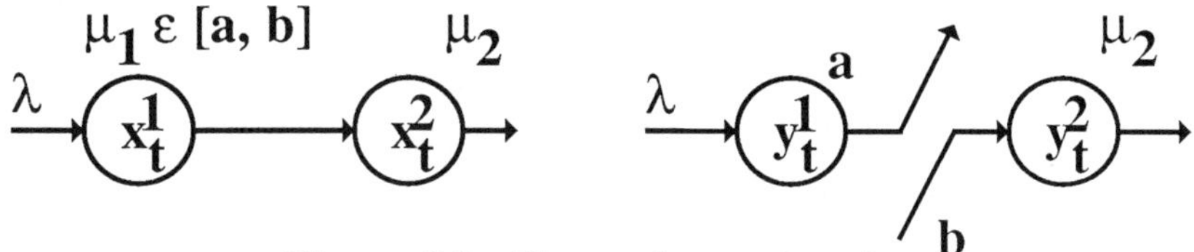

Figure 6.8. Two other networks.

The first queue is no longer M/M/1 in the first network. It is assumed that its service rate (when the queue is nonempty) takes values in the interval $[a, b]$. For instance, the queue could be an M/PH/1-FCFS queue with the following representation for the service times. (See Figure 6.9.)

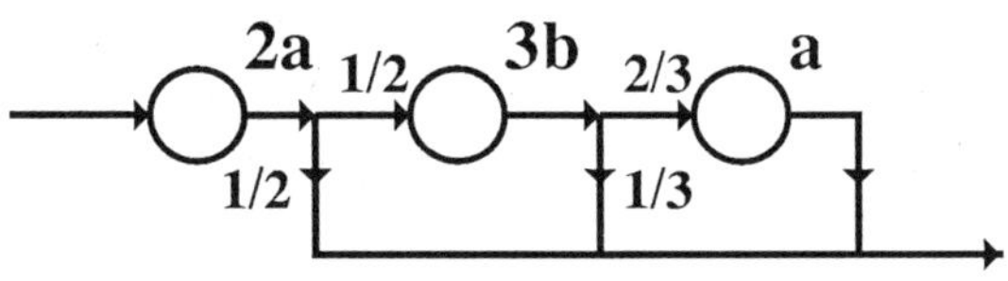

Figure 6.9. The service time distribution.

It will be shown that if $x_0 \underset{stoch}{\leq} y_0$, then $x_t \underset{stoch}{\leq} y_t$ for all $t \geq 0$.

This result extends to more general networks, as we will see below. First we need a simple lemma.

Lemma 6.7.6 [Invariance of stochastic ordering] For $x, y \in \mathbb{R}$ let $x \wedge y := \min\{x, y\}$ and $x^+ := \max\{x, 0\}$. Let $\mathbb{N} = \{0, 1, 2, \ldots\}$.

Let X, Y, M, N be $\mathbb{N}^2$-valued random variables such that $X \underset{stoch}{\geq} Y$, $\{X, Y\}$ and M are independent, and given Y,

$$N = (N^1, N^2) \underset{stoch}{=} (N^2, N^1), \text{ and } (-M^1, M^2) \underset{stoch}{\geq} (-N^1, N^2)\,.$$

Then

$$((X^1 - M^1)^+, X^2 + M^2) \underset{stoch}{\geq} W := ((Y^1 - N^1)^+, Y^2 + N^1 \wedge Y^1)\,.$$

Proof: By conditioning on M one finds that

$$((X^1 - M^1)^+, X^2 + M^2) \underset{stoch}{\geq} ((Y^1 - M^1)^+, Y^2 + M^2) =: V\,.$$

To show $V \underset{stoch}{\geq} W$ one must prove that for all $a^1, a^2 \in \mathbb{N}$,

$$P\{V^1 \geq a^1, V^2 \geq a^2\} \geq P\{W^1 \geq a^1, W^2 \geq a^2\}. \tag{6.8}$$

For $a^1 = 0$, this equation reads $V^2 \underset{stoch}{\geq} W^2$ and follows from

$$Y^2 + M^2 \underset{stoch}{\geq} Y^2 + N^2 \underset{stoch}{=} Y^2 + N^1 \underset{stoch}{\geq} Y^2 + N^1 \wedge Y^1\,.$$

For $a^1 > 0$, the equation (6.8) is implied by

$$(Y^1 - M^1, Y^2 + M^2) \underset{stoch}{\geq} (Y^1 - N^1, Y^2 + N^1)\,. \tag{6.9}$$

To see why (6.9) holds, observe that

$$(Y^1 - M^1, Y^2 + M^2) \underset{stoch}{\geq} (Y^1 - N^1, Y^2 + N^2)\,,$$

so that it suffices to show that

$$(Y^1 - N^1, Y^2 + N^2) \underset{stoch}{\geq} (Y^1 - N^1, Y^2 + N^1)\,.$$

By conditioning on Y it remains only to prove that for all $a^1, a^2 \in \mathbb{N}$

$$P\{N^1 \leq a^1, N^2 \geq a^2\} \geq P\{N^1 \leq a^1, N^1 \geq a^2\}\,.$$

But this last inequality is immediate from $(N^1, N^2) \underset{stoch}{=} (N^2, N^1)$.

Definition 6.7.7 [Stochastic networks] Let $\{x_t, y_t, t \geq 0\}$ be two $\mathbb{N}^d$-valued processes corresponding to the vectors of queue lengths in two networks of d queues.

Alternatively, each $i \in \{1, \ldots, d\}$ could be some pair $(k, c) \in \{1, \ldots, K\} \times \{1, \ldots, C\}$ where k would indicate a node and c a customer class; the components of x_t and y_t indexed to the pair (k, c) would be the number of customers of class c in node k at time t in the two networks.

Assume that those processes admit the following representation. For $t \geq 0$ and $i \in \{1, \ldots, d\}$,

$$\begin{aligned} dx_t^i &= -\sum_{j=0}^{d} 1\{x_{t-}^i > 0\}\, dS_t^{ij} + \sum_{j=0}^{d} dA_t^{ji} \\ dy_t^i &= -\sum_{j=0}^{d} 1\{y_{t-}^i > 0\}\, dR_t^{ij} + \sum_{j=0}^{d} 1\{y_{t-}^j > 0\}\, dR_t^{ji}\,. \end{aligned} \tag{6.10}$$

In these expressions the processes R^{ij}, A^{ij}, S^{ij} are point processes (i.e., piecewise constant with unit jumps) with the following properties. The processes $S^{ij}, A^{ij}, \tilde{A}^{ij}$ (see below) are Poisson processes; $\{S^{ij}, \tilde{A}^{ij}\}, A^{ij}$ for $0 \leq i, j \leq d$ are independent; for every (i, j), the processes A^{ij} and $\tilde{A}^{ij}$ have the same rate; almost surely one has, for $t \geq 0$ and $0 \leq i, j \leq d$,

$$\Delta S_t^{ij} := S_t^{ij} - S_{t-}^{ij} \leq \Delta R_t^{ij} \leq \Delta \tilde{A}_t^{ij} \tag{6.11}$$

Also, the differentials in (6.10) are in the Lebesgue-Stieltjes sense. For instance, $dg(t) = f(t)dA_t^{ij}$ means that

$$g(T) = g(0) + \int_0^T f(t)dA_t^{ij} := g(0) + \sum_n f(T_n)\, 1\{0 < T_n \leq t\}$$

where the T_n are the jump times of A^{ij}. (See Figure 6.10.)

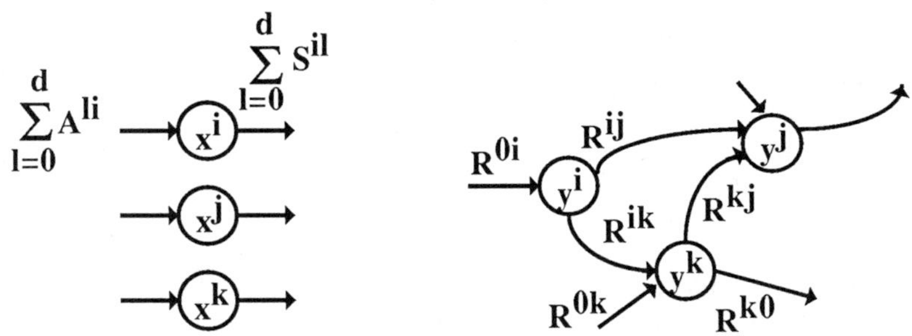

Figure 6.10. The networks described by equation (6.10).

Thus, x_t corresponds to a system of M/M/1 queues in parallel while y_t corresponds to a network of interconnected queues and need not be Markov. The routing and the service rates in y_t may depend on the "state" of the complete network; the basic assumption is that the service rate from queue i to queue j is bounded from above and from below when the queue is not empty.

Proposition 6.7.8 [Stochastic comparison result] Assume that $x_0 \underset{stoch}{\geq} y_0$. Then $x_t \underset{stoch}{\geq} y_t$ for all $t \geq 0$.

Proof: Let $T_1, T_2, \ldots$ be the successive jump times of $\sum_{ij} (S_t^{ij} + A_t^{ij} + \tilde{A}_t^{ij})$ and define $X_n = x_{T_n}, Y_n = y_{T_n}$ for $n \geq 1$. Let also $X_0 = x_0$ and $Y_0 = y_0$. It suffices to show that given $\{T_m, m \geq 1\}$, one has $X_n \underset{stoch}{\geq} Y_n$ for all $n \geq 0$. Assume that this is true for a given n.

Notice that given

$$\Psi := \{X_n, Y_n, T_m, m \geq 1, \Delta A_{T_{n+1}}^{ij} + \Delta \tilde{A}_{T_{n+1}}^{ij} = 1\},$$

the following random variables

$$N^1 = \Delta R_{T_{n+1}}^{ij}, \quad M^1 = \Delta_{T_{n+1}}^{ij}, \quad M^2 = \Delta_{T_{n+1}}^{ij},$$

$$X^1 = X_n^i, \quad X^2 = X_n^j, \quad Y^1 = Y_n^i, Y_n^j$$

satisfy the conditions of Lemma 6.7.6 with N^2 being an i.i.d. copy of N^1 (given Ψ). Hence, given Ψ one has $X_{n+1} \underset{stoch}{\geq} Y_{n+1}$, and this concludes the proof.

Example 6.7.9 For instance, if network Y consists of d queues with exogenous arrival rates λ_i, service rates in $[a^i, b^i]$ in queue i when it is nonempty, and routing probabilities r_{ij}, then y_t is stochastically dominated by the vector of queue lengths of d parallel M/M/1 queues with arrival rates $\lambda_i + \sum_j b_j r_{ji}$ and service rates a_i.

Remarks 6.7.10

(a) The result extends to deterministic, and therefore to arbitrary arrivals, by applying the argument between the arrival times.

(b) The idea of considering the Markov chain $\{Y_n\}$ can be used to show that is a network of M/M/s queues and arbitrary arrivals is started with stochastically more customers, then that ordering is preserved at all times.

Proposition 6.7.11 [Increasing throughput] Consider a closed network of M/M/s queues with random routing and with N customers. Denote by A_t the number of customers who cross a given link (i, j) of the network in $[0, t]$.

Consider now the same network but with $N + 1$ customers and assume that it starts with the same vector of queue lengths, to which one customer was added. Denote by B_t the number of customers who cross the same link (i, j) in $[0, t]$.

Then, the evolution of the processes can be constructed so that

$$P\{A_t \leq B_t, t \geq 0\} = 1 .$$

Proof: Color the N customers **red** and the additional customer **green.** Since only the numbers of customers in the queues matter, one can always reorder the customers in the queues so that the green customer is last. One then sees that the red customers are served at the same rate in each queue of the two networks and at all time. As a consequence, they will contribute the same value A_t to the flow on link (i, j). In the network with $N + 1$ customers, the green customer will contribute an additional number of jumps.

The same argument applies to networks of queues with a service rate that is an increasing function of the queue length.

Remark 6.7.12 [Concavity] The above argument can be used to show that if the service rates in the queues are **increasing and concave** in the queue length, then the expected number of jumps on a link (i, j) in $[0, t]$ (the **throughput**) is **increasing and concave** in the number of customers in the network.

The proof uses the fact that the service rate available to serve the green customer is the excess service rate corresponding to adding one customer in the queue where the green customer is (since the red customers move as before). Moreover, this excess in service rate is decreasing in the number of customers in the network (since the service rates are assumed to be concave). The details are left to the reader.

6.8. EXERCISES

6.1. Diffusion approximations Consider a single GI/GI/1 queue. Particularize the results of section 6.2 to find an approximation to the average queue length in heavy traffic. *Hint:* See Remark 6.2.5.(f).

6.2. Light traffic approximations Find the light traffic approximation for the delay in an M/D/1 queue.

6.3. LCFS Approximations Show that the approximation 6.4.1 agrees with the Pollaczek-Khintchine formula for an M/GI/1 queue. (See Appendix B.)

6.4. Renewal approximations Find an approximation for the delay in a system that consists of an M/D/1 queue followed by an M/M/1 queue in tandem.

6.5. Monotonicity Prove the result of Example 6.5.1 for two GI/GI/s/N queues in tandem.

6.6. SRPT Calculate the average queue length in an M/M/1 queue with the SRPT service discipline.

6.7. FCFS queues Give a formal proof of the last statement of Examples 6.5.3.

6.8. Concavity Provide the details of the discussion of Remark 6.7.12.

6.9. REFERENCES AND COMMENTS

Section 6.2 Good survey of the diffusion approximations are Lemoine (1978) and Flores (1985). Harrison (1985)–(1986), Reiman (1984), and Borovkov (1984) are highly recommended. Key references are Iglehart (1965), Iglehart and Whitt (1970a)–(1970b), Harrison and Reiman (1981), Kobayashi (1974), Gelenbe (1975), Gelenbe and Mitrani (1980), Kingman (1982), Coffman and Reiman (1984). The asymptotic analysis of sojourn times in Jackson networks is due to Reiman (1982). Multiclass results are in Johnson (1983), Peterson (1985), and Harrison (1985), where control applications are also discussed. The existence of stationary distributions for diffusions and their product form is discussed in Harrison and Williams (1987).

Section 6.3 The light traffic approximations were developed in Burman and Smith (1983)–(1986) and in Reiman and Simon (1985a)–(1986).

Section 6.4 This section is a largely inadequate discussion of the vast literature on approximations. The renewal approximations were used in Kuehn (1979), Shantikumar and Buzacott (1981), Whitt (1982)–(1983), Albin (1984), and Bitran and Tirupati (1986).

The blocking approximations were discussed in Altiok (1982), Mitrani (1984), and Kelly (1986a). The LCFS approximation 6.4.1 is in Stoyan (1983).

Section 6.5 Some of the general results of Example 6.5.1 are discussed in van Dijk, et al. (1987).

Section 6.6 The results of this section are from Kelly (1982b). See also Wolff (1982b) and Calo (1982) for related work.

Section 6.7 Proposition 6.7.8 is essentially due to Massey (1984a)–(1984b). The proof is from Tsoucas, et al. (1984b). The arguments in 6.7.11 and 6.7.12 are due to van der Wal (1986). See also Shantikumar and Yao (1986) for a different approach.

CHAPTER SEVEN

STATIONARY QUEUEING SYSTEMS

7.1. INTRODUCTION

A queueing system can be viewed either by an **independent observer** or by a **customer.** This chapter explains the relationship between these two viewpoints. Examples of this relationship were already discussed in Theorem 2.10.6 on the distribution at the jumps of a Markov chain, the **Arrival Theorem** 4.4.1 was a useful particular case. **Little's result** 3.4.1 turns out to be another example of this relationship.

We will also discuss the **stability** of queueing systems: When does the evolution of a queueing system converge to a **stationary** behavior that is **independent of the initial conditions?**

These issues are well addressed in the formalism of **stationary marked point processes.** The generality allows us to emphasize the properties that are essential.

The basic concepts about **marked point processes** are explained in section 7.2. The **Palm distributions** are defined and illustrated by a number of examples. The issues of **stationarity** and **asymptotic stationarity** are explained in section 7.3. The **stability** of simple queueing systems is analyzed in section 7.4. In section 7.5, we take another look at **Little's result** in the light of Palm distributions.

7.2. WAITING FOR A BUS—PALM DISTRIBUTIONS

A simple queueing system is that of customers waiting for a bus at a given bus stop. Say that one bus goes by every hour, on the average. How long can a typical customer, unaware of the bus schedule, be expected to **wait for a bus?** The first answer that comes to mind, half an hour, is not correct. For instance, if the buses arrive in pairs every two hours, then the average waiting time is equal to one hour. In general, the customer is more likely to get to the bus stop during a long interarrival epoch than during a short one. This will be seen to imply that the average waiting time increases with the variance of the interarrival times. This is a simple illustration of a general phenomenon in queueing systems:

fluctuations increase delays.

Assume that some **quantity** is associated with each bus (e.g. the number of customers in the bus). What is the quantity that will be associated with the first bus seen by a typical customer?

The precise answers to these questions are stated in this section. The basic idea is that one can relate in a **one-to-one way** the "bus process" viewed at a **typical time** and the same process viewed from an **arrival time.** The probability distribution of the latter process is called the **Palm distribution** of the former process.

This approach extends to much more general situations and allows us to determine how a **typical customer** views a given queueing process. This question is essential to determine how long a typical customer stays in a network, for instance.

Definition 7.2.1 [Synchronous Marked Point Process] Let $\mathbf{N}_0 = \{(T_n, x_n), n \in \mathbb{Z}\}$ be a sequence of random variables. Assume that

$$\cdots < T_{-2} < T_{-1} < T_0 = 0 < T_1 < T_2 < \cdots$$

and that the x_ns take values in some set X. Typically, $X = \mathbb{R}^d$ for some $d \geq 1$. (See Figure 7.1.) One can think of the T_ns as bus arrival times and of x_n as some quantity associated with bus number n (e.g., the number of customers in the bus or the quantity of fuel in its tank). The subscript 0 in $\mathbf{N}_0$ is used to remind us that $T_0 = 0$. The interarrival times $T_{n+1} - T_n$ may be dependent. The x_ns and the T_ns may also be dependent.

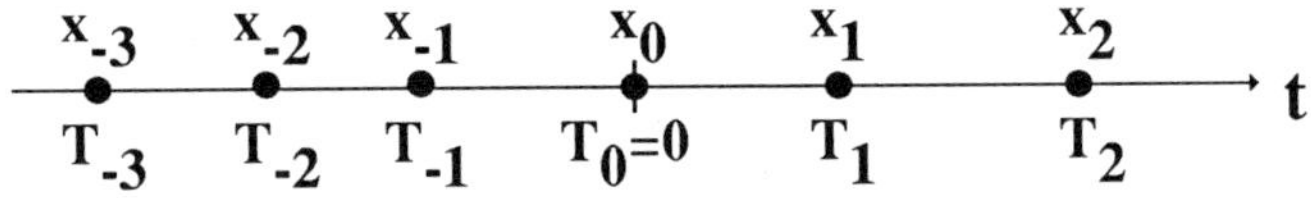

Figure 7.1. Synchronous point processes.

The objective of this section is to determine how the process $\mathbf{N}_0$ is viewed from a **typical** time (in a sense to be made precise below). It will be assumed that the distribution of $\mathbf{N}_0$ does not depend on the specific time T_m from which it is observed. That is, one assumes that the distribution of

$$\{(T_{m+n} - T_m, x_{m+n}), \quad n \in \mathbb{Z}\}$$

does not depend on $m \in \mathbb{Z}$. In particular, the interarrival times $T_{n+1} - T_n$ have the same distribution, as do the marks x_ns.

The process $\mathbf{N}_0$ is then called a **synchronous marked point process.** (The times T_ns are the **points;** the x_ns are the **marks** of the points.)

Definition 7.2.2 [Stationary Point Process] The sequence $\mathbf{N}_0$ viewed from a typical time will be denoted by $\mathbf{N}$. That **marked point process** will be of the form indicated in Figure 7.2.

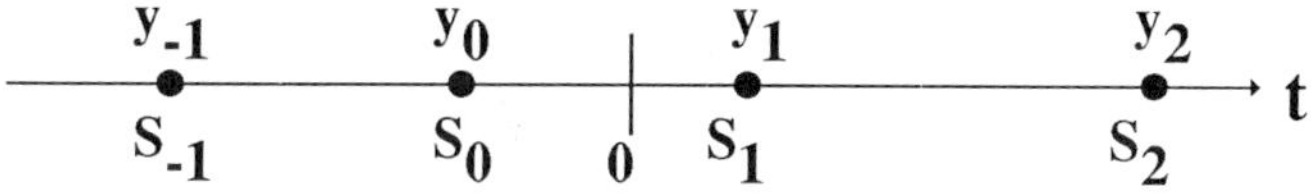

Figure 7.2. Stationary point process.

In Figure 7.2. one considers that the **typical time** is defined to be 0. The quantity y_1 is the first value x_n seen after the typical time and S_1 denotes the time until the first arrival after the typical time. The other quantities S_n and y_n are defined similarly.

Problem 7.2.3 [From $\mathbf{N}_0$ to N] The basic idea used to relate $\mathbf{N}_0$ and N can be described as follows. Let $W := S_1$ and $V := -S_0$. Given $\{V = v, W = w\}$, the distribution of $\{(S_n, y_n), n \in \mathbb{Z}\}$ will be the conditional distribution of $\{(T_n - v, x_n), n \in \mathbb{Z}\}$ given that $\{T_1 = v + w\}$. The justification is that, by the assumption that $\mathbf{N}_0$ looks the same from any point, the only thing that one learns from V and W about $\mathbf{N}_0$ is that the typical time is in an interarrival epoch of duration $V + W$ and that the next arrival will take place after W time units. The above discussion shows that (with $U = V + W$)

$$E[f(\mathbf{N}) \mid V = v, U = u] = E[f(T_n - v, x_n, n \in \mathbb{Z}) \mid T_1 = u]$$
$$= E[f(\mathbf{N}_0 + v) \mid T_1 = u] \qquad (7.1)$$

where one defined

$$\mathbf{N}_0 + v = \{(T_n - v, x_n), n \in \mathbb{Z}\}.$$

(The signs in this definition can be confusing. Remember that $\mathbf{N}_0 + v$ is the process observed by moving the time origin to $+v$.)

To find the distribution of (U, V), one argues as follows. The proportion of interarrival epochs with a duration in $(t, t + dt)$ is given by $P\{T_1 \in (t, t + dt)\}$. Therefore, the fraction of the real line that is covered by such interarrival epochs is proportional to $tP\{T_1 \in (t, t + dt)\}$. Hence,

$$P\{U \in (t, t + dt)\} = \alpha t \, P\{T_1 \in (t, t + dt)\} .$$

The constant α is determined by the fact that this density of U must sum to one. This implies that $\alpha = \lambda := \{ET_1\}^{-1}$. Also, it is reasonable to expect that given $\{U = t\}$ the random variable V is uniformly distributed on $[0, t]$. This leads to

$$P\{U \in (u, u + du), V \in (v, v + dv)\} \tag{7.2}$$

$$= \lambda P\{T_1 \in (u, u + du)\} \, dv \quad \text{for } 0 \le v \le u.$$

Combining (7.1) and (7.2) then gives, with

$$K(du, dv) := P\{U \in (u, u + du), V \in (v, v + dv)\} ,$$

$$E\{f(\mathbf{N})\} = \int_{u=0}^{\infty} \int_{v=0}^{u} E[f(\mathbf{N}) \mid U = u, \ V = v] \, K(du, dv)$$

$$= \int_{u=0}^{\infty} \int_{v=0}^{u} E[f(\mathbf{N}_0 + v) \mid T_1 = u] \, \lambda P\{T_1 \in (u, u + du)\} \, dv$$

$$= \lambda E\{ \int_0^{T_1} f(\mathbf{N}_0 + v) \, dv \} .$$

Summarizing, one has the following result.

Theorem 7.2.4 Let $\mathbf{N}_0$ be a **synchronous marked point process** with $0 < \lambda^{-1} = ET_1 < \infty$.

This process is viewed as $\mathbf{N}$ at a typical time, where $\mathbf{N}$ is such that (provided that the expectations are well defined)

$$E\{f(\mathbf{N})\} = \lambda E\{ \int_0^{T_1} f(\mathbf{N}_0 + v) \, dv \} . \tag{7.3}$$

The distribution of $\mathbf{N}_0$ is called the **Palm distribution** of the process $\mathbf{N}$. One can show that the process $\mathbf{N}$ is **stationary,** i.e., that the distribution of that process observed at time $t \in \mathbb{R}$ does not depend on t. (See Exercise 7.1.)

(Note that the intuitive justification of (7.3) relies implicitly on ergodicity of the point process. Indeed, the statement about the "proportion of interarrival epochs" is an ergodicity statement. However, as the proof in Exercise 7.1 shows, the result is valid for stationary processes, without ergodicity assumption.)

The formula (7.3) can be used to compute quantities about **N** in terms of $\mathbf{N}_0$. The following facts give a few examples. The derivations are outlined in Exercise 7.2.

Facts 7.2.5

(a) The following relations hold (below, h is a function such that the integrals are well defined and $a \wedge b$ denotes the minimum of the real numbers a and b):

$$Eh(y_1) = \lambda ET_1 h(x_1) \tag{7.4}$$

$$Eh(U) = \lambda ET_1 h(T_1) \tag{7.5}$$

$$EW^n = \frac{\lambda}{(n+1)!} E(T_1)^{n+1} \tag{7.6}$$

$$EU = 2EW = ET_1 + \frac{\text{var}(T_1)}{ET_1} \tag{7.7}$$

$$P\{W \leq t\} = \lambda \int_0^t P\{T_1 > s\}\, ds = \frac{E\{T_1 \wedge t\}}{ET_1} \tag{7.8}$$

$$P\{W \leq t\} \equiv P\{T_1 \leq t\}$$

$$\text{if and only if } P\{T_1 \leq t\} \equiv 1 - e^{-\lambda t} \tag{7.9}$$

$$P\{V > v,\ W > w\} = \lambda \int_{v+w}^{\infty} P\{T_1 > t\}\, dt\ . \tag{7.10}$$

(b) Let A be a set of possible realizations of **N**. Then

$$P\{\mathbf{N} \in A\} = \lambda \int_0^{\infty} P\{T_1 > v,\ \mathbf{N}_0 + v \in A\}\, dv\ . \tag{7.11}$$

The distribution of W in (7.8) is the **equilibrium forward recurrence time** corresponding to T_1.

Problem 7.2.6 [From N to $\mathbf{N}_0$] The intuition suggests that $\mathbf{N}_0$ is N conditioned on the fact that the first arrival occurs at time 0. Thus, one expects the following result.

Theorem 7.2.7 For any set of realizations B of $\mathbf{N}_0$ one has

$$P\{\mathbf{N}_0 \in B\} = \lim_{w \downarrow 0} P[\mathbf{N} + S_1 \in B \mid S_1 \leq w] \qquad (7.12)$$

The proof of this intuitively clear result is outlined in Exercise 7.3. (It can be shown that the convergence in (7.12) is uniform in B.)

Consider our typical time and denote by $\lambda(K)$ the rate at which buses go by with an associated quantity y_n that belongs to K (a given subset of X). One might expect that $\lambda(K)\lambda^{-1}$ is equal to the probability of a typical bus having its quantity in K, i.e., equal to $P\{x_0 \in K\}$. The rate $\lambda(K)$ is the number of buses with a quantity in K that go by in (0, 1]. Thus, one expects that

$$P\{x_0 \in K\} = \frac{\lambda(K)}{\lambda} \text{ with } \lambda(K) = E \sum_{n=1}^{\infty} 1\{y_n \in K, \; S_n \leq 1\}. \qquad (7.13)$$

This idea can be generalized as follows. Consider that, in the above argument, y_n is replaced by the **complete evolution** of the process seen from time S_n. That is, consider that y_n is replaced by $\mathbf{N} + S_n$. Applying the above formulas to this case then leads to the following result.

Theorem 7.2.8 For any set B of possible realizations of $\mathbf{N}_0$ one has

$$P\{\mathbf{N}_0 \in B\} = \frac{\lambda(B)}{\lambda} \text{ with} \qquad (7.14)$$

$$\lambda(B) := E \sum_{n=1}^{\infty} 1\{\mathbf{N} + S_n \in B, \quad S_n \leq 1\}.$$

Proof: Choose $\epsilon \ll 1$ and define $f(\mathbf{N}) = 1\{-\epsilon < S_0 < 0, \mathbf{N}_0 \in B\}$. Then

$$E\{f(\mathbf{N})\} = E\,1\{-\epsilon < S_0 < 0, \mathbf{N} + S_0 \in B\}$$

$$= E\,1\{S_1 < \epsilon, \mathbf{N} + S_1 \in B\}$$

$$= E \sum_{n=1}^{\infty} 1\{S_n < \epsilon, \mathbf{N} + S_n \in B\} + o(\epsilon)$$

and

$$E \int_0^{T_1} f(\mathbf{N}_0 + v)\,dv = E \int_0^{T_1} 1\{v \leq \epsilon, \mathbf{N}_0 + v \in B\}\,dv = \epsilon P\{\mathbf{N}_0 \in B\}.$$

(The second equality follows from the stationarity of $\mathbf{N}$.)

Substituting into (7.3) gives

$$E \sum_{n=1}^{\infty} 1\{S_n < \epsilon, \mathbf{N} + S_n \in B\} + o(\epsilon) = \lambda \epsilon P\{\mathbf{N}_0 \in B\} .$$

The stationarity of $\mathbf{N}$ implies that

$$E \sum_{n=1}^{\infty} 1\{S_n < \epsilon, \mathbf{N} + S_n \in B\} = \epsilon E \sum_{n=1}^{\infty} 1\{S_n \le 1, \mathbf{N} + S_n \in B\} .$$

Substituting in the expression above, dividing by ϵ, and letting $\epsilon \to 0$ yield the result.

It is convenient to define the **stationary distribution** P by $P(B) = P\{\mathbf{N} \in B\}$ and the **Palm distribution** P^0 by $P^0(B) = P\{\mathbf{N}_0 \in B\}$. Thus, there is a one-to-one correspondence between P and P^0 given by (7.3) and (7.14).

Corollary 7.2.9 [Campbell's formula] For any given $f : \mathbb{R} \times X \to \mathbb{R}_+$ one has

$$E\{ \sum_{n \in \mathbb{Z}} f(S_n, y_n)\} = \lambda \int_{-\infty}^{\infty} \int_X f(t, x)\, P\{x_0 \in dx\}\, dt . \tag{7.15}$$

Proof: Replacing the interval [0, 1] by the interval $(t, t + dt)$ and the set K by $(x, x + dx)$ in (7.13) gives

$$E\{ \sum_{n \in \mathbb{Z}} 1\{y_n \in (x, x + dx), \quad S_n \in (t, t + dt)\}$$
$$= \lambda P\{x_0 \in (x, x + dx)\}\, dt .$$

Multiplying this equation by $f(t, x)$ and integrating over $\mathbb{R} \times X$ give (7.15).

Example 7.2.10 The simplest example is when $T_n \equiv n$ and $x_n \equiv 0$ for $n \in \mathbb{Z}$. For this example, one finds from (7.8) that W is uniformly distributed on [0, 1], which should be intuitively clear. This distribution of W completely specifies $\mathbf{N}$.

Example 7.2.11 Let the process $\mathbf{N}_0$ be defined by (see Figure 7.3)

$$\mathbf{N}_0 = \begin{cases} \ldots, (-5, 2), (-3, 1), (-2, 2), (0, 1), (1, 2), (3, 1), (4, 2), \\ \qquad \ldots, \text{ with probability } \frac{1}{2} \\ \ldots, (-4, 1), (-3, 2), (-1, 1), (0, 2), (2, 1), (3, 2), (5, 1), \\ \qquad \ldots, \text{ otherwise.} \end{cases}$$

This process corresponds to a sequence of interarrival times $\{\ldots, 2, 1, 2, 1, 2, 1, \ldots\}$. The process $\mathbf{N}_0$ looks the same from any point if one chooses the first interarrival time to be equal to 1 with probability one half, and equal to 2 otherwise.

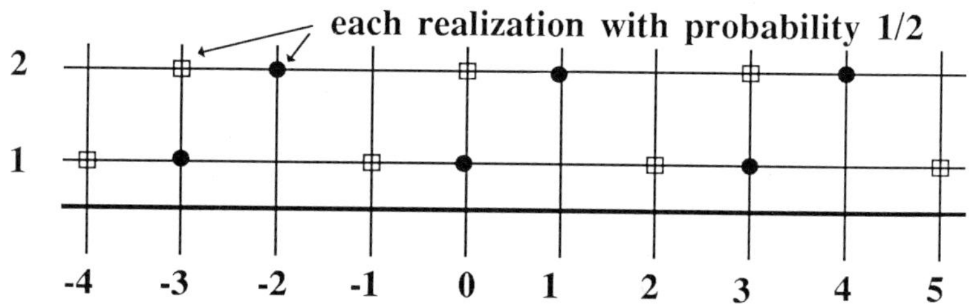

Figure 7.3. The process of Example 7.2.11.

To determine the distribution of $\mathbf{N}$, let us consider (7.3) and choose $f(\mathbf{N}) = 1\{y_1 = 1, S_1 \le t\}$. One finds

$$P\{y_1 = 1, S_1 \le t\} = \lambda E\{\int_0^{T_1} 1\{x_1 = 1, \ T_1 - v \le t\}\, dv$$

$$= \lambda \frac{1}{2} \int_0^2 1\{2 - v \le t\}\, dv = \frac{\lambda}{2}\{2 \wedge t\} = \frac{1}{3}\{2 \wedge t\}$$

where the second equality comes from observing that $x_1 = 1$ with probability one-half and that T_1 is then equal to 2, and the last equality follows from $\lambda^{-1} = ET_1 = \frac{1}{2}\{1 + 2\}$.

Similarly, one finds that

$$P\{y_1 = 2, \ S_1 \le t\} = \frac{1}{3}\{1 \wedge t\}.$$

This completely specifies the distribution of $\mathbf{N}$ since (S_1, y_1) determines the realization of $\mathbf{N}$. In particular, notice that $P\{W \le t\} = \frac{1}{3}\{1 \wedge t\} + \frac{1}{3}\{2 \wedge t\}$ and therefore $EW = \frac{5}{6}$, which is larger than $\frac{1}{2} ET_1 = \frac{3}{4}$. Of course, what is happening is that the typical time is twice as likely to fall in an interarrival epoch of duration 2 than in an interarrival epoch of duration 1. This observation explains the formulas derived above.

Example 7.2.12 A system has M possible states denoted by $\{1, 2, \ldots, M\}$. Given that the system just entered state i, it stays in that state for a random time independent of the past evolution and with distribution F_i and then jumps to state j with probability r_{ij}, for $1 \le i \ne j \le M$, independently of the past evolution. This process is called a **semi-Markov process.** It is illustrated in Figure 7.4.

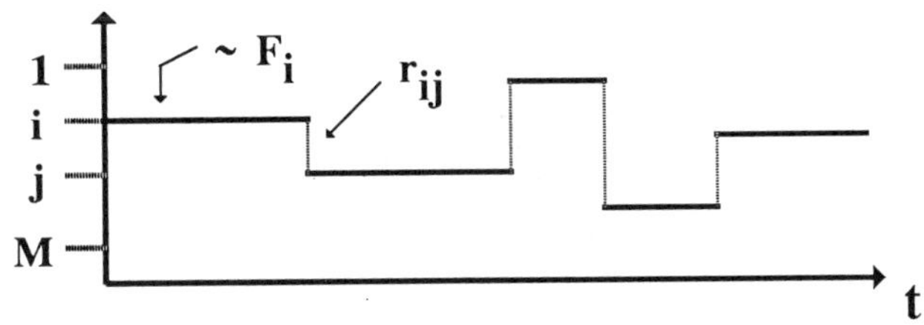

Figure 7.4. The semi-Markov process.

The process $\mathbf{N}_0$ is constructed by choosing x_0 in $\{1, 2, \ldots, M\}$ according to some probability distribution denoted by π. Thus, $\pi(i) = P\{x_0 = i\}$ for $1 \le i \le M$. The precise choice of π will be discussed below. One then decides that $T_{n+1} - T_n$ has distribution F_i if $x_n = i$.
One finds that

$$P\{x_1 = i\} = \sum_{j=1}^{M} P\{x_0 = j, x_1 = i\}$$

$$= \sum_{j=1}^{M} P\{x_0 = j\} P[x_1 = i \mid x_0 = j] = \sum_{j=1}^{M} \pi(j) r_{ji} .$$

For $\mathbf{N}_0$ to be synchronous it is certainly necessary that $P\{x_1 = i\} \equiv P\{x_0 = i\}$, i.e., that

$$\pi(i) = \sum_{j=1}^{M} \pi(j) r_{ji}, \quad i = 1, \ldots, M .$$

Also, this condition is sufficient since knowing x_m determines the law of evolution of the process $\mathbf{N}_0$ viewed from time T_m. In particular, $ET_1 = \sum \pi(i) \lambda_i^{-1}$ where λ_i^{-1} is the mean value of F_i.

Let us calculate $P\{y_0 = i, S_1 \leq t\}$. According to (7.3) with $f(\mathbf{N}) = 1\{y_0 = i,\ S_1 \leq t\}$, one finds (with $\lambda^{-1} := ET_1$)

$$\begin{aligned} P\{y_0 = i, S_1 \leq t\} &= \lambda E \int_0^{T_1} 1\{x_0 = i,\ T_1 - v \leq t\}\, dv \\ &= \lambda E[\{t \wedge T_1\}\, 1\{x_0 = i\}] \\ &= \lambda \pi(i)\, E\{t \wedge \tau_i\} \end{aligned}$$

where τ_i has distribution F_i.

The above calculation shows that $P\{y_0 = i\} = \lambda \pi(i) \lambda_i^{-1}$, which should be intuitively clear. This allows us to write the last equation above as

$$\begin{aligned} P\{y_0 = i,\ S_1 \leq t\} &= P\{y_0 = i\}\lambda_i E\{t \wedge \tau_i\} \\ &= P\{y_0 = i\}\, P\{W_i \leq t\} \end{aligned}$$

where W_i is the equilibrium forward recurrence time corresponding to F_i. See Fact 7.2.5.

Two points can be made about the above calculations. First, the probability that the system is in state i at a typical time (i.e., $P\{y_0 = i\}$) depends on the distributions F_j only through their mean values λ_j^{-1}. This is called an **insensitivity** result. Second, the last formula shows that the insensitivity is accompanied by a form of "product form" stationary distribution. (See section 3.9.)

Example 7.2.13 A bus company has to **dispatch buses** to K bus stops. One bus reaches one of the bus stops every minute, and a fraction p_k of the buses must be sent to stop number k $(1 \leq k \leq K)$. The problem is to find the dispatching policy that will minimize the average waiting time of a typical customer who is equally likely to go to any of the K bus stops.

One policy would be to send each bus to stop number k with probability p_k, independently of the others $(1 \leq k \leq K)$. Let us consider the resulting situation at bus stop number k. The process $\mathbf{N}_0$ that describes the bus arrival process at stop number k is $\mathbf{N}_0 = \{T_n, n \in \mathbb{Z}\}$, since the marks are irrelevant. Here, $T_0 = 0$, and the next arrival time T_1 is such that

$$P\{T_1 = n\} = (1 - p_k)^{n-1} p_k, \quad n \geq 1$$

since $T_1 = n$ if $n-1$ buses are sent to other bus stops and the nth bus is sent to stop number k. Using (7.6), one finds that the average waiting time for a bus at stop number k is given by

$$\frac{\lambda}{2} E(T_1)^2 = \frac{p_k}{2} \frac{2 - p_k}{p_k^2} = p_k^{-1} - \frac{1}{2}.$$

In the particular case where $p_k \equiv \frac{1}{K}$, one finds that the average waiting time is equal to $K - \frac{1}{2}$.

On the other hand, the simpler policy that would send the buses in a **round-robin** manner, i.e., that would send the buses successively to the bus stops $1, 2, \ldots, K, 1, 2, \ldots, K$, etc., corresponds to $T_1 \equiv K$ and therefore to an average waiting time equal to $\frac{K}{2}$. For $K = 2$, the randomized policy yields an average waiting time that is 50 percent larger than the round-robin policy. For K large, the waiting time of the randomized policy is about twice as large as that of the round-robin policy.

Notice that (7.7) shows that EW is minimized by a policy that corresponds to a constant T_1 (since the mean value of T_1 for bus stop number k is known to be p_k^{-1}). This minimum is achieved by the round-robin policy in the case where $p_k \equiv \frac{1}{K}$.

For a general set of fractions $(p_1, p_2, \ldots, p_K)$, it is not possible to realize a constant T_1 simultaneously at all the bus stops. In fact, the optimal dispatching policy is not known in general. However, a good policy can be constructed as follows.

Choose a large integer $M \gg 1$ and let $\{M_1, \ldots, M_K\}$ be positive integers such that $M_k \approx Mp_k$ and $M_1 + \cdots + M_K = M$. For $1 \leq m \leq M$, let a_m denote the decimal part of $m(\frac{\sqrt{5} - 1}{2}) \approx m \times 0.62$. Rank these numbers in increasing values as $(a_{m_1}, a_{m_2}, a_{m_3}, \ldots, a_{m_M})$. The policy then dispatches the buses as follows. Buses $m_1, m_2, m_3, \ldots, m_{M_1}$ are sent to stop number 1, buses $m_{M_1+1}, \ldots, m_{M_1+M_2}$ are sent to stop number 2, ..., buses $m_{M_1+\cdots+M_{K-1}+1}, \ldots, m_M$ are sent to stop number K. Bus number n is dispatched as bus m if $|n - m|$ is a multiple of M.

An example of this policy (called the **Golden Ratio Policy)** is worked out next in the case

$$K = 5, \quad p_1 = 10\%, \quad p_2 = p_3 = 15\%, \quad p_4 = 25\%, \quad p_5 = 35\%.$$

Let $M = 20$, $M_1 = 2$, $M_2 = M_3 = 3$, $M_4 = 5$, $M_5 = 7$. The calculation of the a_m leads to the following sequence of dispatchings:

$$4, 2, 5, 4, 1, 5, 3, 5, 4, 2, 5, 3, 1, 5, 3, 5, 4, 2, 5, 3$$

which is then repeated periodically.

The rationale behind this policy is that the successive interarrival times at any given bus stop will differ very little.

7.3 STATIONARITY

This section has to do with the **typical behavior of a stochastic process.** Say that a given system is modeled by a stochastic process $\mathbf{x} := \{x_t, t \in \mathbb{R}\}$. One is interested in calculating the average value of functions $f(\mathbf{x})$. For instance, x_t could be the "state" of some queueing system at time t and $f(\mathbf{x})$ could be the number of customers in the system at time $t = 0$, or the time spent in the system by the first customer to enter the system after $t = 0$. (The latter is some function of the evolution of the system on $(-\infty, +\infty)$, i.e., of $\mathbf{x}$, provided that the state x_t is suitably defined.)

Denote by $P\{.\}$ the probability distribution of $\mathbf{x}$ and by $E\{.\}$ the expectation with respect to $P\{.\}$. Let also $\mathbf{x} + t := \{x_{t+s}, s \in \mathbb{R}\}$ denote the process $\mathbf{x}$ viewed from time t (for $t \in \mathbb{R}$).

The following three questions are important.

(1) Does $E\{f(\mathbf{x} + t)\}$ depend on t for some function f? In other words, is it essential to set one's clock exactly in order to perform the measurements? If not, the process $\mathbf{x}$ is said to be **stationary.**

(2) Can one determine $E\{f(\mathbf{x})\}$ by observing a single realization of the process $\mathbf{x}$, or does one have to observe the "ensemble" of realizations generated according to $P\{.\}$? The **ergodic** processes have the former desirable property. This has obvious implications in terms of possible simulation experiments, for instance. Thus, if one builds a large number of equivalent ergodic systems and if one of them is being monitored, then it should reveal the characteristics of all the others.

(3) Assuming that the answer to the first question is affirmative, is it the case that $E\{f(\mathbf{x}+t)\}$ approaches some limit? That is, does the process $\mathbf{x}$ approach some "steady-state," independent of the initial value x_0, as time evolves? If the answer is affirmative, then the process is said to be **asymptotically stationary** and it makes sense to investigate the corresponding limiting behavior. The implication is then that there is one behavior of interest, instead of a collection of transient behaviors.

We first consider an example that should make the nature of the answers more intuitive. After discussing these ideas for general processes, we will particularize them to the case of marked point processes.

Example 7.3.1 [Renewal Process] Consider the following simplified model of "waiting for a bus." Buses arrive at times $\mathbf{N} = \{S_n, n \in \mathbb{Z}\}$. As in section 7.2, $\mathbf{N}$ is assumed to be the stationary version of some synchronous process $\mathbf{N}_0 = \{T_n, n \in \mathbb{Z}\}$ with $T_0 = 0$. For simplicity it will be assumed that the interarrival times $T_{n+1} - T_n$ are i.i.d. with some distribution F. It is assumed that $0 < ET_1 < \infty$. A point process $\mathbf{N}_0$ with i.i.d. interarrival times is called a **renewal process.**

As discussed in section 7.2, the distribution P of $\mathbf{N}$ can be interpreted as that of $\mathbf{N}_0$ viewed from some randomly chosen time. Clearly, the process $\mathbf{N}_0$ viewed from time t, i.e., the process $\mathbf{N}_0 + t$ has a distribution that depends on t. For instance, the process $\mathbf{N}_0 + t$ has a point at $-t$. It may be, however, that the distribution of $\mathbf{N}_0 + t$ converges as $t \to \infty$.

Denote by W_t the first positive point of $\mathbf{N}_0 + t$. That is, W_t is the waiting time of a customer who arrives at time t if the bus arrival times are $\mathbf{N}_0$. Let us try to understand when the distribution of W_t can be expected to converge as $t \to \infty$.

First notice that if the distribution F of the interarrival times is **arithmetic** i.e., if it is concentrated on the set multiples of some positive number α, then the distribution of W_t cannot converge. Indeed, in that case, all the T_n must be multiples of α, so that the same will be true of W_t if and only if t is a multiple of α. (For W_t is some $T_n - t$.)

It turns out that if F is nonarithmetic, then the distribution of W_t converges to the equilibrium forward recurrence time distribution corresponding to F. (See Fact 7.2.5 for the definition.) The intuitive meaning is any time $t \gg 1$ is essentially typical, in the sense of section 7.2. Thus, the distribution of $\mathbf{N}_0 + t$ approaches that of $\mathbf{N}$ as $t \to \infty$ if and only if F is nonarithmetic. (For a nice proof, see Lindvall (1977).)

The implication of this result is that the distribution of $\mathbf{N}_0$ started at an arbitrary random time τ (independent of $\mathbf{N}_0$) approaches the distribution of $\mathbf{N}$ when viewed after a long time.

In conclusion, **a renewal process with a nonarithmetic interarrival distribution** that has a finite mean **is asymptotically stationary.**

Definition 7.3.2 [Ergodicity] Let A be a set of trajectories of a random process $\mathbf{x}$. The set A is said to be **invariant** if $\mathbf{x} + t \in A$ for all $t \in \mathbb{R}_+$ whenever $\mathbf{x} \in A$.

For instance, the set

$$A = \{\mathbf{x} \mid x_t \text{ spends a finite time in } V\}$$

is invariant for any set $V \subset X$. Similarly,

$$A = \{\mathbf{x} \mid \lim_{T \to \infty} \frac{1}{T} \int_0^T f(\mathbf{x} + t)\, dt = 0\}$$

is invariant. The set

$$\{\mathbf{x} \mid x_t \in V \text{ for some } t \in [0, 1]\}$$

is not invariant.

The process $\mathbf{x}$ is said to be **ergodic** if any **invariant set** A of realizations is such that

$$P(A) = 0 \text{ or } P(A) = 1 . \tag{7.16}$$

(Notice that a process may be ergodic without being stationary, as is the case for any deterministic process.)

The central result about these processes is **Birkhoff's individual ergodic theorem** stated below.

Theorem 7.3.3 Let $\mathbf{x}$ be a **stationary** process with distribution P and f a function such that $E\{\,|f(\mathbf{x})|\,\}$ is finite.

(a) Then, the following **limit exists** for all possible realizations and defines a random variable:

$$\lim_{T \to \infty} \frac{1}{T} \int_0^T f(\mathbf{x} + t)\, dt$$

(b) Moreover, if the process is also **ergodic,** then

$$\lim_{T \to \infty} \frac{1}{T} \int_0^T f(\mathbf{x}' + t)\,dt = E\{f(\mathbf{x})\} \tag{7.17}$$

for all possible realizations $\mathbf{x}'$ of $\mathbf{x}$.

Similar definitions and considerations apply mutatis mutandis to discrete-time processes, i.e., to random sequences $\{y_n, n \in \mathbb{Z}\}$. The intuitive meaning of ergodicity is that all the possible realizations behave similarly, in the long term.

There is a simple consequence of (7.17) that will be used repeatedly. It is stated next.

Fact 7.3.4 Let $\mathbf{x}$ be an ergodic process and assume that f is such that

$$f(\mathbf{x} + t) = f(\mathbf{x}), \text{ for all } t \in \mathbb{R} \tag{7.18}$$

and for all possible realization of $\mathbf{x}$.

Then

$$f(\mathbf{x}) = C, \text{ for all } \mathbf{x} \text{ for some constant } C\ . \tag{7.19}$$

The point of this fact is that the value of f in (7.18) could a priori depend on the realization, but that the ergodicity makes that impossible.

The proof is obvious since (7.17) implies that $f(\mathbf{x}') = E\{f(\mathbf{x})\}$ for all possible $\mathbf{x}'$. Alternatively, one could argue that $A_c = \{\mathbf{x} \mid f(\mathbf{x}) \le c\}$ is invariant for all $c \in \mathbb{R}$, so that $P\{A_c\} \in \{0, 1\}$ for all $c \in \mathbb{R}$. One concludes by letting $C = \min\{c \mid P\{A_c\} = 1\}$.

Simple examples of functions $f(.)$ with the property (7.18) are (assuming that the limits exist)

$$f(\mathbf{x}) = \lim_{T \to \infty} \frac{1}{T} \int_0^T h(x_t)\,dt$$

$$f(\mathbf{x}) = 1\{x_n \in V \text{ for infinitely many values of } n \in \mathbb{Z}\}\ .$$

Notice that in the latter case, the value of f for an ergodic $\mathbf{x}$ must be either 0 or 1, simultaneously for all the realizations. (Again, this should be clear since f is the indicator of an invariant set.)

The intuition suggests that a process is ergodic if its distribution P cannot be written as a mixture of distinct stationary probabilities. Indeed, if that were possible, then the process would admit distinct long-term averages. This idea is justified below.

Theorem 7.3.5 Let P be stationary for $\mathbf{x}$. P is ergodic, if and only if P is **extreme**, i.e., if it is not possible to write P as

$$P = \beta P_1 + (1 - \beta)P_2 \tag{7.20}$$

for two distinct stationary probabilities P_1 and P_2 and for some $0 < \beta < 1$.

Proof: To see this, first assume that (7.20) holds and that P is ergodic. This implies that P_1 is ergodic. Indeed, if A is invariant, then either $P(A) = 1$ and then $P_1(A) = 1$, or $P(A) = 0$ and then $P_1(A) = 0$. But this implies that for any given A the set A_1 of realizations such that $\alpha(\mathbf{x}) := \lim_{T \to \infty} \frac{1}{T} \int_0^T 1\{\mathbf{x} + t \in A\} = P_1(A)$ must be such that $P_1(A_1) = 1$. Also, the set B such that $\alpha(\mathbf{x}) = P(A)$ is such that $P(B) = 1$, which also implies (by (7.20)) that $P_1(B) = 1$. Thus, for all realizations in $A_1 \cap B$ one has $\alpha(\mathbf{x}) = P(A) = P_1(A)$. Since A was arbitrary, it follows that $P_1 = P = P_2$.
Conversely, assume that P is not ergodic. There must then exist an invariant set A with $0 < P(A) < 1$. Hence one can write, for any B,

$$P(B) = P(A)\, P[B \mid A] + P(A^c)\, P[B \mid A^c]$$

which is of the form (7.20). (It is easily verified that $P_1(.) = P[. \mid A]$ is stationary and similarly for $P_2(.) = P[. \mid A^c]$.) This proves the result.

Proving that a given random process $\mathbf{x}$ is ergodic is usually not very easy. Fortunately, a lot is known about when this happens. One interesting problem, in the framework of service systems, is to show that the ergodicity of the arrivals implies the ergodicity of the state of the system. This will be discussed in the next section.

Let us first indicate that a marked point process is ergodic if and only if its Palm distribution is ergodic.

Theorem 7.3.6. Let $\mathbf{N}$ be a stationary marked point process with Palm version $\mathbf{N}_0$. Then $\mathbf{N}$ is ergodic if and only if $\mathbf{N}_0$ is ergodic, i.e., if and only if the sequence $\{(T_n, x_n), n \in \mathbb{Z}\}$ is ergodic.

Proof: It is shown in Exercise 7.6 that P is extreme if and only if P^0 is extreme. The result then follows from Theorem 7.3.5.

7.4. STABILITY OF QUEUEING SYSTEMS

A queueing system will be said to be **stable** if the distribution of its state process converges to a proper distribution that is independent of the initial condition.

This section will discuss the **stability of a few simple queueing systems.** In order to emphasize the system properties that imply stability, assumptions are kept at a minimum.

Example 7.4.1 [G/G/1 queue] Customers arrive at a single-server FCFS queue at times $\{T_n, n \in \mathbb{Z}\}$. Customer n requires a service time equal to σ_n. Define x_n to be the **waiting time of customer** n (before the start of service).
A moment of reflection shows that

$$x_{n+1} = (x_n + \xi_n)^+ =: f(x_n, \xi_n), \; n \in \mathbb{Z} \tag{7.21}$$

where

$$\xi_n := \sigma_n - (T_{n+1} - T_n) =: \sigma_n - \tau_n . \tag{7.22}$$

Define $\mathbf{N}_0 = \{\tau_n, \sigma_n, n \in \mathbb{Z}\}$. For $m \leq n$ one can solve (7.21) for x_n in terms of x_m and $\mathbf{N}_0$. Let us write the solution as

$$x_n = F(m, n; x_m, \mathbf{N}_0) . \tag{7.23}$$

Thus, $F(m, n; x, \mathbf{N}_0)$ is the value of x_n given that $x_m = x$ and that the "input" is $\mathbf{N}_0$.

Observe that (7.21) implies that $F(m,n; x, \mathbf{N}_0)$ is nondecreasing and left-continuous in $x \in [0, \infty]$ for all $\mathbf{N}_0$ and all $m \leq n$. Indeed, these properties hold for the one-step transition function f in (7.21) and are conserved under composition.

Notice that saying that the sequence $\mathbf{N}_0 := \{\tau_n, \sigma_n, n \in \mathbb{Z}\}$ is stationary is equivalent to saying that the marked point process $\mathbf{N}_0$ is synchronous.

Fact 7.4.2 [Limiting distribution when $x_0 = 0$] Assume that $x_0 = 0$ and that $\mathbf{N}_0$ is stationary. Then the distribution of x_n converges as $n \to \infty$ to that of a random variable ϕ which may be infinite.

Proof: Observe that

$$F(m-1, 0; 0, \mathbf{N}_0) = F(m, 0; f(0, \xi_{m-1}), \mathbf{N}_0)$$
$$\geq F(m, 0; 0, \mathbf{N}_0), \quad m \leq 0 \tag{7.24}$$

and also that

$$F(-n, 0; 0, \mathbf{N}_0) \text{ and } x_n = F(0, n; 0, \mathbf{N}_0) \tag{7.25}$$

have the same distribution for $n \geq 0$.

Indeed, (7.24) follows from the fact that $F(m, 0; x, \mathbf{N}_0)$ is nondecreasing in x. The equality of the distributions of the terms in (7.25) is a consequence of the stationarity of $\mathbf{N}_0$. The inequality (7.24) implies that $F(m, 0; 0, \mathbf{N}_0)$ converges to a random variable ϕ as $m \downarrow -\infty$, with $\phi := \lim_{m \downarrow -\infty} F(m, 0; 0, \mathbf{N}_0)$. Also, (7.25) shows that the distribution of x_n converges to that of ϕ as $n \to \infty$.

Fact 7.4.3 [Stationary distribution] If one chooses $x_0 = \phi$, then $\{x_n, n \geq 0\}$ is stationary.

Proof: It follows from the definition of F in (7.23) that

$$F(m, n; 0, \mathbf{N}_0) = F(0, n; F(m, 0; 0, \mathbf{N}_0), \mathbf{N}_0) \quad \text{for } m < 0 \leq n .$$

Letting $m \downarrow -\infty$ one finds, by left-continuity of F and with $\phi(n, \mathbf{N}_0) := \lim_{m \downarrow -\infty} F(m, n; 0, \mathbf{N}_0)$,

$$\phi(n, \mathbf{N}_0) = F(0, n; \phi, \mathbf{N}_0), \quad n \geq 0 .$$

Hence,

$$\{\phi(n + l, \mathbf{N}_0), 1 \leq l \leq L\}$$
$$= \{F(0, n + l; \phi, \mathbf{N}_0), 1 \leq l \leq L\}. \tag{7.26}$$

Now,

$$\{\phi(n + l, \mathbf{N}_0), 1 \leq l \leq L\} = \{\phi(l, \mathbf{N}_0 + n), \quad 1 \leq l \leq L\}$$

where $\mathbf{N}_0 + n := \{\xi_{m+n}, m \in \mathbb{Z}\}$. Since $\mathbf{N}_0$ is stationary, it follows that the distribution of the left-hand side of (7.26) is stationary (i.e., independent of $n \geq 0$). The equality (7.26) shows that this stationary process is precisely the process $\{x_n, n \geq 0\}$ when $x_0 = \phi = \phi(0, \mathbf{N}_0)$.

Fact 7.4.4 [Finiteness of ϕ] When is ϕ **finite**? A partial answer is given below. Let $\mathbf{N}_0$ be stationary and ergodic and such that the expectation of ξ_n is well defined. Then

$$P\{\phi < \infty\} = \begin{cases} 1 \text{ if } E\{\xi_n\} < 0 \\ 0 \text{ if } E\{\xi_n\} > 0 \,. \end{cases} \tag{7.27}$$

The reader is referred to the literature for a discussion of the "critical" case $E\{\xi_n\} = 0$.

Proof: First, one uses the explicit form of f to verify by induction that

$$F(-n, 0; 0, \mathbf{N}_0) \tag{7.28}$$
$$= \max\{0, \xi_{-1}, \xi_{-1} + \xi_{-2}, \ldots, \xi_{-1} + \cdots + \xi_{-n}\},$$

for $n \geq 1$.
Therefore,

$$\phi = (\sup_{n \geq 1} \sum_{m=1}^{n} \xi_{-m})^{+} \,. \tag{7.29}$$

Second, one uses the ergodicity and stationarity of $\mathbf{N}_0$ to conclude that (with probability one)

$$\lim_{n \to \infty} \frac{1}{n} \sum_{m=1}^{n} \xi_{-m} = E\{\xi_0\} \,.$$

This shows that (with probability one)

$$\lim_{n \to \infty} \sum_{m=1}^{n} \xi_{-m} = \begin{cases} +\infty \text{ if } E\{\xi_0\} > 0 \\ -\infty \text{ if } E\{\xi_0\} < 0 \,. \end{cases} \tag{7.30}$$

Therefore, if $E\{\xi_0\} > 0$, then $P\{\phi = \infty\} = 1$. In the case $E\{\xi_0\} < 0$, (7.30) shows that $P\{\phi < \infty\} = 1$. Indeed, a sequence that converges to $-\infty$ must have a finite supremum.

Theorem 7.4.5 [Stability] Assume that $\mathbf{N}_0$ is ergodic and stationary and that $E\{\xi_n\} < 0$. Then the process $\mathbf{x} := \{x_n, \quad n \geq 0\}$ is **stable.** That is, the distribution of x_n converges as $n \to \infty$ to that of the finite random variable ϕ, independently of the initial condition x_0.

Also, the process $\{x_n, n \geq 0\}$ eventually agrees with the stationary process $\{\phi(n, \mathbf{N}_0), n \geq 0\}$. That is, for all realizations there is a finite n_0 such that $x_n = \phi(n, \mathbf{N}_0)$ for all $n \geq n_0$. Moreover, $\{\phi(n, \mathbf{N}_0), n \in \mathbb{Z}\}$ is the only stationary process with finite values that satisfies (7.21).

Proof: For the first two statements it suffices, in view of Facts 7.4.2–7.4.4, to show the eventual independence of the initial condition. This is done by proving that the system (7.21) is **forgetful,** i.e., that for all $x > 0$ there is a finite $n(x, \mathbf{N}_0)$ such that

$$F(0, n; x, \mathbf{N}_0) = F(0, n; 0, \mathbf{N}_0) \text{ for } n \geq n(x, \mathbf{N}_0) . \quad (7.31)$$

Thus, after a finite number of arrivals, the sequence of waiting times is the same when $x_0 = x$ as when $x_0 = 0$. It follows that the limiting distribution of x_n is independent of x_0.

To show (7.31), it suffices to show that $F(0, n; x, \mathbf{N}_0) = 0$ for some finite $n = n(x, \mathbf{N}_0) > 0$. Indeed, one has $0 \leq F(0, n; 0, \mathbf{N}_0) \leq F(0, n; x, \mathbf{N}_0)$ for all $n \geq 0$.
Let then $x_0 = x$ and assume that $x_n > 0$ for all $n \geq 0$. Then (7.21) shows that

$$x_{n+1} = x_n + \xi_n, \quad n \geq 0 ,$$

so that

$$x_n = x + \sum_{m=0}^{n-1} \xi_m, \quad n > 0 . \quad (7.32)$$

This is clearly impossible, since the right-hand side of (7.32) goes to $-\infty$ as $n \to \infty$. (See the argument leading to (7.30).) Hence it must be that x_n is eventually equal to zero, which concludes the proof of the first two statements.

To prove the uniqueness, assume that $\mathbf{x} := \{x_n, \ n \geq 0\}$ is a finite and stationary process that solves (7.21). Since the system is forgetful, it follows that $x_n = \phi(n, \mathbf{N}_0)$ for all n larger than some finite value. Therefore, $| x_n - \phi(n, \mathbf{N}_0) |$ converges almost surely to zero. However, this random variable has a distribution that does not depend on n. Hence, it must be that $x_n \equiv \phi(n, \mathbf{N}_0)$ for all n, as was to be shown.

Comments 7.4.6 [Summary of arguments]

(a) The **existence** of a stationary process that satisfies (7.21) was shown to be a consequence of the one step transition function f in (7.21) being nondecreasing and left-continuous.

(b) The **finiteness** of that distribution when $E\{\xi_n\} < 0$, i.e., when $E\{\sigma_n\} < E\{\tau_n\}$, follows from the observation that $\phi = \sup\{0, \xi_{-1}, \xi_{-1} + \xi_{-2}, \xi_{-1} + \xi_{-2} + \xi_{-3}, \ldots\}$ and the fact that $\sum_{1}^{n} \xi_{-m}$ converges to $-\infty$ and is therefore bounded from above.

(c) The limiting distribution of x_n does not depend on x_0 when $E\{\xi_n\} < 0$ because x_n must become empty eventually for any value of x_0. Thus, the process x_n is **forgetful.** This also implies the uniqueness of a finite stationary sequence of waiting times.

Example 7.4.7 [G/G/1 queues in parallel] The system illustrated in Figure 7.5 consists of J queues in parallel.

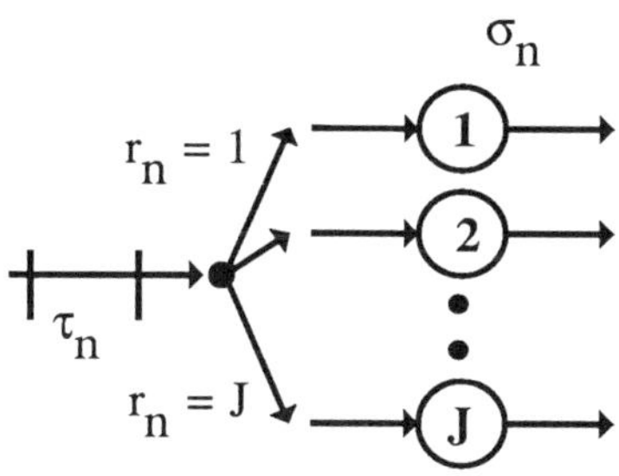

Figure 7.5. G/G/1 queues in parallel.

The arrival process is described by a sequence $\mathbf{N}_0 = \{(\tau_n, r_n, \sigma_n), n \geq 0\}$ where τ_n is the $(n+1)$st interarrival time, r_n indicates the queue to which customer number n is sent, and σ_n is the service time of that customer.

Let $x_n = (x_n^1, \ldots, x_n^J)$ where x_n^i is the **work-to-be-done** in queue i just before the arrival time of customer number n. Thus,

$$x_{n+1}^i = (x_n^i - \tau_n + \sigma_n 1\{r_n = i\})^+, \quad n \geq 0, i = 1, \ldots, J . \tag{7.33}$$

Theorem 7.4.8 [Stability of G/G/1 queues in parallel] Assume that $\{(\tau_n, r_n, \sigma_n), n \geq 0\}$ is ergodic and stationary with

$$E\{\sigma_n 1\{r_n = i\}\} < E\{\tau_n\}, \quad i = 1, 2, \ldots, J. \tag{7.34}$$

Then $\mathbf{x} := \{x_n, n \geq 0\}$ is **stable,** i.e., the distribution of $\{x_{n+l}, l \geq 0\}$ converges to the distribution of a finite stationary process, independently of x_0.

Proof: It is clear from Theorem 7.4.5 that each x_n^i is stable. The convergence of marginal distributions is in general not sufficient to imply the convergence of the joint distribution of x_n. The argument goes as follows.

One can write the equations (7.33) as

$$x_{n+1} = f(x_n, \xi_n), \quad \text{with } \xi_n = (\tau_n, r_n, \sigma_n).$$

The function $f(x, \xi)$ is nondecreasing and left-continuous in x. Hence, if $x_0 = 0$, the distribution of x_n converges to that of some random variable ϕ. We know from Theorem 7.4.5 that ϕ is **finite.**

It remains to show that the system is **forgetful,** i.e., that (7.31) holds for some finite n (with F being the transition function for the system (7.33)). This cannot be done by showing that x_n is eventually 0 for any given x_0. Indeed, this may fail to be the case, as simple examples show. (See Exercise 7.7.)

As seen in Theorem 7.4.5, each queue will eventually go empty. Thus, there is a finite n_i such that

$$F^i(0, n; x, \mathbf{N}_0) = F^i(0, n; 0, \mathbf{N}_0) \tag{7.35}$$

for $n = n_i$ where $F^i(0, n, x, \mathbf{N}_0)$ is the value of x_n^i given that $x_0 = x$. It follows that (7.35) also holds for $n \geq n_i$. Indeed, given $x_{n_i}^i$, the evolution of x_n^i for $n \geq n_i$ is independent of $x_{n_i}^j$ for $j \neq i$.

Therefore,

$$F(0, n; x, \mathbf{N}_0) = F(0, n; 0, \mathbf{N}_0) \text{ for all } n \geq \max\{n_i, 1 \leq i \leq J\},$$

which shows that the system (7.33) is forgetful and concludes the proof of the theorem.

Example 7.4.9 [G/G/1 queue with feedback] In the system in Figure 7.6 the customers arrive at times T_n. Let $\tau_n = T_{n+1} - T_n$ for $n \geq 0$. Customer number n wishes to visit the queue ν_n times, with the successive service times $(\sigma_n^1, \ldots, \sigma_n^{\nu_n})$. Let x_n be the total **work-to-be-done** at time T_n-. (Thus, the system would become empty at time $T_n + x_n$ if no customers were to arrive at times T_m for $m \geq n$.)

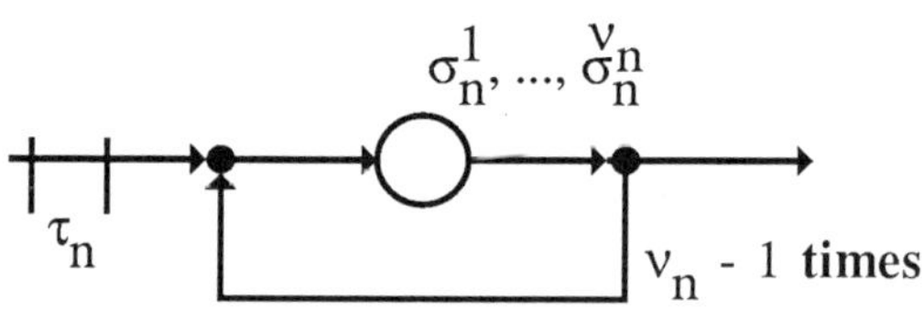

Figure 7.6. The G/G/1 queue with feedback.

Theorem 7.4.10 [Stability of the G/G/1 queue with feedback] Assume that $\{(\tau_n, \nu_n, \sigma_n^1, \ldots, \sigma_n^{\nu_n}), n \geq 0\}$ is ergodic and stationary with $E\{\sum_{m=1}^{\nu_n} \sigma_n^m\} < E\{\tau_n\}$. Then x_n is **stable.**

Proof: Since the work-to-be-done does not depend on the service discipline (as long as the server works at rate 1 while there is work to be done), the work-to-be-done is the same as that in a G/G/1 queue with arrival times T_n and service times $\sigma_n = \sigma_n^1 + \cdots + \sigma_n^{\nu_n}$. The result then follows from Theorem 7.4.5.

Example 7.4.11 [G/G/1 queues in tandem] The input process is $\{(\tau_n, \sigma_n^1, \sigma_n^2), n \geq 0\}$, where τ_n is the interarrival time between the nth and $(n+1)$st customers and σ_n^i is the service time in queue i $(i = 1, 2)$ of customer n. (See Figure 7.7.)

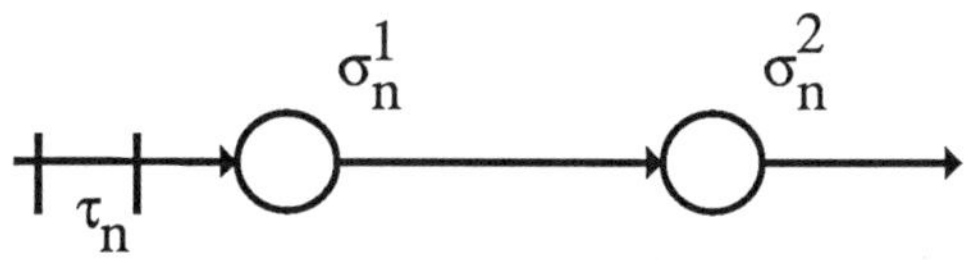

Figure 7.7. G/G/1 queues in tandem.

One might be tempted to conjecture that if the input process is ergodic and if $E\{\tau_n\} > E\{\sigma_n^i\}$ for $i = 1, 2$, then the system will be empty after a finite time. A simple counterexample with $\tau_n = 3$ and $\sigma_n^1 = \sigma_n^2 = 2$ shows that is not necessarily the case.

Define x_n^i to be the **waiting time** of customer n until the start of service in queue i (for $i = 1, 2$).

Theorem 7.4.12 [Stability of G/G/1 queues in tandem] Assume that $\{(\tau_n, \sigma_n^1, \sigma_n^2), \ n \geq 0\}$ is ergodic and stationary with

$$E\{\sigma_n^i\} < E\{\tau_n\}, \quad i = 1, 2 .$$

Then $x_n = (x_n^1, x_n^2)$ is **stable.**

Proof: The convergence to some ϕ, the finiteness of ϕ and the forgetfulness of $\{x_n\}$ are immediate consequences of the second part of Theorem 7.4.5.

Indeed, x_n^1 **eventually agrees with the stationary process** for the first queue, denoted by $\phi^1(n, \mathbf{N}_0)$, so that the **arrival process** into the second queue **eventually agrees** with the stationary and ergodic process with arrival times $S_n := T_n + \phi^1(n, \mathbf{N}_0) + \sigma_n^1$.

Thus, the second queue behaves eventually as a G/G/1 queue with a stationary arrival process. One can then apply Theorem

7.4.5 to that second queue. The only thing to prove is that

$$E\{\sigma_n^2\} \;<\; E\{S_{n+1} - S_n\}\,.$$

Now, it should be intuitively clear that $E\{S_{n+1} - S_n\} = E\{\tau_n\}$. The proof of this fact is left as an exercise.

Remark 7.4.13 The above discussion shows that the process x_n eventually agrees with a stationary process $\phi(n, \mathbf{N}_0)$. Therefore, the result of Theorem 7.4.12 extends to an arbitrary finite number of queues in tandem.

The reader is invited to prove that the same result extends to acyclic (i.e., without loops) networks of G/G/1 queues where the routing is defined as in the case of parallel G/G/1 queues. The same ideas can also be shown to apply to **fork-and-join** networks.

7.5. LITTLE'S RESULT

As we have seen in section 3.4, Little's result relates an average value of quantity observed at a **typical time** to the average value of a quantity observed by a **typical customer.**

One can then expect this property to be a relationship between a stationary process and its Palm version. This section will explain the connection.

Example 7.5.1 [Sojourn time in a G/G/1 queue] Consider the G/G/1 queue of Example 7.4.1 and assume that the conditions of Theorem 7.4.5 are satisfied.

Let L be the long term average number of customers in the queue and W be the average sojourn time per customer. Thus,

$$L \;=\; \lim_{T \to \infty} \frac{1}{T} \int_0^T L_t dt \tag{7.36}$$

where L_t is the number of customers in the queue at time t. Also,

$$W \;=\; \lim_{n \to \infty} \frac{1}{n} \sum_{m=1}^{n} W_m \tag{7.37}$$

where W_m is the sojourn time of customer m.

Theorem 7.5.2 [Little's result for the G/G/1 queue] Under the conditions of Theorem 7.4.5 the limits (7.36) and (7.37) exist and one has

$$L = E\{L_t\} = \lambda W = \lambda E^0\{W_0\} = E^0\{W_n\} \quad (7.38)$$

where $\lambda := (E\{\tau_0\})^{-1}$, $E\{.\}$ denotes the expectation with respect to the stationary queueing system, and $E^0\{.\}$ denotes the expectation with respect to the Palm version of the queueing system at the arrival times.

Before proving this fact we have to explain the terminology.

Comments 7.5.3 [Terminology]

(a) The expectation $E^0\{W_0\}$ is defined as

$$E^0\{W_0\} = E\{\phi + \sigma_0\}$$

where ϕ is the waiting time of customer 0 for the stationary process $\{x_n\} = \{\phi(n, \mathbf{N}_0)\}$. (See Theorem 7.4.5.)

Thus, the sequence of sojourn times $\psi_n := \phi(n, \mathbf{N}_0) + \sigma_n$ is stationary and ergodic. Recall moreover that the sequence of sojourn times W_n corresponding to an arbitrary (finite) initial condition will eventually agree with the stationary sequence. In particular, by ergodicity and stationarity,

$$\lim_{n \to \infty} \frac{1}{n} \sum_{m=1}^{n} W_m = \lim_{n \to \infty} \frac{1}{n} \sum_{m=1}^{n} \psi_m$$

$$= E\{\psi_0\} =: E^0\{W_0\} . \quad (7.39)$$

This shows that the limit exists.

(b) The stationary queueing process can be defined as follows. We know that $\{\phi(n, \mathbf{N}_0)\}$ is the stationary sequence of waiting times. Corresponding to this sequence, customer n arrives at time T_n and leaves at time $T_n + \phi(n, \mathbf{N}_0) + \sigma_n = T_n + \psi_n$. This provides a synchronous view of the queueing process observed at the arrival times T_n. One can consider this process $\mathbf{N}_0^2 := \{T_n, \psi_n\}$ as being the Palm version of a stationary point process $\mathbf{N}^2 = \{S_n, \eta_n\}$, as explained in section 7.2. Corresponding to the process $\mathbf{N}^2$ one can calculate the number L_t of customers in the queue at time t. The distribution of L_t is independent of t since $\mathbf{N}^2$ is stationary. Hence the meaning of $E\{L_t\}$ in (7.38). The equality (7.36) follows by ergodicity.

Proof of Theorem 7.5.2 The starting point is Campbell's Formula (7.15). For the processes $\mathbf{N}_0^2$ and $\mathbf{N}^2$ defined above, this formula reads

$$E\{\sum_{n\in\mathbb{Z}} f(S_n, \eta_n)\} = \lambda \int_{-\infty}^{\infty}\int_{0}^{\infty} f(t,x)\, P\{\psi_0 \in (x, x+dx)\}\, dt \,. \quad (7.40)$$

Let $f(t,x) = 1\{t \le 0, x \ge -t\}$. Then

$$\sum_{n\in\mathbb{Z}} f(S_n, \eta_n) = \sum_{n\in\mathbb{Z}} 1\{S_n \le 0, \eta_n \ge -S_n\} = L_0\,, \quad (7.41)$$

the number of customers in the stationary queue at time 0. Also,

$$\int_{-\infty}^{\infty}\int_{0}^{\infty} f(t,x) P\{\psi_0 \in (x, x+dx)\}\, dt = \int_{-\infty}^{0} P\{\psi_0 \ge -t\}\, dt$$

$$= \int_{0}^{\infty} P\{\psi_0 \ge t\}\, dt = E\{\psi_0\}\,. \quad (7.42)$$

Substituting the expressions (7.41) and (7.42) in (7.40) gives

$$E\{L_0\} = \lambda E\{\psi_0\}\,.$$

Thus, the (time) stationary mean queue length is equal to the (customer) stationary waiting time times the average arrival rate.

Example 7.5.4 [Sojourn times in networks] Consider a stationary ergodic queueing network. Customers arrive at times $\{S_n, n \in \mathbb{Z}\}$. Customer n faces a waiting time η_n and will leave the network at time $S_n + \eta_n$. Let $\mathbf{N}_0 = \{T_n, \psi_n\}$ be the Palm version of the stationary point process $\mathbf{N} = \{S_n, \eta_n, n \in \mathbb{Z}\}$.

The argument given above shows that

$$L = E\{L_0\} = \lambda W = \lambda E\{\psi_0\}$$

where

$$L_0 = \sum_{n} 1\{S_n \le 0, \eta_n \ge -S_n\}\,.$$

Also, (7.36) and (7.37) hold.

A simple example of a stationary and ergodic network is a network described by an ergodic Markov chain. For instance, if the exogenous arrivals are described by Markov modulated Poisson processes (see 5.2.3) which may not necessarily be independent and if the service times are phase-type, then the network is described by a Markov chain. If one can show that the Markov chain is irreducible and positive recurrent, then Little's result applies. Of course, this does not mean that one can calculate L or W.

7.6. EXERCISES

7.1. Stationary point process Prove the stationarity of $\mathbf{N}$ defined by (7.3). *Hint:* Since $\mathbf{N}_0$ looks the same from any point one sees that the interval of integration in (7.3) can be replaced by any $[T_n, T_{n+1}]$. Therefore,

$$E\{f(\mathbf{N})\} = \frac{\lambda}{n} E\{\int_0^{T_n} f(\mathbf{N}_0 + v)\,dv\}, \quad n \geq 1, \tag{7.43}$$

so that

$$\begin{aligned} E\{f(\mathbf{N} + t)\} &= \frac{\lambda}{n} E\{\int_0^{T_n} f(\mathbf{N}_0 + v + t)\,dv\} \\ &= \frac{\lambda}{n} E\{\int_t^{T_n + t} f(\mathbf{N}_0 + v)\,dv\}. \end{aligned} \tag{7.44}$$

Assume that $|f| \leq 1$. Then the right-hand sides of (7.43) and (7.44) differ at most by $\frac{\lambda}{n} 2t$. Letting $n \to \infty$ shows that the left-hand sides must be equal. Hence, the probability distributions of $\mathbf{N}$ and $\mathbf{N} + t$ agree, which proves the stationarity of $\mathbf{N}$.

7.2. Calculations Prove the relations of Facts 7.2.5. *Hints:* All the relations of (a) can be proved by choosing a specific function f in (7.3). For instance, to show (7.8) one chooses $f(\mathbf{N}) = 1\{S_1 \leq t\}$, so that $f(\mathbf{N}_0 + v) = 1\{T_1 - v \leq t\}$ since $T_1 - v$ is the first point of $\mathbf{N}_0 + v$.

To derive (7.11) one starts with the next to last equality before Theorem 7.2.4. Choosing $f(\mathbf{N}) = 1\{\mathbf{N} \in A\}$ in that formula gives

$$\begin{aligned} P\{\mathbf{N} \in A\} &= \lambda \int_{u=0}^{\infty} \int_{v=0}^{u} P\{\mathbf{N}_0 + v \in A,\ T_1 \in (u, u + du)\}\,dv \\ &= \lambda \int_{v=0}^{\infty} \int_{u=v}^{\infty} P\{\mathbf{N}_0 + v \in A,\ T_1 \in (u, u + du)\}\,dv \end{aligned}$$

$$= \lambda \int_0^\infty P\{\mathbf{N}_0 + v \in A,\ T_1 > v\}\, dv$$

which is (7.11).

7.3. Conditioning on a jump Prove Theorem 7.2.7. *Hint:* Choosing $f(\mathbf{N}) = 1\{\mathbf{N} + S_1 \in B, S_1 \leq w\}$ in (7.3) gives

$$P\{\mathbf{N} + S_1 \in B, S_1 \leq w\} = \lambda E \int_0^{T_1} 1\{\mathbf{N}_0 + T_1 \in B, T_1 - v \leq w\}\, dv$$

$$= \lambda E[\{w \wedge T_1\}\, 1\{\mathbf{N}_0 + T_1 \in B\}]$$

where the first equality follows from observing that the integrand is $f(\mathbf{N}_0 + v)$ since $T_1 - v$ is the first positive point of $\mathbf{N}_0 + v$.

The same formula when B is replaced by the set of all possible realizations of $\mathbf{N}_0$ allows us to estimate $P\{S_1 \leq w\}$. Dividing those two expressions then gives

$$P[\mathbf{N} + S_1 \in B \mid S_1 \leq w] = \frac{E[\{w \wedge T_1\}\, 1\{\mathbf{N}_0 + T_1 \in B\}]}{E\{w \wedge T_1\}} .$$

One concludes by observing that the last term converges to $P\{\mathbf{N}_0 + T_1 \in B\} = P\{\mathbf{N}_0 \in B\}$ as $w \downarrow 0$. (For instance, divide numerator and denominator by w and use the fact that $\dfrac{w \wedge T_1}{w}$ converges to 1 almost surely since $T_1 > 0$, use dominated convergence.)

7.4. Exponential case Assume that $P\{T_1 > t\} \equiv e^{-\lambda t}$. Show that V and W are independent and distributed as T_1. *Hint:* Use (7.10).

7.5. Bounds Show that $P\{W \leq t\} \geq P\{T_1 \leq t\}$ for all T if and only if

$$E[T_1 - t \mid T_1 > t] \leq ET_1 \text{, for all } t.$$

Moreover, in that case,

$$P\{W \leq t\} \geq 1 - e^{-\lambda t}, \text{ for all } t \geq 0 .$$

Hint: Since $T_1 - t \wedge T_1 = (T_1 - t)1\{T_1 > t\}$, one finds that

$$E[T_1 - t \mid T_1 > t] = \frac{ET_1 - E\{t \wedge T_1\}}{P\{T_1 > t\}}.$$

One then concludes by (7.8).

7.6. Convex combinations Let P, P_1, P_2 be three stationary distributions of a marked point process **N** with respective intensities λ, λ_1, λ_2. Let P^0, P_1^0, P_2^0 be the corresponding Palm distributions. Show that, for $\beta \in (0, 1)$, one has

$$P = \beta P_1 + (1 - \beta)P_2 \text{ if and only if } P^0 = \beta\frac{\lambda_1}{\lambda}P_1^0 + (1 - \beta)\frac{\lambda_2}{\lambda}P_2^0.$$

Hint: Use (7.3)–(7.4). Note that the intuitive interpretation of this result is immediate.

7.7. Queues in parallel Give an example of two queues in parallel that satisfy the conditions of Theorem 7.4.8 and that are never simultaneously empty.

7.8. Poisson processes Let **N** be a stationary Poisson process and let $\mathbf{N}_0$ be its Palm distribution. Show that **N** has the same distribution as $\mathbf{N}_0$ from which one removes the point at $t = 0$. Show that the same property continues to hold if the marks are independent of the jump times. *Hint:* Use the Facts 7.2.5.

7.9. Poisson arrivals Consider a queueing system such that its state at time t is some functional of the arrivals before time t. Show that the stationary distribution of x_t is the same as the Palm distribution of the state just before the arrivals. (Thus, Poisson arrivals see time averages: PASTA.) *Hint:* Use the result of Exercise 7.8.

7.7. REFERENCES AND COMMENTS

Section 7.2 The standard approach is to first consider a stationary marked point process **N** and then to define $\mathbf{N}_0$ as having the conditional distribution of **N** given that there is a jump time at $t = 0$. This is formula (7.12). We chose instead to start with $\mathbf{N}_0$ and to see how the time origin should be randomized so as to have a stationary process **N**. The heuristic is justified in Exercise 7.1.

The theory of point processes is covered in Baccelli and Brémaud (1987), Franken, et al. (1982), and Neveu (1977).

Section 7.3 For a nice introduction to ergodic theory, see Breiman (1968), Chapter 6 and Doob (1953), Chapter X.

Section 7.4 This section is essentially from Loynes (1962). See also Franken, et al. (1982), Baccelli and Brémaud (1987), Neveu (1983), Loynes (1964), and sections 4.6–4.7 of Borovkov (1984).

CHAPTER EIGHT

DYNAMIC CONTROL OF QUEUEING SYSTEMS

8.1. INTRODUCTION

This chapter is a short introduction to the **optimization of queueing networks.** The emphasis is on the methodology.

The selection of rules of operations, or **control strategies**, is of significant practical importance. Indeed, many examples show that the **performance measures** of a queueing system can often be substantially improved by a careful choice of the control strategy.

Control problems can be classified in terms of the **information** on which the control decisions are based. We will discuss the **complete observation** problems where the **state** of the system is available in real time to the decision maker. Some **open-loop** problems where the decisions have to be taken without observing the evolution of the system will also be considered.

Some problems can be solved by simple **interchange arguments,** as illustrated in section 8.2. Other problems can be solved by a **forward induction** (in time) on the properties of the **state process** that corresponds to the policy that is to be shown optimal. This method is discussed in section 8.3. In principle, the **Dynamic Programming** methodology is sufficient for solving complete observation problems. The basic results of that theory (finite horizon and finite action space) are reviewed in section 8.4 and are illustrated in a number of examples. The infinite horizon problems are explained in section 8.5. For many problems, the calculations required by the dynamic programming may be prohibitive. In section 8.5 one discusses some examples where the **structure** of the optimal policy can be determined without to have to solve the dynamic programming equations. The average cost problems present some technical difficulties that are addressed in section 8.6.

8.2. INTERCHANGE ARGUMENTS

A number of control problems can be solved by simple interchange arguments. The idea is to prove the optimality of a policy by showing that any other strategy can be improved by interchanging action times. This method may require using some general results about optimal strategies, such as the existence of an optimal policy, or the existence of an optimal stationary policy. Simple examples are used to illustrate the method.

A more involved example of interchange argument will be discussed in Chapter Nine (multiarmed bandit problems).

Example 8.2.1 [Nonpreemptive service of N jobs] This is a rather trivial example to introduce the method.

There are N jobs that require independent random service times $S_1, \ldots, S_N$, respectively. These jobs have to be processed by a **single server, nonpreemptively.** That is, once the processing of a job starts, it cannot be interrupted until completion.

The problem is to find in which **order** the jobs should be processed so as to **minimize**

$$E\{\sum_{i=1}^{N} c_i T_i\} \tag{8.1}$$

where c_i is a positive constant and T_i is the time at which the processing of job i is completed ($1 \leq i \leq N$).

The expression (8.1) is the total waiting cost of the jobs given that job i pays c_i units of cost per unit of time until it is completed. The value of (8.1) depends on the order in which the jobs are performed through the T_i.

The objective function (8.1) is called the **expected weighted flowtime** in the scheduling literature. It is the **expected flowtime** if all the c_i are equal to one.

Proposition 8.2.2 [Optimality of the nonpreemptive $c\mu$-rule] The optimal policy is to process the jobs in an order that corresponds to nonincreasing values of $c_i \mu_i$ where $\mu_i := [E\{S_i\}]^{-1}$.

This is called the **nonpreemptive $c\mu$-rule.** Notice that the optimal policy depends on the service timc distributions only through their mean values. It is interesting to note that this property docs not hold for more than one server. (See Exercise 8.1.)

Proof: Let $C(i_1, \ldots, i_N)$ denote the value of (8.1) when the jobs are processed in the order $(i_1, \ldots, i_N)$, an arbitrary permutation of $(1, 2, \ldots, N)$.

Assume that there is some k with $1 \leq k < N$ and $c_{i_k}\mu_{i_k} < c_{i_{k+1}}\mu_{i_{k+1}}$. Interchanging i_k and i_{k+1} would decrease the completion time of job i_{k+1} by S_{i_k}. It would also increase the completion time of job i_k by $S_{i_{k+1}}$. The other completion times would not be affected by this interchange. Thus, one finds that

$$C(i_1, \ldots, i_{k+1}, i_k, \ldots, i_N) - C(i_1, \ldots, i_k, i_{k+1}, \ldots, i_N)$$

$$= E\{c_{i_k} S_{i_{k+1}} - c_{i_{k+1}} S_{i_k}\} = c_{i_k} \mu_{i_{k+1}}^{-1} - c_{i_{k+1}} \mu_{i_k}^{-1} < 0 .$$

This shows that the order $(i_1, \ldots, i_N)$ cannot be optimal unless it corresponds to nonincreasing values of $c_i \mu_i$.

Example 8.2.3 [Preemptive server allocation to parallel queues] The service system operates in discrete time. That is, the arrival times and the service times take values in $\{0, 1, 2, \ldots\}$.

Customers arrive at J queues in parallel at arbitrary and possibly known times. Customers in queue j require geometrically distributed service times with rate μ_j. (Recall that τ is geometrically distributed with rate $\mu \in (0, 1)$ if $P\{\tau = n\} = (1 - \mu)^{n-1}\mu$ for $n \geq 1$.)

All the service times are independent. There is a **single server** who serves the customers in the J queues. At each time, the server serves one customer in one of the queues. Service is **preemptive.** That is, at any time $n \in \{0, 1, 2, \ldots\}$, the server can interrupt the service of a customer and resume it later on. The objective is to **minimize**

$$E\{\sum_{n=0}^{N} \beta^n \sum_{j=1}^{J} c_j\, x_n^j\} \tag{8.2}$$

where the c_j and β are known positive numbers and x_n^j is the number of customers at time n in queue j. (See Figure 8.1.)

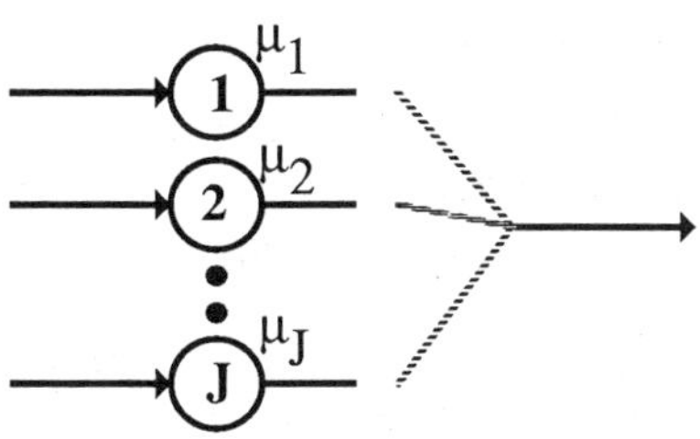

Figure 8.1. Serving parallel queues.

The **interpretation** of (8.2) is that every customer in queue j is charged c_j units of cost per unit of time and that β is the **discount rate** of the cost. (One unit of cost to be paid at time n is worth β^n units of cost at time 0.) The server allocation rule may depend on all the arrival times, but the values of the uncompleted service times are not known.

Proposition 8.2.4 [Optimality of the $c\mu$-rule] The following policy (called the preemptive **$c\mu$-rule)** is optimal: at each time, the server allocates all the service effort to a customer in the first nonempty queue in decreasing order of $c_j \mu_j$ (if any).

Notice that the policy does not depend on the **number** of customers in the nonempty queues. Also, it does not depend on the **arrival times.** This policy is **preemptive.**

Proof: Assume that the $c\mu$-rule is optimal after some time $M \leq N$. (It is certainly optimal after time N.) It will be shown that it is then optimal at time M.

Consider a policy that serves a customer in queue j at time M while there is a customer in queue i at that time, and that agrees with the $c\mu$-rule after time M, where i and j are such that $c_i \mu_i > c_j \mu_j$. Denote by τ the first time after time M that the policy serves a customer in queue i. (Let $\tau = \infty$ if the policy does not serve a customer in queue i during $\{M + 1, \ldots, N\}$.)

Modify the policy by serving a customer in queue i at time M and a customer in queue j at time τ, i.e., by interchanging the actions at times M and τ. (See Figure 8.2.)

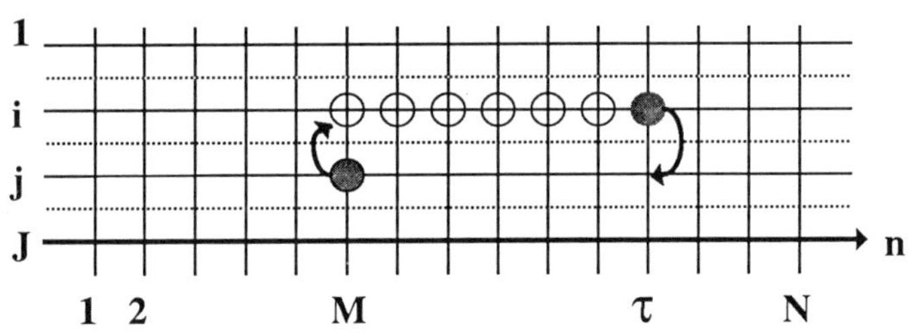

Figure 8.2. The interchange.

The effect of this modification can be calculated as follows. With probability μ_i, the customer of class i is served at time τ in the original policy. With probability μ_i, it will be served at time M in the modified policy. Thus, with probability μ_i, the cost of a customer in queue i is reduced by $\sum_{n=M+1}^{\tau} c_i \beta^n$. Similarly, with probability μ_j, the cost of a customer in queue j is increased by $\sum_{n=M+1}^{\tau} c_j \beta^n$. Thus, the expected reduction in cost is

$$(c_i \mu_i - c_j \mu_j) E\{ \sum_{n=M+1}^{T} \beta^n \}$$

which is positive since $c_i\mu_i > c_j\mu_j$.

This shows that the $c\mu$-rule is optimal.

Example 8.2.5 [Serving Poisson arrivals nonpreemptively with a single server] This example is loosely related to the preceding ones. This is why it is discussed here, even though the proof is not quite an interchange argument.

Jobs of different classes $\{1, 2, \ldots, J\}$ arrive as independent Poisson processes. A job of class i has a mean service time equal to μ_i^{-1} and a waiting cost of c_i per unit of time. All the service times are independent.

The problem is to find the **nonpreemptive** server allocation strategy that minimizes the long-term **average waiting cost per unit of time,** i.e.,

$$K = \sum_{i=1}^{J} c_i N_i \tag{8.3}$$

where N_i denotes the long term average number of customers of class i in the system.

Proposition 8.2.6 [Optimality of the nonpreemptive $c\mu$-rule] The nonpreemptive $c\mu$-rule is optimal. That policy is one that, at any service completion time, starts serving a customer with the largest value of $c_i\mu_i$ among the present customers. (If the queue is empty at a service completion time, then the policy starts serving the next customer to arrive.)

Proof: The proof is based on a **work-conserving property** of nonidling policies. Roughly the argument goes as follows.

First one observes that it suffices to consider nonidling policies, i.e., policies under which the server is never idle when there is a customer to serve. Indeed, one can always consider that an idling policy is in fact serving a class $J + 1$ of customers with $c_{J+1} \approx 0$. If the result is true for nonidling policies, the fact that $c_{J+1}\mu_{J+1} \approx 0$ implies that the class $J+1$ should be served last, i.e., that an optimal policy will be nonidling.

Second, one claims that the value of

$$W = \sum_{i=1}^{J} \mu_i^{-1} N_i \tag{8.4}$$

is the same for all nonidling nonpreemptive policies, where N_i denotes the long term average number of customers of class i in the queue. This is a consequence of Little's result. (See Exercise 8.2.)

Third, assume that $c_1\mu_1 \geq c_2\mu_2 \geq \cdots \geq c_J\mu_J$. Since (8.4) implies that

$$N_1 = \mu_1(W - \sum_{i=2}^{J} \mu_i^{-1} N_i)$$

for all nonidling nonpreemptive strategies, one finds from (8.3) that all such strategies are such that

$$K = c_1 \mu_1 W + \sum_{i=2}^{J} N_i[c_i - c_1 \mu_1 \mu_i^{-1}] . \tag{8.5}$$

The coefficients of N_i in (8.5) are nonpositive for all i (since $c_i\mu_i \leq c_1\mu_1$). It follows that the right-hand side of (8.5) is minimized by a policy that makes the N_i as large as possible. Such a policy must necessarily serve a customer of class 1 whenever it can.

Fourth, consider all the nonidling policies that serve class 1 whenever they can. The set of times available for those policies to to serve the other classes is the same for all the policies. One can then check that all these policies have the same value of $\sum_{i=2}^{J} \mu_i^{-1} N_i$ (by Little's result again). Repeating the above argument shows that among these policies, the ones that minimize K must serve class 2 whenever they can.

Continuing in this way concludes the proof.

Example 8.2.7 [Waiting for a fast server or using a slow one] Customers arrive at a service facility that has two servers. The arrival times form a **Poisson process** with rate λ. The service times are independent and are **exponentially** distributed with rate μ_1 for server 1 and with rate $\mu_2 < \mu_1$ for server 2. Service is **nonpreemptive.** (See Figure 8.3.) The problem is to find the server allocation policy that minimizes

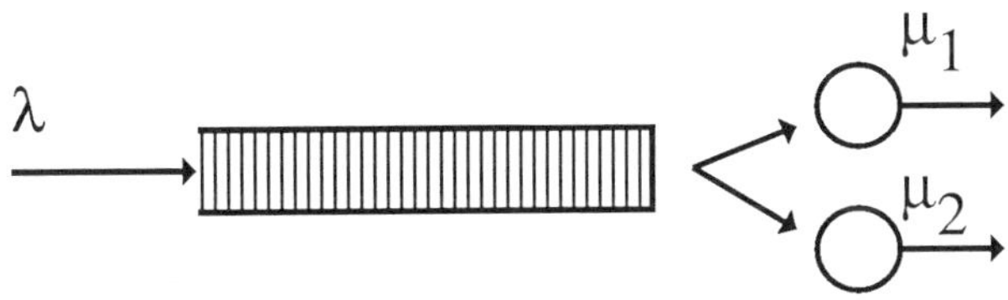

Figure 8.3. System with two servers.

$$E\{\int_0^\infty e^{-\alpha t} x_t \, dt\} \tag{8.6}$$

where α is the positive discount rate and x_t is the number of customers in the system at time t.

This objective function is a measure of the time spent by the customers in the system. The **trade-off** is between waiting for the fast server (server 1) to become available and committing a customer to the slow server.

Proposition 8.2.8 [Optimality of a threshold policy] There is some number N such that the optimal policy is to use the fast server all the time and to send a customer to the slow server (when the latter is available) if and only if the number of customers in the queue at that time is at least N.

Proof: We will rely on the fact that there is a stationary optimal policy. A policy is stationary if the action at a decision time t depends only on the state at that time. Here, the state is the number of customers in the queue and the set of servers that are idle. Decision times are service completion times and arrival times when at least one server is idle. This fact will be discussed in section 8.5.

Consider a stationary policy. We leave to the reader the task of showing that the policy cannot be optimal unless it uses the fast server whenever possible (see Exercise 8.3).

If the stationary policy uses the fast server whenever possible, then it is specified by a subset $\mathbf{A}$ of $\{1, 2, \ldots\}$ with the interpretation that a customer is sent to the slow server at decision times when that server is idle and when the queue length belongs to $\mathbf{A}$. It then only remains to show that the set $\mathbf{A}$ must be of the form $\{N, N+1, N+2, \ldots\}$. This is done by contradiction.

Assume that the set $\mathbf{A}$ contains a "gap." That is, assume that $\{M+1, M+2, \ldots, N-1\} \subset \mathbf{A}^c$ and $M, N \in \mathbf{A}$ for some $M+1 \leq N-1$. Say that at $t = 0$ the fast server is busy, the slow server is idle, and there are $M+1$ customers waiting to be allocated to a server. The policy will then wait until the queue length reaches either M or N before sending a customer to the slow server. That policy can be improved by sending a customer at time 0 to the slow server. To see this, denote by σ the service time of that customer sent at time 0 to the slow server. This service time is known at time σ. Pretend that the policy corresponding to **A** was in fact used, by doing as if the customer had not been sent at time 0 but had been sent only when the queue length hit either M or N, at time τ say, and by pretending that the slow server is busy during $[\tau, \tau + \sigma]$. This shows that the modified policy behaves as the one corresponding to **A**, except that one customer leaves the queue at time σ instead of at time $\tau + \sigma$.

It remains only to show that for all $M \geq 1$ there is some $N > M$ such that $N \in \mathbf{A}$. Again, this can be shown by contradiction. The idea is indicated in Exercise 8.4.

8.3. FORWARD INDUCTION

The **forward induction** method for proving the optimality of a policy can be described as follows.

Denote by x_t the state process that corresponds to that policy, and by y_t the state process corresponding to another arbitrarily chosen policy. Suppose that there is a partial ordering **B** on the set **X** of possible states with the following two properties. First, it should be such that it is possible to prove that x_t **B** y_t implies that x_s **B** y_s for all $s \geq t$. Second, the ordering should imply that the cost corresponding to x_t is not larger than that corresponding to y_t. Then it follows that the policy that corresponds to x_t is optimal.

The examples will show that there is no general method for finding the partial ordering **B**. One usually starts with the intuitive reason why the policy is thought to be optimal (for instance, that it minimizes the number of customers in the system). One then determines **additional properties** that are required to propagate by forward induction (for instance, that the queue lengths are most similar).

Example 8.3.1 [Customer allocations to parallel queues] Customers arrive at arbitrary known times at a system consisting of J identical M/M/1 queues in parallel. (See Figure 8.4.) That is, the service times in all the queues are independent and exponentially distributed with the same rate μ.

The problem is to choose, at each arrival time, **which queue the arriving customer should join** so as to minimize

$$E\{\int_0^\infty e^{-\alpha t} \sum_{j=1}^{J} x_t^j \, dt\} \tag{8.7}$$

where α is a positive constant (the discount rate) and x_t^j is the number of customers at time t in queue j. The information available when the decision is made is the evolution of the vector of queue lengths up to that time and the set of all arrival times.

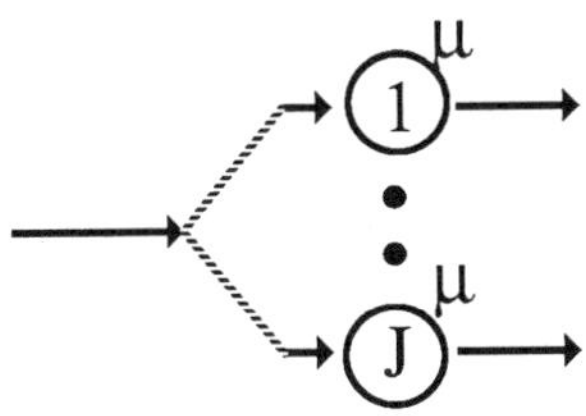

Figure 8.4. Arrivals at parallel queues.

It is assumed that the arrival times are such that (8.7) is finite at least for a policy.

Proposition 8.3.2 [Optimality of "Join-the-shortest-queue"] The policy that sends each arriving customer to the shortest queue achieves the minimum value of (8.7) over all the admissible policies.

It should be noted that this policy is clearly **individually** optimal for each customer, for arbitrary decisions of the customers who arrived before him. However, this does not imply that the policy is optimal **socially,** i.e., in the sense of minimizing (8.7).

Indeed, it is often the case that individuals have to accept sacrifices for the benefit of society at large. Mathematically, each customer should take into account not only the personal cost of that customer (here, the discounted waiting time) but also the **impact of the decision on the other customers** (here, on those who will arrive later).

Proof: One can consider that the potential service completion times occur at the jumps of a Poisson process with rate $J\mu$ and that every service completion is equally likely to take place at each of the J queues. (A potential service completion time is an actual completion time if there is a customer in the queue. By the memoryless property of the exponential distribution, the assumptions on the potential service completion times imply that the service times are exponentially distributed with rate μ.)

For two random variables V and W taking values in $\mathbf{X} = \{0, 1, 2, \ldots\}^J$, write that $V\mathbf{B}W$ if there exist two random variables $\tilde{V}$ and $\tilde{W}$ such that

$$\begin{cases} \tilde{V} \text{ has the same distribution as } V \\ \tilde{W} \text{ has the same distribution as } W \\ P\{S_i(\tilde{W}) \geq S_i(\tilde{V}),\ 1 \leq i \leq J\} = 1 \end{cases} \tag{8.8}$$

where $S_i(\tilde{V})$ denotes the sum of the i largest components of $\tilde{V}$ and similarly for $S_i(\tilde{W})$.

For instance, $W \equiv (2, 4, 3, 2, 0, 0)$ and $V \equiv (1, 3, 2, 0, 3, 1)$ are such that $V\mathbf{B}W$.

Denote by x_t the vector of queue lengths at time $t \geq 0$ corresponding to the JSQ ("Join-the-shortest-queue") policy, and by y_t that corresponding to another arbitrarily chosen policy. The claim is that

$$x_t \mathbf{B} y_t, \quad t \geq 0. \tag{8.9}$$

Since $S_J(x_t) = \sum_{i=1}^{J} x_t^i$ and similarly for y_t, this will imply that

$$E\{\sum_{i=1}^{J} y_t^i\} = \sum_{m=0}^{\infty} P\{S_J(y_t) \geq m\} = \sum_{m=0}^{\infty} P\{S_J(\tilde{y}_t) \geq m\}$$

$$\geq \sum_{m=0}^{\infty} P\{S_J(\tilde{x}_t) \geq m\} = \sum_{m=0}^{\infty} P\{S_J(x_t) \geq m\} = E\{\sum_{i=1}^{J} x_t^i\}$$

and therefore that the cost (8.7) corresponding to x_t is not larger than the cost corresponding to y_t.

Thus, the partial ordering **B** has the second desired property mentioned in the opening paragraph of this section.
(In the above calculation, the first and last equations come from the easily verified fact that any $\{0, 1, 2, \ldots\}$-valued random variable Z is such that $E\{Z\} = \sum_{m=0}^{\infty} P\{Z \geq m\}$. In the second equality, $\tilde{y}_t$ is the random variable that enters in the definition (8.8) of $x_t \mathbf{B} y_t$. Therefore, the second equality comes from the identity of

the distributions of y_t and $\tilde{y}_t$, and similarly for next to last equality. The inequality comes from (8.8).)

To prove (8.9), let $t_1 < t_2 < t_3 < \cdots$ be the values of the arrival and potential service completion times. Assume that $x_t \mathbf{B} y_t$ for some $t \geq 0$. (This property holds for $t = 0$.) Say that $t_{n-1} \leq t < t_n$ for some $n \geq 1$ (with the convention that $t_0 = 0$). It then suffices to show that $x_{t_n} \mathbf{B} y_{t_n}$.

First consider the case when t_n is an arrival time. Define $\tilde{x}_t$ and $\tilde{y}_t$ so that (8.8) holds. Notice that $S_i(\tilde{x}_{t_n}) = S_i(x_t) + 1\{i = J\}$ while $S_i(\tilde{y}_{t_n}) = S_i(y_t) + 1\{i \geq k\}$ if the policy choose to send the customer to the kth largest queue in y_t. This is seen to imply (8.8) at time t_n.

Next consider the case when the event is a potential service completion. The probability that it occurs in the kth longest queue of x_t is J^{-1} (ties are broken arbitrarily in the case of equal queue lengths). The same is true for y_t. Define $\tilde{x}_{t_n}$ and $\tilde{y}_{t_n}$ by deciding that if the potential service completion occurs in the kth longest queue of $\tilde{x}_t$, then the same is true for $\tilde{y}_t$. This modifies the joint distribution but not the marginals, so that the first two statements of (8.8) will hold for $\tilde{V} = \tilde{x}_{t_n}$ and $\tilde{W} = \tilde{y}_{t_n}$.

To verify that the last statement of (8.8) holds, one uses the fact that if the potential service completion occurs in the kth longest queue, then

$$S_i(\tilde{x}_{t_n}) = S_i(\tilde{x}_t) - 1\{S_J(\tilde{x}_t) > S_k(\tilde{x}_t) \text{ and } i \geq k\}$$

and similarly for $\tilde{y}_{t_n}$.

To conclude that $S_i(\tilde{y}_{t_n}) \geq S_i(\tilde{x}_{t_n})$, it suffices to observe that if $S_i(\tilde{y}_{t_n}) = S_i(\tilde{y}_t) - 1$ and $S_i(\tilde{x}_{t_n}) = S_i(\tilde{x}_t)$, then it must be that $i \geq k$ and that the kth longest queue is empty in $\tilde{x}_t$ and not in $\tilde{y}_t$, so that

$$S_i(\tilde{x}_t) = S_{k-1}(\tilde{x}_t) \leq S_{k-1}(\tilde{y}_t) = S_k(\tilde{y}_t) - 1 \leq S_i(\tilde{y}_t)$$

which yields the result.

Example 8.3.3 [Open loop customer allocations to parallel queues] Assume that the customer allocations in Example 8.3.1 must be decided at time 0, prior to any observation. In particular, the queue lengths and the arrival times are not observable.

The objective is again to minimize the cost (8.7).

Proposition 8.3.4 [Optimality of round-robin allocation] Assume that $x_0 = (0, 0, \ldots, 0)$. Then an optimal policy is to allocate the arriving customers successively to the queues $1, 2, \ldots, J, 1, 2, \ldots, J, 1, 2, \ldots$ etc. This policy is called the **round-robin policy.**

The **intuition** is that, given the available information, this is the best way to share the work among the servers and thereby to avoid overloading a server while others are idle.

Proof: The proof is similar to that of Proposition 8.3.2. Let x_t be the vector of queue lengths corresponding to the round-robin policy and y_t be that corresponding to an arbitrary policy.

Since only the **marginal distributions** of the components of x_t and of those of y_t matter, and not the joint distribution of (x_t, y_t) (for the cost (8.7) is additive in the queue lengths and the mean value of a sum only involves the marginal distributions), one can assume that the **potential service completions occur at the same time** in the J queues, at the jump times of a Poisson process with rate μ. (This assumption modifies the joint distribution but not the marginal ones).

The claim is then that $x_t \mathbf{B} y_t$ for all $t \geq 0$ where $\mathbf{B}$ is the partial ordering defined in (8.8). The proof is a simple adaptation of that of Proposition 8.3.2. The details are left to the the reader.

Example 8.3.5 [Preemptive service in a single server queue] Customers arrive at a queueing system at arbitrary **known** times. They are served by a single server according to a preemptive policy. It is assumed that the server knows the service times of the customers as soon as they enter the queue.

The problem is for the server to decide at all times which customer should be served so as to minimize (8.7).

Proposition 8.3.6 [Optimality of SRPT policy] The policy that at all time serves the customer with the **shortest remaining processing time** (SRPT) is optimal.
This policy minimizes the number of customers in the system at all times.

Proof: One can assume that the system is empty at time 0. Let $0 \leq \sigma_t^1 \leq \sigma_t^2 \leq \cdots \leq \sigma_t^{A_t}$ be the residual service times of the A_t customer who arrived in $[0, t]$. Thus, $\sigma_t^n = 0 < \sigma_t^{n+1}$ if the number D_t of departures in $[0, t]$ is equal to n.

One can show by induction on t that the SRPT policy minimizes $\sum_{i=1}^{j} \sigma_t^i$ for all $t \geq 0$ and for all $1 \leq j \leq A_t$. In particular, SRPT maximizes D_t and minimizes x_t for all t. It follows that it minimizes (8.7).

Example 8.3.7 [Processing exponential jobs with precedence constraints] A set of N jobs with i.i.d. and exponentially distributed service times have to be processed by one of two identical processors.

The processing must be done in an order that satisfies **in-tree precedence constraints.**

That is, the N jobs can be thought of as being the nodes of a tree, the interpretation being that the processing of a job cannot start before that of its sons is completed. Thus, the processors can work only on leaves of the tree (jobs without sons, given that completed jobs are removed from the tree). (See Figure 8.5.)

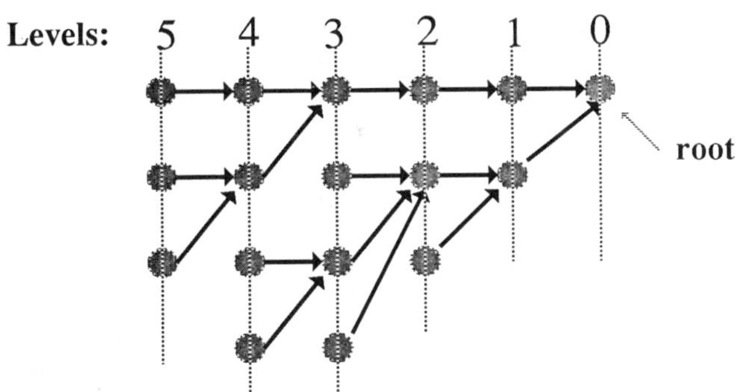

Figure 8.5. Tree of precedence constraints.

The problem is to find a **nonpreemptive** allocation of the processors to the jobs that minimizes the **expected makespan.** (The makespan is the time to complete the processing of all the jobs.)

Proposition 8.3.8 [Optimality of Highest Levels First policies] Define the **level** of a job to be one more than the level of its father, the level of the root (the job without father) being 0.

A policy is **HLF (Highest Levels First)** if it assigns any processor that becomes free to a leave with the highest level among the leaves not yet assigned. Any HLF policy minimizes the expected makespan.

Proof: The decision times for any policy are the job completion times by one of the two processors.

The **state** at a completion time specifies the tree of remaining jobs and the job which was being processed and did not terminate at the decision time. The latter information is needed to guarantee that the policy is nonpreemptive. The value of residual service time required by the job still being processed at a decision time is not useful, in view of the memoryless property of the exponential distribution.

Let then **X** denote the set of possible states. A typical state is of the form $(\mathbf{T}, i)$, where **T** is a tree and i is the index of a leaf in **T**. For $x_1 = (\mathbf{T}_1, i_1)$ and $x_2 = (\mathbf{T}_2, i_2)$ in **X**, one writes that $x_1\mathbf{B}x_2$ if $\mathbf{T}_1$ has fewer jobs than $\mathbf{T}_2$ at levels higher than l for all $l \geq 0$, and if $\mathbf{T}_2$ has at most one leaf at a level higher than that of job i_1 in $\mathbf{T}_1$. (See Figure 8.6.)

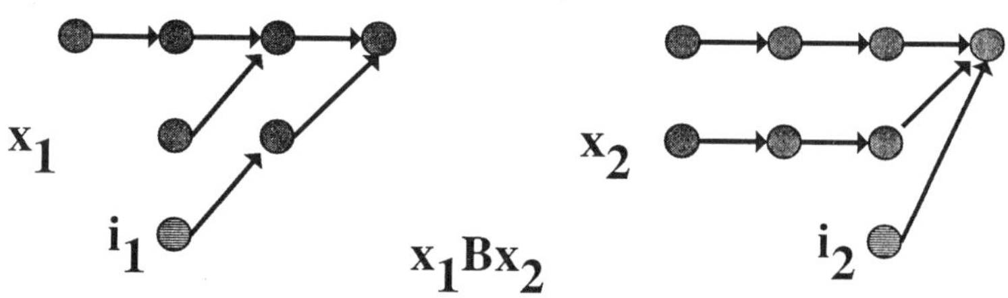

Figure 8.6. The ordering of trees.

Let $\mathbf{T}_1 = \mathbf{T}_2$ be an initial tree of jobs to be processed. Apply an HLF policy to $\mathbf{T}_1$ and another arbitrary policy to $\mathbf{T}_2$. One can assume that the highest level jobs selected by the two policies have the same processing times, and that the same is true of the other two jobs initially selected by the two policies. Denote by x and y the states at the next completion time. Notice that $x\mathbf{B}y$.

Let now $x\mathbf{B}y$ in **T**. Apply an HLF policy to both x and y. Match the processing times as in the previous paragraph. It is again immediate to see that the next state x' corresponding to x and the next state y' corresponding to y are such that $x'\mathbf{B}y'$.

It follows that one can construct the sequences $\{x_n, n \geq 1\}$ and $\{y_n, n \geq 1\}$ of states at the successive decision times corresponding respectively to using an HLF policy at times $\{1, 2, \ldots\}$ and to using an arbitrary policy at time 1 and an HLF policy at times $\{2, 3, \ldots\}$ in such a way that

$$P\{x_n\mathbf{B}y_n, n \geq 1\} = 1 . \tag{8.10}$$

Indeed, the transitions from the initial tree $\mathbf{T}_1 = \mathbf{T}_2$ to x_1 using an

HLF policy and to y_1 using an arbitrary policy is such that $x_1 \mathbf{B} y_1$ by the first construction. By the Strong Markov Property (see 2.9.8), one can then restart the evolution afresh from x_1 and y_1 and use the second construction. It follows that $x_2 \mathbf{B} y_2$, etc.

This proves that if HLF policies minimize the makespan whenever there are fewer than N jobs to process, then they also do when there are $N+1$ jobs to process. Indeed, the only policies that need to be compared are such that they use an HLF policy after the first job completion, and among those the HLF policies achieve the smallest makespan, since they minimize the number of jobs in the system at each decision time, by (8.10).

The result follows from the fact that HLF policies are certainly optimal when there is only one job to process. It should be noted that the same argument proves the optimality of HLF policies when all the processing times are equal to a **common constant.** It is known that HLF policies are not generally optimal in the case of 3 processors. (See Exercise 8.6.)

8.4. DYNAMIC PROGRAMMING: FINITE CASE

The central idea of **Dynamic Programming** is to consider the **optimal cost-to-go** starting from an arbitrary state at an arbitrary time. This **value function** is shown to solve recursive equations in the number of steps to go (called the **Dynamic Programming Equations).** Also, the optimal strategy is easily characterized in terms of the value function.

These ideas are illustrated in a number of examples.

Example 8.4.1 [Guessing a diamond] This is a simple example that illustrates the basic method. A deck of 52 cards is to be turned over one at a time. Before each card is turned we are given the opportunity to say whether or not it will be a diamond. We are allowed to say that a card will be a diamond only once. The objective is to maximize the probability of being correct.

Proposition 8.4.2 [Indifference] All the decision rules that select at least one card before all the diamonds are turned over are optimal.

Proof: Denote by $V_n(m)$ the maximum probability of guessing a diamond when there are n cards left to be turned and when m of those cards are known to be diamonds.

Thus, $V_m(m) = 1$ for $1 \le m \le 13$ and $V_n(0) = 0$ for $n \ge 1$. The claim is that

$$V_n(m) = \max\{\frac{m}{n}, \frac{m}{n}V_{n-1}(m-1) + \frac{n-m}{n}V_{n-1}(m)\}, \quad (8.11)$$

for $n \geq 2$ and $1 \leq m \leq \min\{n, 13\}$.

To see this, notice that the first term in the maximization is the probability of being correct if the decision is to declare that the next card is a diamond. The second term gives the maximum probability of being correct if the decision is to not declare that the next card is a diamond. Indeed, in the latter case, there are two possibilities. With probability $\frac{m}{n}$, the next card is a diamond, so that there remain $m - 1$ diamonds in the deck among $n - 1$ cards, with a maximum probability of being correct being equal to $V_{n-1}(m-1)$, by definition. With probability $\frac{n-m}{n}$, the next card is not a diamond, so there are then still m diamonds among $n - 1$ cards, which corresponds to the maximum probability of being correct equal to $V_{n-1}(m)$.

Also, a policy is optimal if and only if it achieves the maximum in (8.11) for all possible values of (n, m) that can be reached. Indeed, this is clearly necessary. This is also sufficient, since all the policies that achieve the maximum are seen to have the same probability of guessing correctly and since the others have a smaller probability.

Now, solving (8.11) shows that

$$V_n(m) = \frac{m}{n}, \quad \text{for } n \geq 1 \text{ and } 0 \leq m \leq \min\{n, 13\}.$$

But this implies that the two terms in the maximization of (8.11) are equal for $n \geq 2$ and $1 \leq m \leq \min\{n, 13\}$. Hence all the policies are optimal, as long as they select at least one card before all the diamonds are turned over.

Comments 8.4.3

(a) In $V_n(m)$, n is the number of steps to go and m can be thought of as being the state of the game. The function $V_n(.)$ is called the **value function.** The interpretation is that $V_n(m)$ is the value of state m when there are n steps to go.

(b) The equations (8.11) are the **Dynamic Programming Equations** (DPEs).

(c) **A policy is optimal if achieves the maximum in the DPEs**, for all possible states and values of the time-to-go.

(d) In some cases it is possible to solve analytically the DPEs and thereby to derive the optimal policy. In most cases, however, this cannot be done. One may then attempt to calculate, either exactly or approximately, the value functions by computer. It may also be possible to derive properties of the optimal policy without having to solve the DPEs. Examples below illustrate that possibility.

Example 8.4.4 [Processing a set exponential jobs on parallel machines] A set of N jobs with independent and exponentially distributed service times with rates $\mu_1 \leq \mu_2 \leq \cdots \leq \mu_N$ have to be processed by one of J identical servers. (See Figure 8.7.)

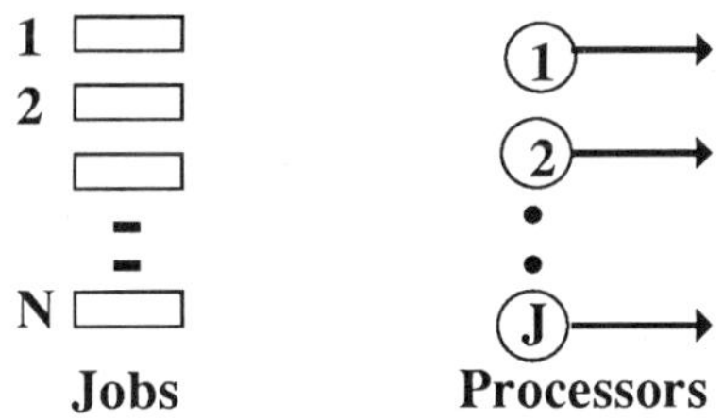

Figure 8.7. N jobs and J parallel machines.

The N jobs are ready to be processed at time 0 (thus, there are no arrivals). Two objective functions will be considered:

$$FT := E\{\sum_{n=1}^{N} T_n\} \tag{8.12}$$

and

$$MS := E\{\max(T_1, \ldots, T_N)\} .$$

In these expressions, T_n is the completion time of job n. FT is called the **expected flowtime** and MS is the **expected makespan** of the jobs.

Proposition 8.4.5 [SEPT and LEPT]

(a) FT is minimized by the SEPT policy that, at each service completion time and at time 0, allocates the jobs to available servers in the order $N, N-1, \ldots, 1$ of shortest expected processing times first.

(b) MS is minimized by the LEPT policy that allocates the jobs in the order $1, 2, \ldots, N$ of longest expected processing times first.

Proof of the optimality of LEPT [outline]: The proof is given in the case of two processors, to simplify the notation. The reader is invited to write the general proof.

Assume that the LEPT is known to be optimal whenever there are fewer than N jobs to process. It will be shown to be optimal for N jobs. Let V^i denote the minimum expected makespan for the jobs $\{1, 2, \ldots, N\} - \{i\}$. By the hypothesis, this minimum expected makespan is achieved by the LEPT policy. The minimum expected makespan V for the jobs $\{1, 2, \ldots, N\}$ is then, by conditioning on the first of the two jobs initially processed to complete,

$$V = \min_{i \neq j} \frac{1 + \mu_i V^i + \mu_j V^j}{\mu_i + \mu_j} \tag{8.13}$$

Simple algebra shows that (8.13) is equivalent to

$$0 = \min_{i \neq j} \{1 + \mu_i(V^i - V) + \mu_j(V^j - V)\} \tag{8.14}$$

and the minimum in (8.14) is achieved by the same pair as in (8.13).

Also, an optimal policy is one that achieves the minimum in (8.13). Thus, to show that LEPT is optimal when there are N jobs to process, one has to show that the minimum in (8.14) is achieved by choosing $\{i, j\} = \{1, 2\}$. It follows that a sufficient condition for the optimality of LEPT is that $\mu_1 \leq \mu_2 \leq \cdots \leq \mu_N$ and $\mu_i < \mu_j$ imply

$$D_{ij} = V_i - V_j \geq 0 \quad \text{whenever } \mu_i < \mu_j \tag{8.15}$$

and

$$\frac{d}{d\mu_1} D_{12} \geq 0 \, . \tag{8.16}$$

with $V_i := \mu_i(V^i - V)$. (The property (8.16) is necessary only to be able to prove (8.15) by induction.)

To start the induction one verifies that when $N = 2$, $D_{ij} = \frac{\mu_i}{\mu_j} - \frac{\mu_j}{\mu_i}$, so that (8.15) and (8.16) hold.

Assume now that (8.15)–(8.16) hold when there are fewer than N jobs. Some algebra shows that

$$D_{12} = \frac{\mu_1}{\mu_1 + \mu_2} D_{32}^1 + \frac{\mu_2}{\mu_1 + \mu_2} D_{13}^2 \tag{8.17}$$

and

$$D_{2i} = \frac{\mu_1}{\mu_1 + \mu_2} D^1_{2i} + \frac{\mu_2}{\mu_1 + \mu_2} D^2_{3i}, \quad i = 3, 4, \ldots, N \tag{8.18}$$

where D^k_{ij} is the value of D_{ij} that corresponds to the set of jobs $\{1, 2, \ldots, N\} - \{k\}$.

From the induction hypothesis (8.15) it follows that $D^2_{13} \leq 0$ and $D^1_{23} \leq 0$. Similarly, the induction hypothesis (8.16) implies that $\frac{d}{d\mu_1} D^2_{13} \geq 0$. By integrating this inequality with respect to μ_1 one finds $D^2_{13} \leq D^1_{23} = -D^1_{32} \leq 0$. Since $\mu_1 \leq \mu_2$, this implies that (8.17) is nonpositive. The induction hypothesis also implies that $D^1_{2i} \leq 0$ and $D^2_{3i} \leq 0$, so (8.18) is nonpositive. This shows (8.15) for N jobs.

To show (8.16) for N jobs, one differentiates the right hand side of (8.17) with respect to μ_1 and one uses the induction hypothesis to check that every term is nonnegative.

Proof of the optimality of SEPT [outline]: Once again we consider the case of two processors. Considering the minimum expected flowtime V for the jobs $\{1, 2, \ldots, N\}$ and V_i for $\{1, 2, \ldots, N\} - \{i\}$, one finds that (8.15) becomes

$$0 = \min_{i \neq j} \{n + \mu_i(V^i - V) + \mu_j(V^j - V)\}. \tag{8.19}$$

It suffices then to show that $\mu_1 \geq \mu_2 \geq \cdots \geq \mu_N$ and $\mu_j > \mu_i$ imply

$$-1 \leq D_{ij} \leq 0 \tag{8.20}$$

and

$$\frac{d}{d\mu_1} D_{12} \leq 0. \tag{8.21}$$

This is shown by induction. For $N = 2$ one has $D_{12} = 0$. Algebra gives, instead of (8.17)–(8.18),

$$D_{12} = \frac{\mu_1}{\mu_1 + \mu_2}(D^1_{32} - 1) + \frac{\mu_2}{\mu_1 + \mu_2}(D^2_{13} + 1)$$

and

$$D_{2i} = \frac{\mu_1}{\mu_1 + \mu_2} D^1_{2i} + \frac{\mu_2}{\mu_1 + \mu_2}(D^2_{3i} - 1), \quad i = 3, 4, \ldots, N.$$

These equations can be used to show that if (8.20)–(8.21) hold for fewer than N jobs, then they continue to hold for N.

Example 8.4.6 [Serving returning customers] The system illustrated in Figure 8.8 consists of one server having to serve N returning customers. That is, once customer i is served, he leaves the server and comes back after an exponential time with rate λ_i $(1 \le i \le N)$. Every service time of customer i is exponentially distributed with rate μ_i. All those times are independent.

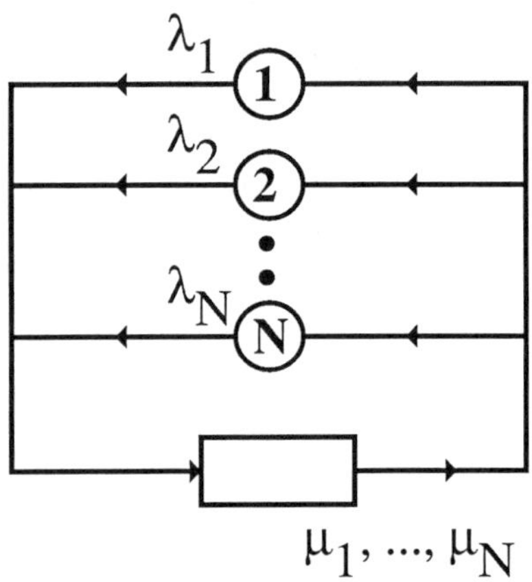

Figure 8.8. Returning customers.

The problem is to find the preemptive service policy that maximizes the long-term average fraction of time that the server is idle. A policy is a rule that chooses, for each set $x \subset \{1, 2, \ldots, N\}$ of customers waiting for service, the customer $i \in x$ to serve. Service can be preemptive since the policy may decide to choose another customer when one returns.

Proposition 8.4.7 [Optimality of slowest-to-return policy] The policy that always serves the customer with the **smallest** value of λ_i (among the customers waiting for service) **maximizes** the long-term average fraction of time that the server is **idle.**

Notice that this policy, called the **slowest-to-return,** is independent of the service rates μ_i. In particular, if all the λ_i are equal, then the fraction of server idle time is independent of the policy.

In fact, this policy **maximizes stochastically** the time from any state x to the state ϕ corresponding to no customer waiting for service. (See 6.7.1 and 6.7.2.)

Conversely, one can show that the **fastest-to-return policy,** that always serves the customer with the largest value of λ_i, **minimizes** the fraction of time that the server is idle.

Proof of Proposition 8.4.7: Denote by x_t the set of customers waiting to be served at time $t \geq 0$. Let ϕ be the state corresponding to an empty server.

Assume without loss of generality that $\lambda_1 \leq \lambda_2 \leq \cdots \leq \lambda_N$. Let Q^* denote the rate matrix of $\mathbf{x} := \{x_t, \quad t \geq 0\}$ when the policy claimed to be optimal is used. That is,

$$q^*(x, y) = \lambda_i \, 1\{y = x + i, i \notin x\} + \mu_{m(x)} 1\{y = x - m(x)\}$$
$$- [\sum_{i \notin x} \lambda_i + \mu_{m(x)}] \, 1\{y = x\}$$

where $m(x) := \min\{i \mid i \in x\}$.

With $\alpha := \sum_{i=1}^{N} (\lambda_i + \mu_i)$ one finds that $\mathbf{x} := \{x_t, \quad t \geq 0\}$ has the same law as $\{y_{N_t}, t \geq 0\}$, where $\{N_t, t \geq 0\}$ is a Poisson process with rate α and $\{y_n, n \geq 0\}$ is a discrete-time Markov chain with transition matrix

$$P^* := I + \alpha^{-1} Q^* . \tag{8.22}$$

(This is the uniformization technique of Exercise 2.12.)

Another control strategy would correspond to some rate matrix Q and therefore to some transition matrix $P = I + \alpha^{-1} Q$. It will be shown that the Markov chain $\mathbf{y}$ with transition matrix P^* takes longer to reach ϕ than with any other feasible transition matrix P. To do this, one modifies P and P^* by making the state ϕ absorbing. This corresponds to setting $P(\phi, \phi) = P^*(\phi, \phi) = 1$.

For $x \in \mathbf{X}$ (the collection of subsets of $\{1, 2, \ldots, N\}$) let e_x denote the row vector indexed by $\mathbf{X}$ with components $e_x(y) = 1\{x = y\}$. The probability that the Markov chain $\mathbf{y}$ with modified transition matrix P^* reaches ϕ in not more than n steps is then given by

$$e_x P^{*n} e_\phi^T$$

if it starts in state x (the vector e_ϕ^T is the vector e_ϕ transposed).

It suffices then to show that

$$e_x [P - P^*] P^{*n} e_\phi^T \geq 0 \tag{8.23}$$

for all $n \geq 0$ and for all feasible matrix P. Indeed, this will show that if one assumes that the strategy * is optimal when there are n steps to go, then it is also optimal when there are $n + 1$ steps to go.

Denote by R a feasible value of $P - P^*$. To show (8.23) it is sufficient to prove the existence of nonnegative matrices A and B and of a matrix H such that

$$\begin{aligned} He_\phi^T &\geq 0 \\ R &= AH \\ HP^* &= BH\,. \end{aligned} \tag{8.24}$$

Indeed, the last equation then implies that $HP^{*n} = B^n H$ so that $RP^{*n}e_\phi^T = AHP^{*n}e_\phi^T = AB^n He_\phi^T \geq 0$.

Now one can verify that the rows of R must be positive linear combinations of vectors $T(ij, x)$ where $i < j \in x$ and

$$T(ij, x) := S(i, x) - S(j, x) \quad \text{with } S(i, x) := \mu_i[e_x - e_{x-i}]\,.$$

Some calculations show that if $i = m(x) < j$ and $k = \max\{l \mid l \in x\}$, then

$$S(i, x)\,[Q^* + (\lambda_i + \mu_i + \sum_{l \notin x} \lambda_l)I]$$

$$= \mu_i\, S(k, x-k) + \sum_{l \notin x} \lambda_l\, S(i, x+l)$$

and

$$T(ij, x)\,[Q^* + (\lambda_i + \mu_i + \sum_{l \notin x} \lambda_l)I]$$

$$= (\lambda_j - \mu_i)\, S(i, x) + \mu_i\, T(kj, x-i) + \sum_{l \notin x} \lambda_l\, T(ij, x+l)\,.$$

Similarly, if $k = m(x) < i < j \in x$, then

$$S(i, x)\,[Q^* + (\lambda_i + \mu_k + \sum_{l \notin x} \lambda_l)]$$

$$= (\lambda_j - \lambda_i)\, S(i, x) + \mu_k\, T(ij, x-k) + \sum_{l \notin x} \lambda_l\, T(ij, x+l)\,.$$

This shows that one can choose H to be the matrix with row $S(i, x)$ and $T(ij, x)$ for all $i < j \in x$.

8.5. INFINITE HORIZON

The previous section dealt with **finite horizon problems,** i.e., problems that involved only a finite number of steps. The method for constructing an optimal policy was summarized in Comments 8.4.3. This section considers **infinite horizon problems** when the cost is additive over the infinite number of steps.

Definition 8.5.1 [Controlled Markov chain] Let **X** be a countable set and **U** a subset of some $\mathbb{R}^d$. The random sequence $\mathbf{x} = \{x_n, n \geq 0\}$ with values in **X** is called a **controlled Markov chain** with transition matrix $[P_{ij}(u), i, j \in \mathbf{X}, u \in \mathbf{U}]$ if

$$P[x_{n+1} = j \mid x_n = i, u_n = u, x_{n-1}, \ldots, x_0, u_{n-1}, \ldots, u_0] = P_{ij}(u) \quad (8.25)$$

for all $i, j \in X$ and $u \in U(i)$, where $U(i) \subset \mathbf{U}$ is the set of admissible control values when the Markov chain is in state i.

Definition 8.5.2 [Total cost problem] Let **x** be as in Definition 8.5.1. The **total discounted cost problem** with horizon N and discount factor β is the following. Find $u_n = \gamma_n(x_0, \ldots, x_n)$ that achieves

$$V_N^\beta(i) = \min E_i\{\sum_{n=0}^{N-1} \beta^n c(x_n, u_n)\} \quad (8.26)$$

where $N \in \{0, 1, \ldots, \infty\}$, $\beta \in (0, 1]$, and $c(.,.)$ are given. Functions that achieve (8.26) are called **optimal policies** (or optimal control strategies).

The **interpretation** is that $c(i, u)$ is the cost of using the control value u when the state is i.

Proposition 8.5.3 [Dynamic Programming Results]
(a) One has

$$V_N^\beta(i) = \inf_{u \in U(i)} \{c(i, u) + \beta \sum_j P_{ij}(u) V_{N-1}^\beta(j)\}, \; N \geq 1 \quad (8.27)$$

with $V_0(.) := 0$.
(b) If $u = \gamma_N(i)$ achieves the minimum in (8.27), then $u_n = \gamma_{N-n}(x_n)$, $0 \leq n \leq N$ is optimal for (8.26) with $N < \infty$.

(c) Assume that

$$\begin{cases} 0 < \beta < 1 \\ \mathbf{U} \text{ is compact} \\ c(i, u) \text{ is bounded and continuous in } u \text{ for each } i \\ P_{ij}(u) \text{ is continuous in } u \text{ uniformly in } j . \end{cases} \tag{8.28}$$

Then

$$V_N^\beta(i) \rightarrow V^\beta := V_\infty^\beta(i) \text{ as } N \rightarrow \infty \tag{8.29}$$

and V^β is the unique bounded solution of

$$V^\beta(i) = \min_{u \in \mathbf{U}} \{c(i, u) + \beta \sum_j P_{ij}(u) V^\beta(j)\} . \tag{8.30}$$

(d) Finally, under (8.28), if $u = \gamma(i)$ achieves the minimum in (8.30), then $u_n = \gamma(x_n)$ is optimal for (8.26) with $N = \infty$. This optimal policy is then **stationary**.

Comments 8.5.4

(a) The statements (a) and (b) of Proposition 8.5.3 are as discussed in Comments 8.4.3.

(b) There **may not exist** a minimizer u in (8.27). By choosing a policy that achieves the infimum in (8.27) within ϵ for all N and all i one sees that the cost of that policy over N steps is at most $V_N^\beta(i) + \epsilon \sum_{n=0}^{N-1} \beta^n$. This shows the existence of an ϵ-**optimal policy.**

(c) The **continuity-compactness** assumption of (8.28) is sufficient to guarantee the existence of a minimizing u. Indeed, any continuous function on a compact set achieves its minimum and (8.28) implies that the expression to minimize in (8.27) is continuous in u.

(d) The result of part (c) of Proposition 8.5.3 can be explained **intuitively** from the fact that if c is bounded, then for any ϵ one can choose N large enough so that the cost after time N contributes at most ϵ to the total cost over the infinite horizon. The problem is then (up to ϵ) a finite horizon problem.

A **formal proof** of that result is based on the observation that the mapping $V_N^\beta = TV_{N-1}^\beta$ defined by (8.27) is a **contraction** with respect to the supremum norm.

Extensions 8.5.5 [Weakening the assumptions (8.28)] The assumption that c is bounded is too strong in many applications. It can be replaced by the assumption A1 or A2 below.

A1: There exist constants $\{B_i, i \in \mathbf{X}\}$ and k such that

$$E[\,|c(x_n, u_n)|\,|\,x_0 = i] \leq B_i n^k \text{ for all policies and } n \geq 0. \tag{8.31}$$

A2: The problem is such that

$$c(.,.) \geq 0 \text{ and } \mathbf{U} \text{ is finite}. \tag{8.32}$$

Other conditions are also known to be sufficient.

Remark 8.5.6 [Structural properties] Proposition 8.5.3 provides a method for deriving **structural properties** of optimal policies. Assume that (8.28) holds (or with the variation (8.31) or (8.32)) and that $\mathbf{H}$ is a set of functions from $\mathbf{X}$ into $\mathbb{R}$ that contains the function that is identically zero and is such that

$$f \in \mathbf{H} \Rightarrow g(i) := \inf_{u \in \mathbf{U}} \{c(i, u) + \beta \sum_j P_{ij}(u) f(j)\} \in \mathbf{H} \tag{8.33}$$

and

$$\mathbf{H} \text{ is closed under pointwise limits} \tag{8.34}$$

(i.e., if $f \in \mathbf{H}$ for all n and $f_n(i) \to f(i)$ for all $i \in \mathbf{X}$, then $f \in \mathbf{H}$). Then $V^\beta \in \mathbf{H}$.

This result, which is immediate from (8.29), may be used to find properties of the optimal strategy for the infinite horizon, i.e., of the minimizer in (8.30).

Examples 8.5.7 [Convexity, supermodularity,...]

(a) Let $\mathbf{X} = \{0, 1, 2, \ldots\}$. Assume that $h(x)$ is convex and that α is a positive constant. Then

$$f(x) := \min\{\alpha + h(x), h(x+1)\} \tag{8.35}$$

is convex. (See Exercise 8.7.)

(b) Let $\mathbf{X} = \{0, 1, 2, \ldots\}^2$. Define

$$A_1 x = (x^1 + 1, x^2) \tag{8.36}$$

$$A_2 x = (x^1, x^2 + 1) \tag{8.37}$$

$$D_1 x = ((x^1 - 1)^+, x^2) \tag{8.38}$$

$$D_2 x = (x^1, (x^2 - 1)^+). \tag{8.39}$$

Let $\mathbf{H}$ be the set of functions $h : \mathbf{X} \to \mathbb{R}_+$ that are such that

$$\begin{cases} h(x) \text{ is increasing in } x^1 \text{ and } x^2 \\ h(x+y) - h(x) \text{ is decreasing in } x^1 \text{ and } x^2 \text{ for all } y \geq 0 \\ h(A_2x) - h(A_1x) \text{ is decreasing in } x^1 \text{ and increasing in } x^2 . \end{cases}$$

Then the function f defined by

$$f(x) = c_1x^1 + c_2x^2 + \mu_1 h(D_1x) + \mu_2 h(D_2x) + \lambda \min\{h(A_1x), h(A_2x)\} \tag{8.40}$$

belongs to **H** whenever h does.

These examples will be used to determine the structure of optimal policies.

Example 8.5.8 [Admissions into an M/M/1 queue] Customers arrive at a service facility as a Poisson process with rate λ. They are to be served by a single server. The service times are independent and exponentially distributed with rate μ. (See Figure 8.9.)

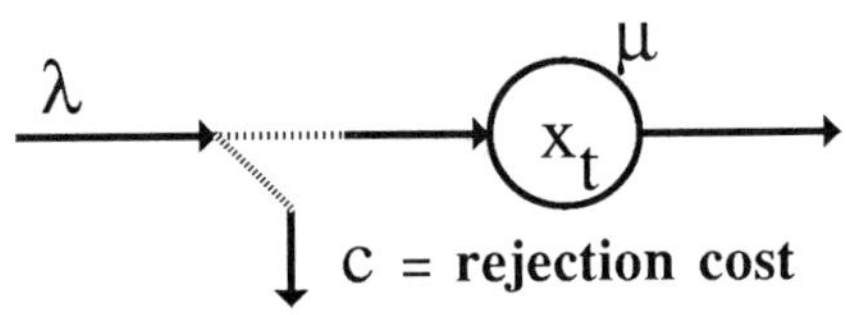

Figure 8.9. Control of arrivals.

One has the option of **refusing entry** to each customer, on the basis of the queue length at the arrival time of that customer. However, a cost c must be paid for each rejection. Rejected customers never come back.

The objective is to minimize the total discounted cost

$$E\{\int_0^\infty e^{-\alpha t} x_t \, dt + \sum_{n=1}^\infty c e^{-\alpha T_n}\}$$

where the T_n are the rejection times and x_t is the queue length at time t. Thus, the optimal policy has to compromise between congestion (as measured by the queue length) and the cost of rejecting customers.

Proposition 8.5.9 [Optimality of a threshold policy] The optimal policy is to reject customers who arrive when the queue length **exceeds some threshold** value N.

Proof: This is a very simple example, but the methodology is similar to that used to solve many other problems. We will therefore give all the details.

The proof consists in the following steps: (1) the problem is shown to be **equivalent to a discrete-time problem;** (2) the **value function** for the discrete-time problem with N steps to go is shown by induction on N to be **convex;** (3) one concludes that the value function for the **infinite horizon** is also **convex;** (4) one shows that this latter result implies that the **optimal policy is of the threshold type.**

Step 1 [Discrete-time] The system evolves at arrival and service times. We will call **potential** transitions all the arrival times and all the service completions times that would occur if the queue were never empty. Thus, these potential transitions occur at the jump times of a **Poisson process** with rate $\lambda + \mu$. It should be noted that these potential transition times remain Poisson with rate $\lambda + \mu$ even given the sequence of queue lengths. This idea is similar to the **uniformization** method of Exercise 2.12.

It will prove convenient to consider the discounting as resulting from **killing** the process after an exponentially distributed random variable with rate α. If one decides that the cost of running the process vanishes as soon as the process is killed, then the cost to be paid at time t is the original cost provided that the system is still alive. Thus, the average cost at time t given the evolution of the system is equal to $e^{-\alpha t}$ times the original cost. As a consequence, the discounted cost for the original problem is the **undiscounted** cost for the killed process.

One will then consider that potential transitions occur at the jump times of a Poisson process with rate $\lambda + \mu + \alpha$ and that a transition is a potential arrival with probability $\lambda(\lambda + \mu + \alpha)^{-1}$, a potential departure with probability $\mu(\lambda + \mu + \alpha)^{-1}$ and is the killing of the process with probability $\alpha(\lambda + \mu + \alpha)^{-1}$.

By scaling the time one can assume that $\lambda + \mu + \alpha = 1$. Denote by τ_n the nth potential transition time, with $\tau_0 := 0$. Let also $y_n := x_{\tau_n}$. The cost can be written as

$$E[\sum_{n=0}^{\infty} \int_{\tau_n}^{\tau_{n+1}} x_t dt + \sum_{n=1}^{\infty} c1\{\text{rejection at time } \tau_n\}]$$

$$= E[\sum_{n=0}^{\infty} y_n(\tau_{n+1} - \tau_n) + \sum_{n=1}^{\infty} c1\,\{\text{rejection at time } \tau_n\}]$$

$$= x_0 + E[\sum_{n=1}^{\infty} (y_n + c1\,\{\text{rejection at time } \tau_n\})]$$

Now, the process $\{y_n, n \geq 0\}$ is a controlled Markov chain. Denote by $u_n = 1$ a rejection at time τ_n and by $u_n = 0$ the fact that an arrival is accepted if it occurs. The transition probabilities $[P_{ij}(u)]$ are then equal to (for $i > 0$)

$$P_{ij}(u) = \begin{cases} \lambda 1\{u = 0\}, & \text{if } j = i + 1 \\ \mu, & \text{if } j = i - 1 \\ \alpha, & \text{if } j = \partial \\ \lambda 1\{u = 1\}, & \text{if } j = i \end{cases}$$

where ∂ denotes the state of the **killed** process. The Dynamic Programming Equations read

$$V_N^{\alpha}(i) = i + \mu V_{N-1}^{\alpha}((i-1)^+)$$
$$+ \lambda \min\{V_{N-1}^{\alpha}(i+1), c + V_{N-1}^{\alpha}(i)\}. \quad (8.41)$$

Indeed, the first term is the waiting cost, the term with the factor μ is the future minimum cost if the next transition is a departure, and the term with the factor λ is the future minimum cost if the next transition is an arrival. In the latter case, one has the option of rejecting the arrival, so that the next state keeps the value i and the cost c is incurred, or of accepting the arrival and not paying the rejection cost.

Step 2 [Convexity] The claim is that $V_N^{\alpha}(.)$ is convex and nondecreasing for all N. To see this, assume that that is the case for $N-1$. The first of the Examples 8.5.7 then shows that the minimum in (8.41) is a convex nondecreasing function. The other terms in (8.41) are also convex and nondecreasing, so that these properties hold for N. Since they hold trivially for $N = 0$, this proves the claim.

Step 3 [Convexity of V_∞^α] To prove that V_∞^α is nondecreasing and convex it suffices to show that one can use Remark 8.5.6. This is the case since the conditions (8.32) hold.

Step 4 [Threshold] The optimal policy is given by part (d) of Theorem 8.5.3. For this problem, the equations (8.30) are (see (8.41))

$$V^\alpha(i) = i + \mu V^\alpha((i-1)^+) \tag{8.42}$$

$$+ \lambda \min\{V^\alpha(i+1), c + V^\alpha(i)\} .$$

The optimal policy is the function γ that achieves the minimum in this equation. That is,

$$\gamma(i) = \begin{cases} 0, & \text{if } V^\alpha(i+1) \le c + V^\alpha(i) \\ 1, & \text{otherwise.} \end{cases}$$

Since the function V^α is convex, it follows that $V^\alpha(i+1) - V^\alpha(i)$ is nondecreasing in i. Therefore, there is a minimum value i_0 of i such that this difference exceeds c. Hence,

$$\gamma(i) = 1\{i \ge i_0\}$$

which shows that the optimal policy is of the threshold type.

Remarks 8.5.10

(a) The above exercise was intended to illustrate the general procedure. However, it should be clear that the result of Proposition 8.5.9 is **without interest.**

Indeed, a more natural formulation of the problem is the average cost where one tries to minimize the average cost per unit of time of running the system with a given control policy. One can show that this problem admits a stationary optimal policy (see section 8.6). Say that $x_0 = 0$ and let i be the minimum queue length such that the optimal policy is to reject the new arrival. Then one may assume that the policy is of the threshold type, with threshold value i. Indeed, the queue length will evolve on $\{0, 1, \ldots, i\}$.

(b) Calculating the optimal threshold for the average cost problem is left as an easy exercise. (See Exercise 8.8.)

Example 8.5.11 [Routing to parallel M/M/1 queues] This example is an extension of Example 8.3.1 to the case of queues with **unequal service rates.**

Thus, customers arrive as a Poisson process with rate λ at a system of two queues in parallel. The service times are independent and exponentially distributed with rate 1. The server in queue i works at rate μ_i ($i = 1, 2$).

Equivalently, one may consider that the service times are exponentially distributed with rate μ_i in queue i. (See Figure 8.10.)

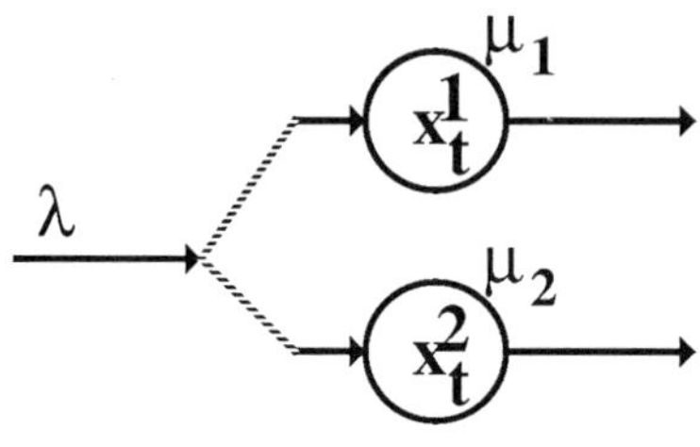

Figure 8.10. Routing to parallel M/M/1 queues.

The problem is to decide to which queue each arriving customer should be sent so as to minimize

$$E\{\int_0^\infty e^{-\alpha t}[c_1 x_t^1 + c_2 x_t^2]\,dt\}$$

where α, c_1, c_2 are positive constants and x_t^i is the number of customers at time t in queue i.

Proposition 8.5.12 [Optimality of switch-over strategy] There is a nondecreasing function $s(.) : \{0, 1, 2, \ldots\} \to \{0, 1, 2, \ldots\}$ such that the optimal policy sends a customer who arrives when the queue lengths are x^1 and x^2 to queue 1 if $x^2 \geq s(x^1)$ and to queue 2 otherwise. (See Figure 8.11.)

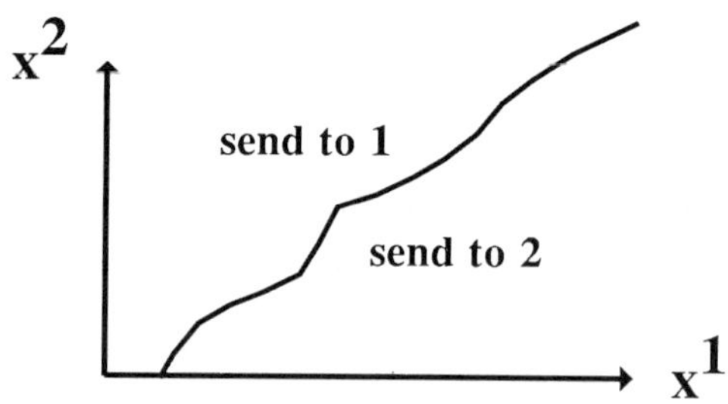

Figure 8.11. Switch-over strategy.

Proof: The proof consists in the following steps: (1) reduction to a discrete-time problem; (2) structural properties of the value function for horizon N; (3) convergence and structural properties for the value function for infinite horizon; (4) optimality of switch-over strategy.

Step 1 [Discrete-time] A potential transition is either an arrival or a potential service completion, or the killing of the process (at rate α). Let the V_n be the minimum cost over n steps. Then

$$V_n(x) = c_1x^1 + c_2x^2 + \mu_1 V_{n-1}(D_1x) + \mu_2 V_{n-1}(D_2x)$$
$$+ \lambda \min\{V_{n-1}(A_1x), V_{n-1}(A_2x)\}, \quad n \geq 1$$

where $D_1x = (x^1 - 1, x^2)$ if $x^1 > 0$ and $D_1x = x$ otherwise, $A_1x = (x^1 + 1, x^2)$, and similarly for A_2x and D_2x. The initial value is $V_0(x) \equiv 0$. We assumed $\alpha + \mu_1 + \mu_2 + \lambda = 1$.

Step 2 [Structural properties] It follows from the second of Examples 8.5.7 that

$V_n(x)$ is increasing in x^1 and x^2

$V_n(x + y) - V_n(x)$ is decreasing in x^1 and x^2 for all $y \geq 0$

$V_n(A_2x) - V_n(A_1x)$ is decreasing in x^1 and increasing in x^2.

Step 3 [Limit] One uses Remark 8.5.6 to conclude that the above properties hold for the value function V_∞.

Step 4 [Switch-over] The optimal policy is the one that achieves the minimum in (8.30), i.e., the optimal decision is to send the customer to queue 1 if and only if $V_\infty(A_1x) \leq V_\infty(A_2x)$.

Assume that it is optimal to send the arriving customer to queue 1 when the state is x. This implies that $V_\infty(A_1x) \leq V_\infty(A_2x)$. The last of the properties given in step 2 then implies that

$$V_\infty(A_2A_2x) - V_\infty(A_1A_2x) \geq V_\infty(A_2x) - V_\infty(A_1x) \geq 0.$$

This implies that the same decision is again optimal when there is an additional customer in the second queue. Similarly, one can see that if the optimal decision is to send an arriving customer to queue 2, then the decision remains optimal when there are more customers in the first queue. This proves the optimality of a switch-over strategy.

Proposition 8.5.13 [Bang-Bang Principle] Consider a control problem where the set of control values is some interval $[a, b]$ of the real line. Under suitable assumptions, the optimal policy will take only the **extreme** values a and b.

For instance, say the problem is to control the rate matrix $Q(u) = [q_{ij}(u), i, j \in \mathbf{X}]$ of a continuous-time Markov chain so as to minimize a total cost

$$E\{\int_0^T e^{-\alpha t} c(x_t, u_t)\, dt\}$$

subject to the condition that $u_t = \gamma(t, x_t) \in [a, b]$ for all $t \geq 0$.

Assume that $c(x, u) = f(x) + ug(x)$ and that $Q(u) = Q(0) + Au$. Thus, both c and Q are assumed to be linear in u. Then the optimal policy takes values in $\{a, b\}$. This is the **Bang-Bang Principle.**

Proof (Outline): One expects Dynamic Programming Equations of the form

$$V_t(i) = \inf_u \{c(i,u)\epsilon + \sum_j [1\{i = j\} + \epsilon\, q_{ij}(u)] V_{t-\epsilon}(j)\} + o(\epsilon).$$

The minimizing value of this expression (which is linear in u) will therefore be either a or b.

Example 8.5.14 [Controlling parallel processors] *Jobs consisting of two parts* each arrive as a Poisson process at a processing facility. Each part is processed by a **dedicated processor,** and the parts corresponding to the same job are then **reassembled** after processing. (See Figure 8.12.)

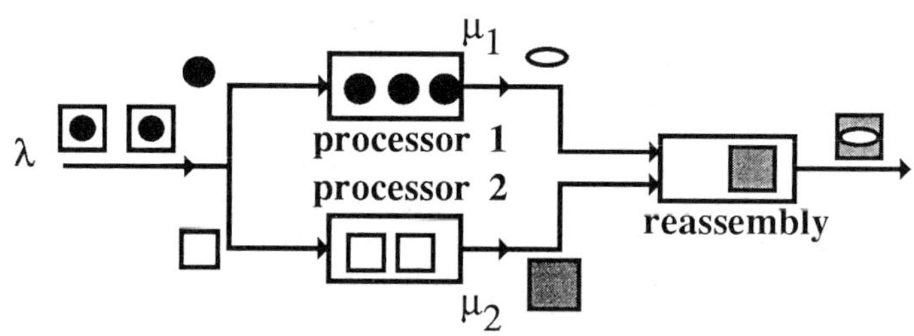

Figure 8.12. The parallel processing system.

The problem is to **control the processing rates** so as to minimize a measure of the **inventory cost** in the processors and the reassembly queue.

The rate μ_i of processor i is allowed to depend on the number of parts queued at each processor and in the reassembly queue. It is assumed that $\mu_i \in [0, a]$. The cost to be minimized is

$$E\{\int_0^\infty e^{-\alpha t}[c_1(y_t^1 + y_t^2) + c_2 y_t^3]\,dt\}$$

where α is the positive discount rate and $c_2 > c_1 > 0$ are the inventory costs per unit of time. The assumption that $c_2 > c_1$ reflects the fact that the processed parts are more valuable than the others.

In the above expression, y_t^i is the number of parts queued at processors $i = 1, 2$ and y_t^3 is the number of parts in the reassembly queue. It is assumed that reassembly is essentially instantaneous (compared to processing). Therefore, $y_t^3 = |y_t^1 - y_t^2|$.

Proposition 8.5.15 [Optimality of switch-over policy] The optimal policy (μ_1, μ_2) is such that

(a) if $y_t^1 > y_t^2$, then $\mu_1(y_t^1, y_t^2) = a$;

(b) if $y_t^1 \le y_t^2$, then $\mu_1(y_t^1, y_t^2)$ is as indicated in Figure 8.13;

(c) $\mu_1(y_t^1, y_t^2) = \mu_2(y_t^2, y_t^1)$.

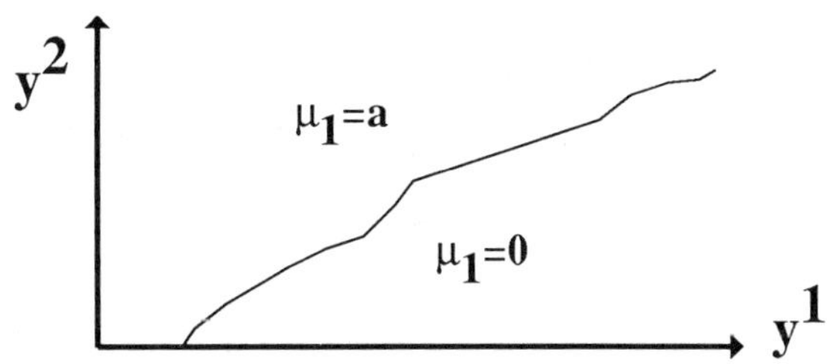

Figure 8.13. The optimal policy.

Proof (outline): As usual, one discretizes the time by using the fact that the optimal policy is bang-bang and by observing that serving at rate 0 is equivalent to serving at maximum rate and then deciding not to accept the service completion.

The dynamic programming equations for the discrete-time problem are seen to be given by

$$\begin{cases} V_{n+1} = d'x + \lambda V_n(Ax) + a\,V_n(Dx) \\ \qquad + a \min\{V_n(x), V_n(Tx)\}, \qquad \text{if } x^2 > 0 \\ V_{n+1} = d'x + \lambda V_n(Ax) \\ \qquad + 2a \min\{V_n(x), V_n(Tx)\}, \qquad \text{if } x^2 = 0 \end{cases}$$

where

$$\begin{aligned} x^1 &= \min\{y^1, y^2\}, \quad x^2 = y^3 \\ d'x &= d_1x^1 + d_2x^2 = 2c_1x^1 + (c_1 + c_2)x^2 \\ Ax &= (x^1 + 1, x^2) \\ Dx &= (x^1, x^2 - 1) \\ Sx &= (x^1 - 1, x^2 + 1) \quad \text{if } x^1 > 0 \text{ and } Sx = x \text{ otherwise}. \end{aligned}$$

One then shows by induction on n that V_n has the following properties:

$$\begin{aligned} &V_n(Sx) - V_n(x) \text{ is nonincreasing in } x^1 \text{ for } x^2 \ge 0 \\ &V_n(Sx) - V_n(x) \text{ is nondecreasing in } x^2 \text{ for } x^1 \ge 0 \\ &V_n(x) \text{ is increasing in } x^1 \text{ and } x^2 \\ &V_n(x + y) - V_n(x) \text{ is nondecreasing in } x^1, x^2 \text{ for all } y \ge 0 \\ &V_n(x^1, 1) - V_n(x^1 + 1, 0) \ge V_n(x^1, 0) - V_n(x^1, 1) \text{ for } x^1 \ge 0. \end{aligned}$$

The first two properties are then sufficient to conclude that the optimal policy is of the switch-over type. (See Exercise 8.9.)

8.6. AVERAGE COST

In many applications one desires to design a system that yields a satisfactory reward per unit of time, or that corresponds to a small operating cost per unit of time. The mathematical formulation of such problems is discussed in this section.

Definition 8.6.1 [Average cost] For a given policy $u_n = \gamma_n(x_0, \ldots, x_n)$, define the **average cost per unit of time** by

$$\bar{V}_\gamma(i) = \limsup_{N \to \infty} \frac{1}{N} E_i\{\sum_{n=0}^{N-1} c(x_n,u_n)\} . \tag{8.43}$$

Let also

$$\bar{V}(i) = \inf_\gamma \bar{V}_\gamma(i) . \tag{8.44}$$

The problem is to find a policy that achieves $\bar{V}$. It will be assumed that $c \geq 0$.

Remarks 8.6.2 [Difficulties]

(a) An optimal policy **may fail to exist.** To see this, consider a situation where $u_n \in \{0, 1\}$, $c(x_n, u_n) = 1$, and $x_n = n$ until the first value of n such that $u_n = 0$, at which time x_n freezes at its current value i, say, and generates the constant cost $(1 + i)^{-1}$ at each time thereafter. An average cost $(1 + i)^{-1}$ can than be achieved for any i, but the average cost 0 cannot be achieved.

(b) The optimal policy **may not be stationary**, i.e., of the form $u_n = \gamma(x_n)$. To see this, consider the case where $x_{n+1} = x_n + u_n$ and $c(x_n, u_n) = u_n + (1 - u_n)(1 + x_n)^{-1}$. In this example, $x_0 = 0$ and $u_n \in \{0, 1\}$. It is easily checked that the sequence $\{u_n, n \geq 0\}$
$= \{1, 0, 1, 0, 0, 1, 0, 0, 0, 1, 0, 0, 0, 0, 1, \ldots\}$
yields the optimal average cost which is 0, while any policy of the form $u_n = \gamma(x_n)$ corresponds to $\bar{V}_\gamma(0) = (1 + i)^{-1}$ where $i = \min\{j \mid \gamma(j) = 0\}$. If $\gamma(.) \equiv 1$, then the average cost is 1.

(c) One might be tempted to conjecture that $\bar{V}(i) = \lim_{N \to \infty} \frac{1}{N} V_N^1(i)$, and perhaps also that the optimal policies for V_N^1 converge to the optimal policy for $\bar{V}$. To see that this may fail to be the case, consider the following situation. Let $u_n \in \{1, 2, 3, \ldots\}$ and $x_0 = 0$. If $u_0 = u$, then $c(x_n, u_n) = 0$ for $n = 0, 1, \ldots, u - 2$, and $c(x_n, u_n) = u^2$ for $n = u - 1$, and $x_u = 0$. This can readily be made into a controlled Markov chain model. For this system, $V_N^1(0) = 0$ is achieved by $u_0 = N+1$ for all $N \geq 0$, but $\bar{V}(0) = 1$ and is achieved by $u_n \equiv 1$. The main idea of the example is that costs can be postponed as long as desired, but have to be paid ultimately.

(d) The example above also shows that the optimal policies for V^β **may fail to converge** to the optimal policy for $\overline{V}$ as $\beta \rightarrow 1$.

Sufficient condition 8.6.3 To avoid the lack of regularity of Remarks 8.6.2.(c), one can impose "uniform" recurrence conditions. One example is given next.

Assume that there is a state i_0 and some finite K such that

$$| V^\beta(i) - V^\beta(i_0) | \leq K \quad \text{for all } \beta \in (0, 1) \text{ and } i, \tag{8.45}$$

then there is an optimal policy of the form $u_n = \gamma(x_n)$ with

$$c \equiv \overline{V}_\gamma(i) \equiv \lim_{\beta \rightarrow 1} (1 - \beta) V^\beta(i_0) .$$

Moreover, there is a bounded function $h(i)$ such that

$$\begin{aligned} c + h(i) &= \min_u [c(i,u) + \sum_j P_{ij}(u) h(j)] \\ &= c(i,\gamma(i)) + \sum_j P_{ij}(\gamma(i)) h(j) \end{aligned} \tag{8.46}$$

and

$$h(i) = \lim_n [V^{\beta_n}(i) - V^{\beta_n}(i_0)]$$

for some sequence $\beta_n \rightarrow 1$.

This last result is used to infer some properties of $\overline{V}$ from properties of V^β. The resulting properties for $\overline{V}$ can be used to obtain properties of γ.

It is known that (8.45) holds if there is some i_0 such that the mean time from i to i_0, using the optimal policy for V^β is bounded over all i and $\beta \in (0, 1)$. This is an example of a recurrence condition. The intuition is that it makes it impossible to postpone at will the payment of costs.

Another sufficient condition 8.6.4 It turns out that the condition (8.45) is not satisfied for a large class of queueing system control problems. Fortunately, the condition discussed below is usually satisfied and guarantees the existence of a optimal stationary policy that is given by the dynamic programming equation (8.46).

Assume that $[P_{ij}(u), u \in \mathbf{U}(i), i, j \in \mathbf{X}]$ and $\{c(i, u) \equiv c(i), u \in \mathbf{U}(i), i \in \mathbf{X}\}$ satisfy the following conditions:

(a) For all $M \geq 0$ there is a finite set $\mathbf{Y} \subset \mathbf{X}$ such that

$$c(i) \geq M \quad \text{for } i \notin \mathbf{Y} \text{ and } u \in \mathbf{U}(i); \tag{8.47}$$

(b) For all $i \in \mathbf{X}$ there exists a finite set $\mathbf{X}_i \subset \mathbf{X}$ such that

$$P_{ij}(u) = 0 \quad \text{for } j \notin \mathbf{X}_i \text{ and } u \in \mathbf{U}(i); \tag{8.48}$$

(c) There is some $u_n \equiv \gamma(x_n)$ such that $\{x_n, n \geq 0\}$ is positive recurrent with a single communicating class and a finite average cost per unit of time.

Then there is an **optimal stationary policy** $u_n \equiv \gamma(x_n)$. Moreover, **the equation (8.46) holds.**

Condition (a) means that **the cost of far away states is growing without bounds.** Condition (b) says that **each state communicates with a finite set of neighbors.** Condition (c) guarantees that the problem is **not degenerate.** (That is, there is at least one policy with a finite cost.)

The objective of these conditions is again to avoid the arbitrary postponement of costs.

Example 8.6.5 [Counterexample to 8.6.4] One can verify that the example given in 8.6.2(c) fails to satisfy the conditions 8.6.4.

Example 8.6.6 [Routing to parallel M/M/1 queues] Consider the problem of Example 8.5.11. Assume that objective function is now the average cost per unit of time.

The claim is that the conditions 8.6.4 are satisfied for discrete-time version of the problem provided that $\lambda < \mu_1 + \mu_2$. This is obvious for (8.47) and (8.48). To see why the last condition is satisfied one can use the fact that the policy that sends arriving customers to the shortest queue (for instance) is ergodic and corresponds to finite average queue lengths.

Example 8.6.7 [Controlling parallel processors] Assume that $\lambda < \min\{\mu_1, \mu_2\}$ in the system of Example 8.5.14. Then the conditions 8.6.4 are satisfied.

To see this, one has to find a stationary policy that corresponds to an ergodic system with finite average queue lengths. One can check that the policy that always serves at the maximum rates has the required properties.

Comments 8.6.8 [Properties of h] The result of 8.6.4 does not tell us how to derive **structural properties** of h, and therefore of the optimal policy γ obtained by (8.46). One method for doing this is to approximate h by the limit of V_n^1. This method is outlined below.

Proposition 8.6.9 [From finite time to average cost] Assume that there is a state $i_0 \in \mathbf{X}$ and a policy such that

$$M(i) := E[\sum_{n=0}^{\tau} c(x_n,u_n) \mid x_0 = i] < \infty, \quad i \in \mathbf{X} \qquad (8.49)$$

where $\tau = \min\{n \mid x_n = i_0\}$.

Assume also that (with $V_n(i) := V_n^1(i)$)

$$V_n(i) - V_n(i_0) \text{ is nondecreasing in } n \geq 0 \qquad (8.50)$$

and

$$E[c(x_n, u_n) \mid x_0 = i_0] \leq E[c(x_n, u_n) \mid x_0 = i] \qquad (8.51)$$

for all $i \in \mathbf{X}$ and all policy.
Then

$$h(i) = \lim_{n \to \infty} h_n(i) := V_n(i) - V_n(i_0) \qquad (8.52)$$

exists, is finite, and is the minimum nonnegative solution to (8.46) which determines the optimal policy.

Proof: Consider a policy that agrees with the one such that (8.49) holds until time τ and then (if $\tau < n$) switches to the optimal policy for horizon n and initial state i_0. The cost of this policy starting from state i is at most $M(i) + V_n(i_0)$ so that

$$V_n(i) \leq M(i) + V_n(i_0).$$

Hence, $h_n(i) \leq M(i) < \infty$.

Denote by $V_{n+1}(i) = TV_n(i)$ the dynamic programming equations for the finite horizon. Then, by definition of h_n and h_{n+1},

$$h_{n+1} = Th_n - [V_{n+1}(i_0) - V_n(i_0)]. \qquad (8.53)$$

As $n \to \infty$, one has $h_{n+1} \to h$ and $Th_n \to Th$ so that $V_{n+1}(i_0) - V_n(i_0)$ must converge to some value c.
Now, using (8.51) one finds that

$$E\{\sum_{m=0}^{kn} c(x_m, u_m)\} \geq kV_n(i_0). \qquad (8.54)$$

Indeed, the right-hand side assumes that one restarts the Markov chain in state i_0 every n steps and that one uses the optimal policy for each of the periods of n steps.

The equation (8.54) shows, after dividing by kn and letting n tend to infinity, that the limit c of $V_{n+1}(i_0) - V_n(i_0)$ must be such that

$$c = \sup_{n \geq 0} \frac{1}{n} V_n(i_0) \leq \bar{V}(i), \quad i \in \mathbf{X}. \tag{8.55}$$

Hence, (8.53) implies that $h = Th - c$, so that (8.46) follows.

Assume that γ is given by (8.46). One finds from (8.46) that

$$h(i) = E_i^\gamma \sum_{m=0}^{n-1} c(x_m, u_m) - nc + E_i^\gamma h(x_n)$$

and therefore

$$\frac{1}{n} E_i^\gamma \sum_{m=0}^{n-1} c(x_n, u_n) = \frac{1}{n} h(i) + c - \frac{1}{n} E_i^\gamma h(x_n).$$

Since $h \geq 0$ it follows that

$$\limsup_{n \to \infty} \frac{1}{n} E_i^\gamma \sum_{m=0}^{n-1} c(x_m, u_m) \leq c.$$

Combined with (8.55), this equation shows that

$$\lim_{n \to \infty} \frac{1}{n} E_i^\gamma \sum_{m=0}^{n-1} c(x_m, u_m) = c = \bar{V}(i), \quad i \in \mathbf{X}$$

which proves the optimality of γ.

Example 8.6.10 [Routing to parallel M/M/1 queues] Consider the problem of Example 8.6.6. One can show that the equations (8.49)–(8.52) are satisfied with $i_0 = (0, 0)$ and with the control policy of section 8.6. The reader is referred to the literature for the details.

The same considerations apply to Example 8.6.7. These results then show the optimality of switch-over policies for the average cost problem in Examples 8.6.6 and 8.6.7.

Comments 8.6.11

(a) In principle, one can solve an average cost problem by solving the dynamic programming equations (8.46). These equations can be formulated as a linear programming problem to which efficient algorithms can be applied. However, the size of the state space is usually very large, so approximation (e.g. truncation) methods have to be used.

(b) A numerical solution, useful as it may be in a concrete application, may not give much insight into how a system should be controlled. It is the objective of the papers on structural results to contribute to that insight.

(c) A very small number of optimal control problems have been solved for multiple queue systems.

8.7. EXERCISES

8.1. cμ-rule Show by an example that the result of Proposition 8.2.2 does not extend to more than one server.

8.2. Work-conservation Show that the value of (8.4) does not depend on the nonidling nonpreemptive policy.

8.3. Two servers Show that in Example 8.2.7 a policy cannot be optimal unless it uses the fast server whenever possible. *Hint:* Show that another policy can be improved. Do this by comparing the realizations of the processes under two policies.

8.4. Two servers (cont.) Show that, in the proof of Proposition 8.2.8, there must be queue lengths such that the slow server is used. *Hint:* The intuition is that if the queue length is very large, then it is very likely that a customer at the end of the queue would be served by the slow server before the fast server could become available.

8.5. SRPT Provide the details of the proof of Proposition 8.3.6.

8.6. HLF policies Show by an example that HLF policies may fail to be optimal for more than two processors (see Proposition 8.3.8).

8.7. Convexity Prove the result stated as the first of Examples 8.5.7.

8.8. Optimal threshold Give an algorithm for calculating the optimal threshold for the problem of Example 8.5.8 with an average cost criterion. *Hint:* One can evaluate the cost that corresponds to a given threshold by solving the balance equations for the invariant distribution.

8.9. Switch-over policies Give the details on the last statement of the proof of Proposition 8.5.15.

8.8. REFERENCES AND COMMENTS

Section 8.2 Interchange arguments are standard in stochastic scheduling. See Cox and Smith (1961) for 8.2.6. See also Baras, et al. (1985) for 8.2.4 (the proof is from Buyukkoc, et al. (1984)). Proposition 8.2.8 is due to Lin and Kumar (1984).

Section 8.3 The forward induction proof of Proposition 8.3.4 is from Ephremides, et al. (1980). The proof of Proposition 8.3.2 is a simple variation. Proposition 8.3.6 is due to Schrage (1968). Proposition 8.3.8 is due to Chandy and Reynolds (1977). When all the jobs have equal processing times, the optimality of the HLF policies was first noticed by Hu (1961). The same result continues to hold when the jobs have identical distributions that have log-concave increasing hazard rates, as shown in Ko (1985). Also, HLF policies remain optimal for exponentially distributed processing times with a rate that depends on the level of the job. See Pinedo and Weiss (1984). It is also known that HLF policies are asymptotically optimal for i.i.d. processing times and an arbitrary number of processors when the number of jobs tends to infinity, see Papadimitriou and Tsitsiklis (1985).

Section 8.4 Example 8.4.1 is from Ross (1983b), which is a highly recommended introduction to stochastic dynamic programming. Proposition 8.4.5(a) is due to Glazebrook (1979). Part (b) is due to Bruno, et al. (1981). The proof presented here is from Weber (1982b). See also Weiss (1982) for a survey of stochastic scheduling results. The result of Proposition 8.4.5(a) has been extended to jobs with stochastically ordered processing times; see Weber, et al. (1986). Proposition 8.4.7 is due to Weber (1980). The proof is from Nash and Weber (1982).

Section 8.5 The basic result of Proposition 8.5.3 is proved in Chapter 1 of Bertsekas (1987), Chapter 8 of Kumar and Varaiya (1986), Chapter 5 of Ross (1983b), and in Chapter 23 of Whittle (1983a). The Extensions 8.5.5 are also discussed in those references. The method introduced in 8.5.6 for obtaining structural properties of optimal policies has been used by a number of authors. It is also explained in Chapter 15 of Whittle (1982) and in Section 8.4 of Kumar and Varaiya (1986). Example 8.5.7(b) is from Hajek (1984). The idea of converting a continuous-time problem into a discrete-time one has been used by many authors. See, for instance, Serfozo (1979)–(1981). Proposition 8.5.11 is a particular case of Hajek (1984). See also Rosberg, et al. (1982) for a similar result. Example 8.5.14 and Proposition 8.5.15 are from

Ko (1985). See also Weber and Stidham (1987) for the control of service rates in queues in a cycle. Recently, Bartroli and Stidham (1987) proposed a new method for proving results such as those of Examples 8.5.7.

Section 8.6 The average cost problem is discussed in the references given for section 8.5 on dynamic programming. We have used Ross (1983b) extensively. The conditions 8.6.4 are due to Borkar (1983). The method explained in 8.6.9 is from Hajek (1984).

CHAPTER NINE

MULTIARMED BANDIT PROBLEMS

9.1. INTRODUCTION

This chapter discusses a class of problems that are frequently encountered in the control of queueing systems. These problems concern the **optimal allocation of a single server** to a number of queues or customers. The terminology "multiarmed bandit problems" comes from the case where the tasks to be performed are plays of **slot machines.** (A slot machine is often called a one-armed bandit, for obvious reasons.) The basic method of analysis is the **interchange argument.** This method was already encountered in Chapter Eight, section 8.2.

Simple examples are discussed in section 9.2. **Gittins' theory** of Markov bandits is given in section 9.3. In section 9.4 one presents the case of **arm-acquiring bandits.** The optimal control of a **single server queueing network** is discussed in that section where the theorem of Klimov is explained.

9.2. SIMPLE EXAMPLES

Example 9.2.1 [Deterministic sequences] Consider three sequences of nonnegative numbers $\{x_n^i,\ n \geq 0\}$ for $i = 1, 2, 3$. Let $\beta \in (0, 1)$ be such that $\sum_{n=0}^{\infty} \beta^n x_n^i < \infty$ for $i = 1, 2, 3$.

At each time, one selects one of the three sequences. The **interpretation** is that there are three **slot machines** and that x_n^i is the reward obtained the $(n+1)$st time that machine i is played. Denote by r_n the reward at time n, for $n \geq 0$. The problem is to find the **optimal order** in which the machines are to be played so as to maximize

$$R = \sum_{n=0}^{\infty} \beta^n r_n .$$

Proposition 9.2.2 [Optimality of index rule] Consider sequence i and assume that it has already been selected n times, so that the next reward from this sequence is x_n^i. Define

$$\gamma_n^i = \sup_{\tau > n} \frac{\sum_{m=n}^{\tau - 1} \beta^m x_m^i}{\sum_{m=n}^{\tau - 1} \beta^m} \tag{9.1}$$

where the maximization is over all the integers τ larger than n.

The interpretation is that γ_n^i is the **maximum reward per unit of time** (both discounted) that can be obtained from the remainder of sequence i. The number γ_n^i is called the **Gittins index** of the sequence after it has been operated n times. Then the policy that at each time selects the sequence with the largest index value is optimal. This policy is called the **index rule.**

Since this result is a particular case of a stochastic version, the proof will not be given at this time.

Notice that the calculation of the indices is done sequence by sequence. This results in a **decomposition** of the original problem.

Example 9.2.3 [Numerical example] Let

$$x^1 = (3, 2, 4, 1, 0, 0, \ldots),$$
$$x^2 = (2, 3, 2, 0, 0, \ldots),$$
$$x^3 = (2, 1, 4, 0, 0, \ldots).$$

Let also $\beta = 0.5$.
One finds that

$$\gamma_0^1 = \max\{3, \frac{3 + 2(0.5)}{1 + 0.5}, \frac{3 + 2(0.5) + 4(0.5)^2}{1 + 0.5 + (0.5)^2},$$
$$\frac{3 + 2(0.5) + 4(0.5)^2 + 1(0.5)^3}{1 + 0.5 + (0.5)^2 + (0.5)^3}\} = 3.$$

This procedure allows us to verify that

$$\gamma_0^1 = 3, \quad \gamma_1^1 = \frac{8}{3}, \quad \gamma_2^1 = 4, \quad \gamma_3^1 = 1, \quad \gamma_4^1 = 0, \ldots$$

$$\gamma_0^2 = \frac{7}{3}, \quad \gamma_1^2 = 3, \quad \gamma_2^2 = 2, \quad \gamma_3^2 = 0, \ldots$$

$$\gamma_0^3 = 2, \quad \gamma_1^3 = 2, \quad \gamma_2^3 = 4, \quad \gamma_3^3 = 0, \ldots .$$

The index rule then corresponds to the following sequence of machine selections:

$$1, 1, 1, 2, 2, 3, 3, 3, 2, 1 .$$

Example 9.2.4 [Nonincreasing rewards] For $1 \leq i \leq J$ let $\{x_n^i, n \geq 0\}$ be a sequence of nonincreasing random variables. That is,

$$P\{x_{n+1}^i \leq x_n^i, \ n \geq 0, \ 1 \leq i \leq J\} = 1 . \tag{9.2}$$

As in Example 9.2.1, x_n^i is the reward obtained when slot machine i is played the $(n+1)$st time.

Assuming that the reward to be obtained next by each slot machine is known, what is the optimal sequence in which the slot machines should be played one at a time so as to maximize

$$E\{\sum_{n=0}^{\infty} \beta^n r_n\}$$

where $0 < \beta \leq 1$ and r_n is the reward obtained at the $(n+1)$st play?

Proposition 9.2.5 [Optimality of myopic strategy] Under the assumption (9.2), the optimal policy is to play the slot machine with the largest next reward.

Proof: This intuitively obvious result can be proved by a straightforward interchange argument. The details are left to the reader.

9.3. MARKOV BANDITS

This section presents the basic result on multiarmed bandit problems in the case of Markov bandits.

Example 9.3.1 [When to retire] Let $\mathbf{x} = \{x_n, n \geq 0\}$ be a Markov chain on $\mathbf{X}$ with transition matrix P. The interpretation is that the values of x_n are the successive states of a slot machine. It is assumed that the states are observable (i.e., that the machine is open, say). If the machine is played when its state is x_n, then it yields the reward $f(x_n)$.

This situation will be summarized by saying that one is given a slot machine $(\mathbf{x}, f(.))$. At each time $n \geq 0$, one is given the option of retiring forever, and then collect a constant reward R every unit of time after retiring, or to continue to play. Denote by σ the random retirement time. This time must be a stopping time of $\mathbf{x}$ for the decision to be physically feasible.

The problem is to find the stopping time σ that maximizes

$$E_i\{\sum_{n=0}^{\sigma-1} \beta^n f(x_n) + \sum_{n=\sigma}^{\infty} \beta^n R\}. \tag{9.3}$$

where $0 < \beta < 1$ is a discount factor and $E_i\{.\} := E[. \mid x_0 = i]$. To guarantee that this reward is well defined it will be assumed that

$$E_i\{\sum_{n=0}^{\infty} \beta^n \mid f(x_n) \mid\} < \infty, \quad i \in \mathbf{X}. \tag{9.4}$$

Proposition 9.3.2 [Optimal Stopping] For each $i \in \mathbf{X}$ define

$$\gamma(i) = \sup_{\tau > 0} \frac{E_i\{\sum_{n=0}^{\tau-1} \beta^n f(x_n)\}}{E_i\{\sum_{n=0}^{\tau-1} \beta^n\}} \tag{9.5}$$

where the maximization is over all stopping times $\tau > 0$ of $\mathbf{x}$.

The quantity $\gamma(i)$ is called the **Gittins index** of the machine $(\mathbf{x}, f(.))$ in state i. The stopping time σ that maximizes (9.3) is then given by

$$\sigma = \min\{n \geq 0 \mid \gamma(x_n) \leq R\} = \min\{n \geq 0 \mid x_n \in S(R)\} \tag{9.6}$$

where $S(R) = \{i \in X \mid \gamma(i) \leq R\}$.

Moreover, the stopping time τ that achieves $\gamma(i)$ in (9.5) is given by σ in (9.6) with R replaced by $\gamma(i)$. Thus, the supremum in (9.5), which is well defined by (9.4), is also achieved by some stopping time. An algorithm is given in Exercises 9.1 and 9.2.

Proof: First consider the case when $R = 0$. If $\gamma(i) \le 0$, then the value of (9.3) is nonpositive for any positive stopping time σ. It is zero for $\sigma \equiv 0$. Thus it is optimal to retire in a state i with $\gamma(i) \le 0$. Similarly, it is not optimal to retire in a state i with $\gamma(i) > 0$. Thus, σ given by (9.6) is optimal when $R = 0$.

For $R \ne 0$, observe that replacing $f(.)$ by $f(.) - R$ reduces (9.3) by $R(1 - \beta)^{-1}$ for any σ and replaces every $\gamma(i)$ by $\gamma(i) - R$. The problem is then reduced to the case $R = 0$.

To prove the statement about τ, notice that when $R = \gamma(i)$ the objective function (9.3) can be written as

$$E_i\{\sum_{n=0}^{\sigma-1} \beta^n f(x_n)\} - \gamma(i)\, E_i\{\sum_{n=0}^{\sigma-1} \beta^n\} + (1-\beta)^{-1}\gamma(i)\,.$$

This shows that the stopping time σ that maximizes this expression is the same as the one that maximizes (9.5).

Proposition 9.3.3 [No idle time if $R \ge 0$] Consider the same situation as in Example 9.3.1, but assume that one is given the additional option of inserting **random idle times** between plays. Thus, instead of playing the machine continuously and then retiring forever, one has the possibility of pausing for random durations between the plays, until one retires forever.
The claim is that the optimal policy does not insert idle times if $R \ge 0$.

Proof: Assume that it is optimal to pause for ϵ seconds once the slot machine reaches state i. Since the situation is the same after ϵ seconds, it must be optimal to keep on pausing. Hence, if it is optimal to pause in state i, then it is optimal to pause forever after reaching that state, i.e., to forego any further rewards. Obviously, this is not useful if $R \ge 0$.

Example 9.3.4 [B slot machines] For $1 \le b \le B$, $(\mathbf{x}^b, f_b(.))$ is a slot machine as in Example 9.3.1. At each time n one must play one of the B slot machines. Let r_n be the reward obtained at time n. Thus, $r_n = f_b(x_m^b)$ if machine b is played at time n for the $m+1$st time.

The objective is to maximize

$$E\{\sum_{n=0}^{\infty} \beta^n r_n\}\,. \tag{9.7}$$

Proposition 9.3.5 [Optimality of index rule] The optimal strategy is to play at each time n the slot machine with the largest index value.

Proof: Assume, without loss of generality, that $\nu := \gamma_1(x_0^1) \geq \gamma_b(x_0^b)$ for $2 \leq b \leq B$. ($\gamma_b(x_n^b)$ is the index of slot machine b in state x_n^b, calculated as in (9.5).) By subtracting ν from each $f_b(.)$, one can also assume that $\nu = 0$.

It will be shown that there is an optimal policy that selects machine 1 at time 0. One can repeat the argument at subsequent times, and this will prove the theorem. The basic idea is to show, by an interchange argument, that an arbitrary policy can be improved by a modification such that machine 1 is selected at time 0.

Let then π be an arbitrary policy and assume that its sequence of rewards is as follows:

$$\pi : [\ldots]_1 \, f_1(x_0^1) \, [\ldots]_2 \, f_1(x_1^1) \ldots f_1(x_{\tau-1}^1) \, f_b(x_k^b) \, [\ldots]_l$$

where $[\ldots]_1, [\ldots]_2, \ldots [\ldots]_{l-1}$ indicate rewards obtained when machine 1 is not played and τ is the stopping time of x^1 that achieves $\gamma_1(x_0^1) = 0$. Denote by T the time when the reward $f_1(x_{\tau-1}^1)$ is received.

Define the policy $\tilde{\pi}$ so that it corresponds to the rewards below:

$$\tilde{\pi} : f_1(x_0^1) \, f(x_1^1) \ldots f_1(x_{\tau-1}) \, [\ldots]_1 \, [\ldots]_2 \ldots f_b(x_k^b) \, [\ldots]_l \, .$$

By construction, the two sequences of rewards agree after time T. Denote by R and $\tilde{R}$, respectively, the expected rewards of these two policies up to time T. It suffices to show that $\tilde{R} \geq R$.

To do this, denote by t_n^b [resp. s_n^b] the time when b is selected for the $(n+1)$st time by the policy π [resp. $\tilde{\pi}$]. Observe that $t_n^b - n$ and $t_n^b - s_n^b$ are nondecreasing in n for $b \neq 1$. Now,

$$\tilde{R} - R = E\{\sum_{n=0}^{\tau-1} \beta^n f_1(x_n^1)\} - E\{\sum_{n=0}^{\tau-1} \beta^{t_n^1} f_1(x_n^1)\}$$
$$+ \sum_{b=2}^{B} E\{\sum_{n=0}^{m_b} (\beta^{s_n^b} - \beta^{t_n^b}) \, f_b(x_n^b)\} \qquad (9.8)$$

where $m_b + 1$ is the number of times that machine b is selected up to time T by the policies π and $\tilde{\pi}$.

Notice that the first term on the right-hand side of (9.8) is equal to zero. The second term is nonpositive since it corresponds to the expected reward obtained by playing **x** with inserted idle times (see Proposition 9.3.3).

It remains only to show that the term corresponding to each $b = 2, \ldots, B$ is nonnegative. To do this, one first observes that

$$E\{\sum_{n=0}^{m_b} (\beta^{t_n^b} - \beta^{s_n^b}) f_b(x_n^b)\} = E\{\sum_{n=0}^{m_b} \beta^{\sigma_n^b} f_b(x_n^b)\} \tag{9.9}$$

where the σ_n^b are random times such that $\sigma_{n+1}^b \geq \sigma_n^b + 1$. To verify this fact, define σ_n^b by $\beta^{\sigma_n^b} = \beta^{t_n^b} - \beta^{s_n^b}$ and notice that the construction of the policy $\tilde{\pi}$ is such that $\beta^{\sigma_n^b - n} = \beta^{t_n^b - n}(1 - \beta^{s_n^b - t_n^b})$ is nonincreasing in n, so that $\sigma_n^b - n$ must be nondecreasing in n.

Also, the value of σ_n^b is a function of $\mathbf{x}^b$ up to time n and of the evolution of the other slot machines; it is therefore a stopping time of $\mathbf{x}^b$ by the independence of the machines. See Definition 2.9.7.

Thus, one may view (9.9) as the expected reward of playing $\mathbf{x}^b$ with inserted idle times, the idle times being randomized by the evolution of the other slot machines. Since these idle times cannot improve the maximum expected reward which is nonpositive, it follows that (9.9) is nonpositive. This completes the proof.

Example 9.3.6 [Emptying a queueing network] Consider a queueing network with J queues and a single server. The routing is Markov. That is, a customer who leaves queue i is sent to queue j with probability r_{ij} for $1 \leq j \leq J$, and leaves the network otherwise. There are no arrivals and the service times are equal to one unit of time.

The problem is to decide at each time which customer should be served so as to minimize

$$E\{\sum_{n=0}^{\infty} \beta^n \sum_{i=1}^{J} c(i)[y_n^i - 1\{s_n = i\}]\} \tag{9.10}$$

where y_n^i denotes the number of customers at time n in queue i ($n \geq 0$ and $1 \leq i \leq J$), $0 < \beta < 1$, and s_n indicates the queue being served at time n. The interpretation of the cost (9.10) is that each customer is charged a waiting cost equal to $c(i) \geq 0$ for each unit of time he **waits** (and is not being served) in queue i ($1 \leq i \leq J$).

That is, (9.10) is the total discounted waiting cost.

This problem can be reformulated as in Example 9.3.4. To do this, say that there are B customers at time 0. Each customer b will be viewed as a bandit modeled by a Markov chain on $\{0, 1, \ldots, J\}$ with transition matrix $R = [r_{ij}, 0 \le i, j \le J]$ with $r_{0j} = 1\{j = 0\}$. The reward of playing (serving) customer b when it is in queue i is defined to be

$$f_b(i) = c(i) - \beta \sum_j r_{ij}\, c(j), \quad 1 \le i \le J \tag{9.11}$$

and $f_b(0) = 0$.

To see the equivalence between the original problem and that of Example 9.3.4, denote by t_n^b the $(n+1)$st time that customer b is served $(1 \le b \le B)$. Then, the waiting cost in (9.10) paid by customer b is seen to be equal to

$$c(x_0^b) \sum_{m=0}^{t_0^b - 1} \beta^m + c(x_1^b) \sum_{m = t_0^b + 1}^{t_1^b - 1} \beta^m + \cdots .$$

This cost can be rewritten as

$$c(x_0^b) \frac{1 - \beta^{t_0^b}}{1 - \beta} + c(x_1^b) \frac{\beta^{t_0^b + 1} - \beta^{t_1^b}}{1 - \beta} + \cdots$$

$$= \frac{c(x_0^b)}{1 - \beta} - \frac{1}{1 - \beta} \sum_{n=0}^{\infty} \beta^{t_n^b} [c(x_n^b) - c(x_{n+1}^b)] .$$

Therefore, the average value of the waiting cost paid by customer b is equal to

$$\frac{c(x_0^b)}{1 - \beta} - \frac{1}{1 - \beta} E\{ \sum_{n=0}^{\infty} \beta^{t_n^b} f_b\,(x_n^b)\}$$

where $f_b(.)$ was defined in (9.11). This shows that minimizing (9.10) is equivalent to maximizing (9.7).

As a consequence, we know that the optimal policy is to serve the customer (bandit) with the largest index value. Since the states of the customers are the queues, this means that the index rule is in fact a priority ordering of the queues. That is, there is an ordering $(i_1, i_2, \ldots, i_J)$ and a number $K \le J$ such that the optimal policy is to always serve the first nonempty queue in the order $(i_1, \ldots, i_K)$ and never to serve the other queues. To determine the optimal ordering one has to calculate the indices (9.5). The number K is such that the indices of the queues $\{i_{K+1}, \ldots, i_J\}$ are negative. In view of (9.11), these indices become

$$\gamma(i) = \max_{\tau > 0} \frac{E_i\{\sum_{n=0}^{\tau-1} \beta^n [c(x_n) - \beta c(x_{n+1})]\}}{E_i\{\sum_{n=0}^{\tau-1} \beta^n\}}$$

$$= \max_{\tau > 0} \frac{c(i) - E_i\{\beta^\tau c(x_\tau)\}}{E_i\{\sum_{n=0}^{\tau-1} \beta^n\}} . \tag{9.12}$$

The interpretation of the right-hand side of (9.12) is that $\gamma(i)$ is the maximum rate of reduction of waiting cost per unit of time that can be achieved by serving a customer who is initially in queue i.

Variation 9.3.7 [Total undiscounted cost] Consider once again the model of Example 9.3.6 but assume that $\beta = 1$.

A policy that minimizes (9.10) with $\beta = 1$ must be nonidling. That is, such a policy should keep serving as long as the network is nonempty. Indeed, if the policy stops serving when the network is in some state, then it will never serve again (for the state stops changing — there are no arrivals), and the total cost will be infinite, which cannot be optimal.

Observe that the cost

$$E\{\sum_{n=0}^{\infty} \sum_{i=1}^{J} c(i)\, 1\{s_n = i\}\} = E\{\sum_{n=0}^{\infty} c(s_n)\}$$

is the same for all nonidling policies. As a consequence, the minimization of (9.10) when $\beta = 1$ is equivalent to the minimization of

$$E\{\sum_{n=0}^{\infty} \sum_{i=1}^{J} c(i)\, y_n^i\} ,$$

i.e., of the total waiting cost including the cost while being served.

Examples 9.3.6 and 9.3.7 are finite problems, in the sense of Chapter Eight. As a consequence, the optimal policy for Example 9.3.6 converges to the optimal policy for 9.3.7 as $\beta \to 1$. Therefore, the optimal policy for 9.3.7 is an index rule where the indices are obtained as in (9.12) but with $\beta = 1$. This gives

$$\gamma(i) = \max_{\tau > 0} \frac{c(i) - E_i\{c(x_\tau)\}}{E_i\{\tau\}} . \tag{9.13}$$

An algorithm can be found in Theorem 9.4.9 with $P(i, j) := r_{ij}$ and $m_i := 1$.

Variation 9.3.8 [Random service times] Consider the problems of Examples 9.3.6 and 9.3.7 but assume now that the service times are all independent and are distributed as σ_i in queue i, with $m_i = E\{\sigma_i\} \in (0, \infty)$.

One can once again consider the problems as multiarmed bandit problems. The only difference is that the stopping times used to play the bandits are restricted. More specifically, one can get a Markov model of each bandit (customer) by using a discrete-time phase-type approximation for the service times. Once a customer starts being served, the server can only switch customers after a number of service completions. The service completions correspond to a subset of states of the Markov chain that models the evolution of the customer.

The discussions of 9.3.1–9.3.5 can then be reproduced with this restriction on the admissible stopping times. It follows that the optimal policy is an index rule where the indices are calculated as in (9.5) except that the stopping times must correspond to service completion times.

Consider the problem 9.3.7 with random service times. The cost to be minimized is then

$$E\{\int_0^\infty \sum_{i=1}^J c(i)\, y_t^i \, dt\} \tag{9.14}$$

where y_t^i is the number of customers at time t in queue i. The optimal policy is the index rule (i.e., priority rule) based on the indices (9.13), where the stopping times must be service completion times.

An algorithm for finding the optimal priority rule will be given in Theorem 9.4.9. It will be explained there that this priority rule also minimizes the average waiting cost per unit of time when customers arrive into the queues as independent Poisson processes.

Example 9.3.9 There are B jobs with independent service times $\sigma_1, \ldots, \sigma_B$. All these random variables takes values in $\{1, 2, 3, \ldots\}$. The problem is to find the preemptive policy that maximizes

$$E\{\sum_{b=1}^{B} c_b \beta^{T_b}\} \tag{9.15}$$

where T_b denotes the completion time of job b, β is a discount factor in (0, 1), and the c_bs are positive constants.

Let $F_b(t) = P\{T_b \leq t\}$, for $1 \leq b \leq B$ and $t \geq 0$. Define x_n^b by

$$x_n^b = \begin{cases} n \text{ if } T_b > n \\ \partial_0 \text{ if } T_b = n \\ \partial_1 \text{ if } T_b < n \end{cases}$$

Let also $f_b : X^b := \{\partial_0, \partial_1\} \cup \{0, 1, 2, \ldots\} \to R$ be defined by $f_b(x) = c_b 1\{x = \partial_0\}$.

One then verifies that each x^b is a Markov chain on X^b and that the problem is now reformulated as a multiarmed bandit problem. It remains only to calculate the indices (9.5) for these bandits. This is left as an exercise. See Exercise 9.3.

Example 9.3.10 Consider the previous problem for $\beta \approx 1$. Notice that $\beta^T = [1 - (1 - \beta)]^T \approx 1 - (1 - \beta)T$. As a consequence, the reward to be maximized in (9.15) is almost equal to

$$E\{\sum_{b=1}^{B} c_b[1 - (1 - \beta)T_b]\} = \sum_{b=1}^{B} c_b - (1 - \beta) E\{\sum_{b=1}^{B} c_b T_b\}.$$

Therefore, this problem is almost equivalent to that of minimizing the waiting cost

$$E\{\sum_{b=1}^{B} c_b T_b\}. \tag{9.16}$$

This leads to suspect that the above cost will be minimized by the index rule corresponding to the indices of 9.3.9 calculated for β equal to one. This conjecture is correct, and can be proved directly by applying the interchange argument to the cost (9.16). This is left as an exercise. (See Exercise 9.5.)

Example 9.3.11 [Tax problem] Consider the setup of Example 9.3.4. Assume that the problem is now to minimize

$$E\{\sum_{n=0}^{\infty} \beta^n c_n\} \tag{9.17}$$

where $c_n = \sum_{b \neq b_n} f_b(x_{n_b}^b)$ if machine b_n is selected at time n when the states of the other machines are $x^b_{n_b}$ for $b \neq b_n$.

The interpretation is that if machine b is not selected when its state is x, then it incurs a "waiting cost" $f_b(x)$. The problem is to minimize the total discounted waiting cost.

The argument of 9.3.6 shows that this can be reformulated as a bandit problem with rewards

$$\tilde{f}_b(x) = f_b(x) - \beta E\{f_b(x_1^b) \mid x_0^b = 0\} . \qquad (9.18)$$

This then permits to solve the problem by an index rule.

9.4. ARM-ACQUIRING BANDITS

This section deals with the situation where new bandits arrive continuously. It will be shown that, under suitable assumptions on the arrival process, the optimal policy is again given by an index rule.

Problem 9.4.1 [Arriving bandits] The bandits are again described by Markov chains as in Example 9.3.4. If bandit b is played at time n when it is in state $i \in \mathbf{X}$, then

(a) a reward $r_n = f(i)$ is received;

(b) bandit b moves to state j with probability $P(i, j)$ for $j \in \mathbf{X}$;

(c) the other bandits present at time n do not change state;

(d) N_j new bandits arrive with state j for $j \in \mathbf{X}$.

It is assumed that the random vector $\{N_j, j \in \mathbf{X}\}$ is independent of the evolution of the bandits up to time n. Also, those arrivals occur even if no bandit is played at time n.

The problem is to find, at each time, which bandit should be played in order to maximize $E\{\sum_{n=0}^{\infty} \beta^n r_n\}$.

Theorem 9.4.2 [Index rule] There is an ordering $(i_1, i_2, \ldots)$ of the states in $\mathbf{X}$ and a number K so that the optimal policy is, at each time, to play the bandit with the first state in the order $(i_1, \ldots, i_K)$ that is present and never play bandits in other states.

The state i_1 does not depend on the distribution of the random variables $\{N_j, j \in \mathbf{X}\}$. The other states and K generally depend on that distribution.

The proof will be outlined in Comments 9.4.4. First, we need a new concept.

Definition 9.4.3 [Dominating policy] Assume that at time 0 there is a single bandit available and that the bandit is in state i. At time 0 one plays that bandit and new bandits arrive. At time 1 one can choose either to play the initial bandit or one of the bandits that arrived subsequently. The game continue at later times. Denote by $\{r_n^\pi, n \geq 0\}$ the sequence of rewards that is obtained by using a given policy π.

The policy π is said to be **dominating** if, for all other policy $\tilde{\pi}$,

$$R(\pi, a) := \max_{\tau > 0} E\{\sum_{n=0}^{\tau - 1} \beta^n [r_n^\pi - a]\} \geq R(\tilde{\pi}, a) \text{ for all } a \in \mathbb{R}. \quad (9.19)$$

In (9.19), the maximization is over all the positive stopping times of the evolution of the bandits.

Comments 9.4.4 [Outline of the proof of Theorem 9.4.2]

(a) The interpretation of a dominating policy is that it is preferable to any other policy even if one is given the option of retiring and earn a constant reward a thereafter, for any value of a. It can be shown that there exists a dominating policy.

Thus, the dominating policy π is such that the corresponding sequence of rewards, denoted by $\{r_n^i, n \geq 0\}$ when i is the initial state of the bandit, is the best possible, regardless of future retirement options being presented to the player. This leads to suspect that this will also be the optimal way of playing the initial bandit if other bandits are available. That intuition can be shown to be correct. The proof is similar to that of Proposition 9.3.5.

(b) Assume now that at time 0 there are two bandits with states i and j, respectively. Which bandit should be played initially? The choice can be viewed as choosing how to play the two sequences of random rewards $\{r_n^i, n \geq 0\}$ and $\{r_n^j, n \geq 0\}$. To see this, notice that one can view one possibility of the game as playing the initial bandit in state i with its subsequent arrivals and then playing the initial bandit in state j with its own subsequent arrivals. Indeed, the arrivals available after j is played do not not depend on when that sequence is played. (See Figure 9.1.)

As a consequence, the optimal choice is to play initially the bandit with the largest Gittins index of $\{r_n^i, n \geq 0\}$ or

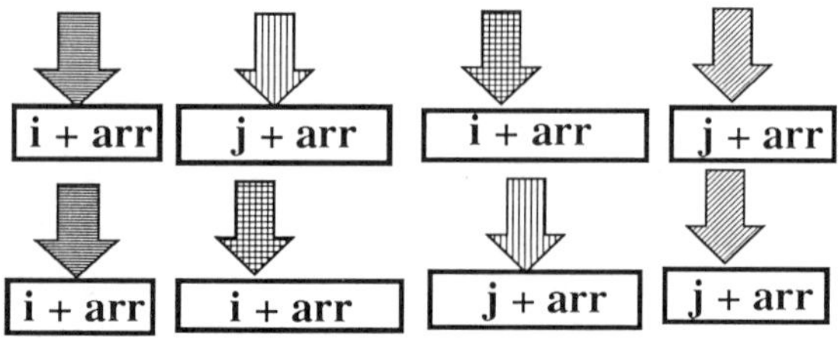

Figure 9.1. The sequences of rewards with arrivals.

$\{r_n^j, n \geq 0\}$.

The argument extends to an arbitrary number of bandits and proves Theorem 9.4.2. The ordering $(i_1, i_2, \ldots)$ corresponds to decreasing values of the Gittins indices of the sequences $\{r_n^i, n \geq 0\}$ for $i \in \mathbf{X}$. Also, the value K is the minimal value of k such that the Gittins index of $\{r_n^{k+1}, n \geq 0\}$ is negative.

Variation 9.4.5 [Semi-Markov bandits] The result of Theorem 9.4.2 continues to be true when the bandits are modeled by semi-Markov processes. That is, one may assume that the duration of a play of a bandit in state i is random. Moreover, it can be assumed that the duration of a play in state i and the number of arrivals during that play are not independent.

The problem is to maximize

$$E\{\sum_{n=1}^{\infty} e^{-\beta T_n} r_n\} \tag{9.20}$$

where T_n is the time when the nth reward r_n is received ($n \geq 1$).

This can be viewed by considering that the bandits are Markov but that the optimization is over a restricted set of stopping times that correspond to "complete plays." This is similar to the discussion of Variation 9.3.8.

Particular case 9.4.6 [Poisson arrivals] The calculations of the optimal ordering will be detailed in the case of semi-Markov bandits with Poisson arrivals.

The model of each bandit is as follows. The possible states are $\{0, 1, 2, \ldots, J\}$. When the bandit is played in state $i \in \{1, \ldots, J\}$, the reward R_i is obtained at the **end** of the play, the play has a duration σ_i that is independent of the past evolution, and the next state is j with probability $P(i, j)$ for $j \in \{0, 1, \ldots, J\}$ independently of σ_i and of the past evolution. When the bandit is in state 0, it remains in that state and no reward is obtained by playing it.

New bandits arrive with initial states in $\{1, 2, \ldots, J\}$ as independent Poisson processes with rate λ_i for state i.

Comments 9.4.7 [Single-server queueing network] One can view the model of 9.4.6 as being a network of J queues and with Poisson arrivals into the queues. The service times in all the queues are independent and are distributed as σ_i in queue i. There is a single server in the network. (See Figure 9.2.)

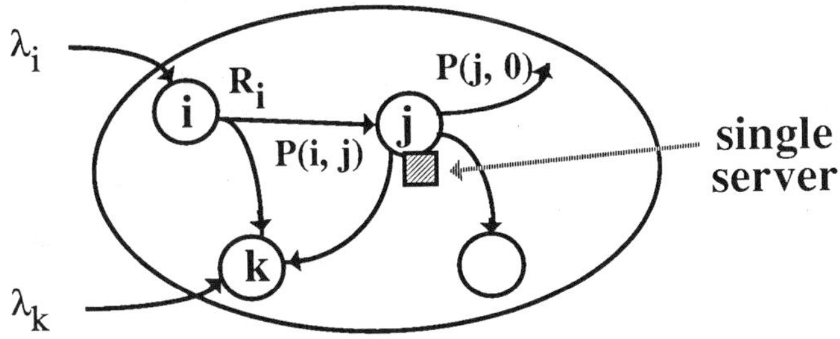

Figure 9.2. The single-server network.

When a customer is served in queue i, a reward R_i is received, and the customer is sent to queue j with probability $P(i, j)$ for $1 \leq j \leq J$ and leaves the network otherwise.

The problem is then to find, at each service completion time, which customer should be served nonpreemptively so as to obtain the maximum total discounted reward.

From the discussion of 9.4.5 we know that the optimal policy corresponds to an ordering $(i_1, \ldots, i_J)$ of the queues and to a number $K \in \{1, \ldots, J\}$ such that at each service completion time, the server should start serving the first nonempty queue in the order $(i_1, \ldots, i_K)$ and should never serve the other queues.

The reader is referred to the literature for the calculation of the optimal ordering. The case of the average cost per unit of time turns out to be substantially simpler to treat. The result is given next.

Problem 9.4.8 [Average reward per unit of time] The formulation is the same as in 9.4.7 except that the objective is to minimize

$$\limsup_{T \to \infty} \frac{1}{T} E\{\int_0^\infty \sum_{i=1}^J c_i\, x_t^i\, dt\} \tag{9.21}$$

over the **nonpreemptive** service policies. In (9.21), x_t^i is the number of customers in queue i at time t and the positive constants c_i represent the cost per unit of time for a customer in queue i $(1 \le i \le J)$.

Klimov's Theorem 9.4.9 [Optimality of index rule] The optimal policy corresponds to an **ordering** of the queues such that the server always chooses to serve a customer in the first nonempty queue in that order.

The optimal order is calculated as follows.

Step 1 [Highest priority queue] The highest priority queue is the one that achieves

$$\max_{1 \le i \le J} \frac{c_i - \sum_j P(i, j)\, c_j}{m_i}$$

where $m_i := E\{\sigma_i\}$.

Order the queues so that the highest priority queue is queue 1.

Step 2 [Remove queue 1] Modify the network by "removing" queue 1. To do this, calculate

$$\begin{aligned} P'(i, j) &:= P(i, j) + P(i, 1)P(1, j), \quad 2 \le i,\ j \le J \\ m'_j &= m_j + P(j, 1)m_1, \quad 2 \le j \le J \\ c'_j &= \frac{c_j m_j + P(j, 1)c_1 m_1}{m'_j}, \quad 2 \le j \le J . \end{aligned}$$

Step 3 [Next highest priority queue] Consider the network with $J - 1$ queues defined in Step 2. The highest priority queue for this network is obtained as in Step 1. It is the second highest priority queue for the original network.

The procedure then continues until all the priorities have been determined.

Comments 9.4.10

(a) The priority ordering does not depend on the arrival rates. It is the same as the index rule that minimizes the total waiting cost until the network is empty when there are no arrivals.

(b) The interpretation of Step 1 is that the highest priority queue is the one that reduces the waiting cost the fastest. It can be justified as follows. Say that the arrival rates are so large that the queues never become empty. If the highest priority queue is i, then only that queue will be served. By serving a customer in queue i, one reduces the waiting cost per unit of time of all the customers by $a_i := c_i - \sum_j P(i, j)\, c_j$. Consider the cost over a large time interval $[0, T]$. Say that $T \approx Km_i$. Serving queue i will then reduce the cost over $[0, T]$ of the first customer served by $a_i(K - 1)m_i$, that of the second served customer by $a_i(K-2)m_i$, and that of the next to last customer served by $a_i m_i$. Thus, the total reduction of cost is roughly equal to $a_i m_i K^2 \approx a_i m_i^{-1} T^2$.

(c) The justification for Step 2 is that the customers will always be served when they are in queue 1 (the highest priority queue). Thus, one may consider that the server will follow a customer who starts in some queue i until queue 1 is departed (if the customer happened to go from queue i to queue 1). The motion of that customer is then the same as in the modified network. The mean service time is calculated by considering the mean time until the customer completes the initial service and leaves queue 1. The average waiting cost per unit of time is also modified accordingly.

9.5. EXERCISES

9.1. Algorithm for indices Consider the situation of Example 9.3.1 and Proposition 9.3.2. Assume that $\mathbf{X} = \{1, 2, \ldots, N\}$ for a finite N. Show that the following algorithm orders the states in nonincreasing values of their Gittins indices. That is, it produces $\mathbf{X} = \{i_1, \ldots, i_N\}$ such that $\gamma(i_1) \geq \gamma(i_2) \geq \cdots \geq \gamma(i_N)$.

Step 1 [State i_1] A state i_1 with the highest index value is the one that achieves

$$\max_i f(i) .$$

Step k [State i_k] A state with the kth largest index value is a state i_k that achieves

$$\max_{i \notin A_k} \frac{E_i\{\sum_{n=0}^{\tau_k} \beta^n f(x_n)\}}{E_i\{\sum_{n=0}^{\tau_k} \beta^n\}}$$

where $A_k = \{i_1, \ldots, i_{k-1}\}$ and

$$\tau_k := \min\{n > 0 \mid x_n \notin A_k\}.$$

Hint: This follows from the property of σ stated in Proposition 9.3.2.

9.2. Calculations Propose a procedure to calculate the expressions needed in step k of the algorithm given in Exercise 9.1. *Hint:* Consider the case $k = 2$. One keeps playing a bandit when it is in state i_1. Thus, one can consider the bandit as taking values in $\mathbf{X} - \{i_1\}$ by watching it when it leaves the value i_1. This becomes a bandit on $\mathbf{X} - \{i_1\}$ with parameters (where $i, j \neq i_1$)

$$P'(i, j) = P_i\{x_{\tau_1} = j\}$$

$$f'(i) = \frac{E_i\{\sum_{n=0}^{\tau_1 - 1} \beta^n f(x_n)\}}{E_i\{\sum_{n=0}^{\tau_1 - 1} \beta^n\}}$$

where τ_1 was defined in Exercise 9.1.

One then applies Step 1 of the algorithm to find the state in $\mathbf{X} - \{i_1\}$ with the largest index value. That is, one finds that i_2 is the value of $i \neq i_1$ that achieves the maximum value of $f'(i)$. The procedure then continues by updating the parameters so as to define a new bandit on $\mathbf{X} - \{i_1, i_2\}$, etc.

The calculation of $f'(i)$ is done by first step equations. Denote by $a(i)$ the value of the numerator in $f'(i)$. Then

$$a(i_1) = f(i_1) + \beta P(i_1, i_1)\, a(i_1)$$
$$a(i) = f(i) + \beta P(i, i_1)\, a(i_1), \quad i \neq i_1.$$

This allows us to calculate $a(i)$ for $i \neq i_1$. The denominator of $f'(i)$ is obtained by the same equations with $f \equiv 1$.

9.3. Preemptive server allocation Calculate the indices required in Example 9.3.9. *Hint:* Say that one is serving job b, which requires a service time equal to σ_b. After having worked on that job for n steps, the only information available is whether the job is finished or not, i.e., whether $\sigma_b \leq n$. It follows that a stopping time τ is of the form $\tau = n \wedge \sigma_b$. (One could introduce a randomization, but this would not help, as explained in the next exercise.) It follows that the index of job b after it has been served $n+1$ times is given by

$$\gamma(n) = \max_{m > n} \frac{c_b E[\beta^{\sigma_b} 1\{\sigma_b \leq m\} \mid \sigma_b > n]}{E[\sum_{k=n}^{m \wedge \sigma_b} \beta^k \mid \sigma_b > n]} .$$

9.4. Randomization does not help In the preceding exercise we used the fact that randomizing the stopping time does not help. The intuition is that if there is a best decision, then it should be made with probability one, and it does not help to choose a worse decision with some positive probability. Give a formal proof of this observation. *Hint:* Let X be an independent random variable whose outcome decides the choice of a policy $\pi(X)$. Denote by $R(\pi)$ the average reward of a policy π. The average reward of the randomized policy $\pi(X)$ is then $E\{R(\pi(X))\}$. This value cannot be larger than $R^* := \sup_\pi R(\pi)$. (For the average value of a random variable cannot be larger than its maximum value.) As a consequence, if π^* achieves R^*, then the expected reward of $\pi(X)$ cannot exceed that of π^*.

Note that randomization sometimes helps when there is a constraint of the form $E\{C(\pi(X))\} \leq A$ for some $C(.)$ and A.

9.5. Total waiting cost Prove the result of Example 9.3.10.

9.6. Single-server network Find the optimal priority list for the network below.

The model is as in Comments 9.4.7 and the average service times and routing probabilities are as indicated in the figure.

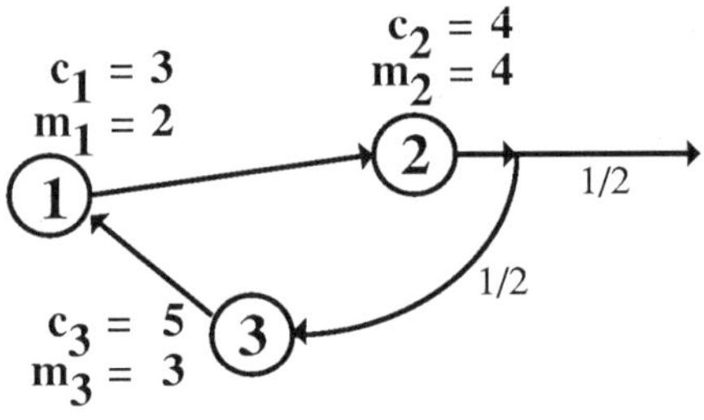

Figure 9.3. A single-server network.

9.6. REFERENCES AND COMMENTS

Sections 9.2 and 9.3 The multiarmed bandit problem was solved, after puzzling researchers for thirty years, by Gittins and Jones (1974). See Gittins (1979) for an informative discussion of the problem and its applications.

The interchange argument used to prove Proposition 9.3.5 is from Varaiya, et al. (1985). A nice proof based on dynamic programming is due to Whittle (1980). See Whittle (1982), Ross (1983b), Kumar and Varaiya (1986), Glazebrook (1983), Katehakis and Veinott (1985), Mandelbaum (1985), Berry and Fristedt (1985), and Wetherill and Glazebrook (1986). See Lai and Robbins (1985) for interesting recent developments of the theory of adaptive resource allocation.

Section 9.4 The discounted cost problem for semi-Markov bandits with Poisson arrivals was first solved in a particular case (queues in parallel) by Harrison (1975a)–(1975b) and then in the form presented in Theorem 9.4.2 by Tcha and Pliska (1977). Those papers present the algorithms for calculating the optimal ordering and the cut-off queue K.

The optimality of an index policy with Poisson arrivals was given a dynamic programming proof by Whittle (1981).

The average cost per unit of time (Theorem 9.4.9) is due to Klimov (1974). See also Meilijson and Weiss (1977). The algorithm given in Theorem 9.4.9 is from Varaiya, et al. (1985). The algorithm of Klimov (1974) finds the states in the order of increasing priorities.

CHAPTER TEN

STATISTICAL ISSUES

10.1. INTRODUCTION

This chapter investigates some **statistical issues** concerning queueing systems.

The **nonlinear filtering** theory is explained in section 10.2. The objective is to compute the conditional distribution of the state of a queueing network given a collection of jump times on certain links. This is used to prove the interchangeability of M/M/1 queues and to determine whether certain flows are Poisson. **Hypothesis testing** problems are analyzed in section 10.3. The typical question addressed in that section is how to test whether a queueing network is operating as it is supposed to. The detection of parameter changes (called **disorder problem**) is also discussed. In section 10.4 one attempts to understand how the population builds up in a queueing network. The practical motivation is to determine the probability of **rare events** that might be undesirable. The approach, based on the theory of **large deviations,** leads to a **quick simulation method.**

10.2. FILTERING

In abstract terms, the **filtering problem** is that of extracting (filtering) relevant information from observed data. More specifically, there are two dependent stochastic processes, only one of which is observed. The problem is to calculate the best estimate (in the least-mean-squared-error sense) of the value of the unobserved process given the past values of the other.

For instance, the unobserved process could be the vector of queue lengths of a Jackson network and the observed process the departure times from the network.

Filtering formulas can be used to calculate the **stochastic intensity** of a point process. The importance of that quantity is that it determines the law of the point process. The formulas lead to methods for verifying that a given point process is Poisson, or to compare point processes.

Other applications are decision problems with incomplete observations, such as hypothesis testing and stochastic control.

Definition 10.2.1 [Basic model] Let $\mathbf{x} = \{x_t, t \geq 0\}$ be a Markov chain on $\mathbf{X}$ with rate matrix Q and initial distribution π_0. One is given a subset $\mathbf{J}$ of $\{(i, j) \in \mathbf{X}^2 \mid i \neq j\}$. Let then

$$J_t = \sum_{0<s\leq t} 1\{(x_{s-}, x_s) \in \mathbf{J}\} . \tag{10.1}$$

Thus, J_t is the number of jumps in $\mathbf{J}$ made by $\mathbf{x}$ in $[0, t]$. This model is the same as that of 2.10.4, but the questions to be investigated are different.

Theorem 10.2.2 [Nonlinear filtering formulas] The **conditional distribution** (considered as a row vector)

$$\psi_t = \{\psi_t(i) := P[x_t = i \mid J_s, s \leq t], \quad i \in \mathbf{X}\} \tag{10.2}$$

can be computed as

$$\psi_t = \frac{\nu_t}{\nu_t 1} , \tag{10.3}$$

i.e., $\psi_t(i) = \nu_t(i)[\sum_j \nu_t(j)]^{-1}$, where the nonnegative vector ν_t is determined by the equations

$$\begin{cases} \nu_0 = \pi_0 \\ \dfrac{d}{dt}\nu_t = \nu_t Q_J, \; t \neq T_n \quad \text{for } n \geq 1 \\ \nu_t = \nu_{t-} Q^J, \; t = T_n \quad \text{for some } n \geq 1 \end{cases} \tag{10.4}$$

where the T_n denote the jump times of $\{J_t, t \geq 0\}$ and Q^J and Q_J are the matrices indexed by $\mathbf{X}$ defined as

$$q^J(i, j) = q(i, j)\, 1\{(i, j) \in \mathbf{J}\}, \quad Q_J = Q - Q^J . \tag{10.5}$$

Comments 10.2.3

(a) The Theorem states that ν_t is an **unnormalized** version of the conditional distribution ψ_t, and that it can be calculated recursively by the linear equations (10.4). **Recursively** means that ν_t can be determined for $t > s$ from ν_s and the values of the jump times of J_u during $(s, t]$. The point here is that it is not necessary to keep track of the values of the jump times prior to time s if ν_s is known.

(b) The **size** of the filter is the dimension of ν_t, i.e., the cardinality of **X**. This is the number of real numbers that have to be updated in carrying out the calculations specified by (10.4).

(c) It may be the case that when **X** is infinite, the formulas (10.4) are equivalent to a finite dimensional set of equations. Similarly, when **X** is finite, there may be a set of equations equivalent to (10.4) but of a smaller dimensionality than (10.4).

We will prove Theorem 10.2.2 when Q is uniform. The general case is outlined in Exercise 10.1. We will need the following simple result.

Lemma 10.2.4 One has, for $0 \leq t < t + \epsilon$ and $i \in \mathbf{X}$,

$$P\{T_1 \in (t, t+\epsilon), x_{T_1} = i\} = \epsilon \pi_0 e^{Q_J t} Q^J(i) + o(\epsilon) \quad (10.6)$$

and

$$P\{T_1 > t, x_t = i\} = \pi_0 e^{Q_J t}(i) . \quad (10.7)$$

Proof: Let $\mathbf{y} = \{y_t, t \geq 0\}$ be the Markov chain $\mathbf{x}$ **killed** after T_1. That is,

$$y_t = \begin{cases} x_t, & \text{if } t < T_1 \\ \delta, & \text{if } t \geq T_1 \end{cases}$$

where δ is a point not in **X** (the coffin state).

It should be clear that $\mathbf{y}$ is a Markov chain on $\mathbf{X} \cup \{\delta\}$. Also, the rate matrix of $\mathbf{y}$ restricted to **X** is seen to be equal to Q_J. It follows that the distribution of y_t on **X** is given by $\pi_0 e^{Q_J t}$. This proves (10.7). Now,

$$\begin{aligned} &P\{T_1 \in (t, t+\epsilon), x_{T_1} = i\} \\ &= \sum_j P\{y_t = j\} P[y_{t+\epsilon} = i \mid y_t = j] \, 1\{(j, i) \in \mathbf{J}\} + o(\epsilon) \\ &= \sum_j \pi_0 e^{Q_J t}(j) \, q^J(j, i) \epsilon + o(\epsilon) , \end{aligned}$$

by definition of Q^J and Q_J (see (10.5)). This concludes the proof.

The following lemma will then imply the theorem.

Lemma 10.2.5 For $0 < t_1 < \cdots < t_n < t, i \in \mathbf{X}$, and $0 < \epsilon \ll 1$, let $\Delta_k := (t_k, t_k + \epsilon)$ and define

$$\nu_{t_n}(t_1, \ldots, t_n, i)\epsilon^n = P\{T_i \in \Delta_i, 1 \le i \le n, x_{T_n} = i\} + o(\epsilon^n)$$

and

$$\nu_t(t_1, \ldots, t_n, i)\, \epsilon^n$$

$$= P\{T_i \in \Delta_i, 1 \le i \le n, T_{n+1} > t, x_t = i\} + o(\epsilon^n).$$

Then, considering ν_t as a row vector and omiting the arguments $(t_1, \ldots, t_n)$ for simplicity, one has

$$\nu_{T_n}(i) = \nu_{T_n-}\, Q^J(i)$$

and

$$\nu_{t+s} = \nu_t\, e^{Q_J s}(i), \quad \text{if } T_n < t < t + s < T_{n+1}\,.$$

(To see that this implies Theorem 10.2.2, notice that ν_t satisfies the equations (10.3)–(10.5).)

Proof: This result follows from Lemma 10.2.4 by induction on n.

Example 10.2.6 [Quasi-reversibility] Assume that $\pi_0 = \pi$, with $\pi Q = 0$, i.e., that $\mathbf{x}$ is stationary. Then

$$\psi_t \equiv \pi \text{ if and only if } \pi Q^J = \lambda\pi \text{ for some } \lambda > 0. \quad (10.8)$$

Indeed, the condition is sufficient since (10.4) then implies that

$$\nu_t = \pi e^{-\lambda t} \lambda^{J_t},$$

so that $\psi_t = \pi$. (Since $\pi Q = 0$, the assumption implies that $\pi Q_J = -\lambda\pi$.)

To see that the condition is necessary, observe that $\psi_t \equiv \pi$ implies that $\nu_t = \alpha_t \pi$, for some scalar α_t (by (10.3)). It then follows from the last equation of (10.4) that $\alpha_t \pi = \alpha_{t-} \pi Q^J$, i.e., that $\pi Q^J = \lambda\pi$ for some $\lambda > 0$, as claimed.

The condition $\pi Q^J = \lambda\pi$ of (10.8) can be seen to be identical to the condition $\pi_S = \pi$ of Theorem 2.10.6. That condition was used to define **quasi-reversible** queues in section 3.2. Part (d) of Theorem 2.10.6 is immediate from (10.8).

Example 10.2.7 [Markov-modulated Poisson source] A machine produces J_t objects in $[0, t]$, for $t \geq 0$. The machine alternates between the two states 0 and 1 according to some continuous-time Markov chain with a known rate matrix Q. When in state i, the objects are produced at the Poisson rate λ_i.

The problem is to determine, for $t \geq 0$ and $k \in \{0, 1\}$,

$$\phi_t(k) = P[\text{machine is in state } k \mid J_s, s \leq t].$$

To cast this model into the framework of Definition 10.2.1, let $\mathbf{X} = \{0, 0', 1, 1'\}$ and let x_t be the Markov chain with the transition diagram in Figure 10.1.

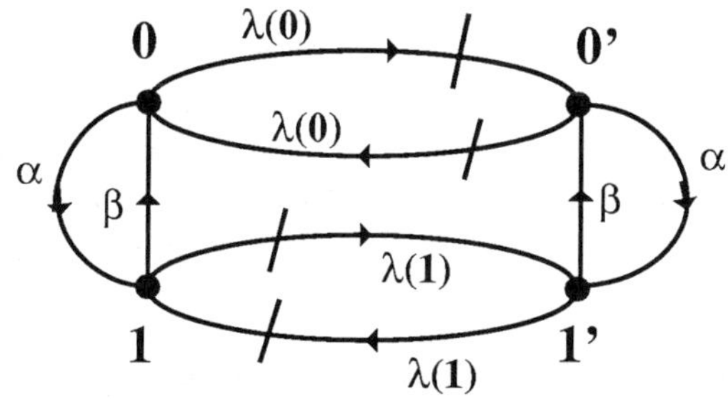

Figure 10.1. The process of Example 10.2.7.

In this diagram, $\alpha = q(0, 1)$ is the rate of transition of the machine from state 0 to state 1. Similarly, $\beta = q(1, 0)$. The interpretation of x_t is that the machine is in state 0 when $x_t \in \{0, 0'\}$ and similarly for state 1. An object is produced when $(x_{t-}, x_t) \in \mathbf{J}$ where $\mathbf{J} = \{(0, 0'), (1, 1'), (0', 0), (1', 1)\}$.

According to Theorem 10.2.2, one can then derive

$$\psi_t(i) = P[x_t = i \mid J_s, s \leq t]$$

by solving the 4-dimensional linear equations (10.4) and then normalizing by (10.3). One then obtains ϕ_t by

$$\phi_t(0) = \psi_t(0) + \psi_t(0') \text{ and } \phi_t(1) = \psi_t(1) + \psi_t(1').$$

The initial condition π_0 can be chosen as

$$\pi_0(0) = \pi_0(0') = \frac{1}{2} P\{\text{machine is in state 0 at time 0}\}$$

and similarly for the states 1 and 1'.

It should be clear by symmetry that, with the above choice of the initial condition, one has

$$\psi_t(0) = \psi_t(0') = \frac{1}{2}\,\phi(0), \quad t \geq 0$$

and similarly for 1 and 1′. This can be verified from (10.4). This fact leads to the simplified equations

$$\phi_t = \frac{\mu_t}{\mu_t 1}$$

$$\frac{d}{dt}\mu_t = \mu_t \begin{bmatrix} -\alpha - \lambda_0 & \alpha \\ \beta & -\beta - \lambda_1 \end{bmatrix}, \quad \text{if } t \neq T_n$$

$$\mu_t = \mu_{t-} \begin{bmatrix} \lambda_0 & 0 \\ 0 & \lambda_1 \end{bmatrix}, \quad \text{if } t = T_n .$$

Of course, these simplified equations could have been derived without having to introduce the additional states 0′ and 1′. The reader is invited to do it. Also, the formulas extend to an arbitrary number of states for the machine.

Definition 10.2.8 [Stochastic Intensity] Let $\mathbf{x}$, Q_J, Q^J, ψ_t be as in Theorem 10.2.2. Define

$$\lambda_t = \psi_t\, Q^J 1 . \tag{10.9}$$

Thus, λ_t is some functional of $\{J_s, s \leq t\}$, called the **stochastic intensity** of the point process $\{J_t, t \geq 0\}$. One can verify that

$$P[J_{t+\epsilon} = J_t + 1 \mid J_s, s \leq t] = \lambda_t \epsilon + o(\epsilon), \tag{10.10}$$

so that λ_t is the conditional intensity of the point process at time t given its past up to time t.

The stochastic intensity can be calculated from Theorem 10.2.2 and from (10.9). The importance of that functional lies in the following result.

Theorem 10.2.9 [The stochastic intensity determines the law] The functionals $\lambda_t(J_s, s \leq t)$ for $t \geq 0$ determine the law of the point process $\{J_t, t \geq 0\}$.

Proof: [Outline] It follows from the definition of $\lambda_t = \lambda_t(t_1, \ldots, t_n)$ that $P[T_{n+1} > t + \epsilon \mid T_1 = t_1, \ldots, T_n = t_n]$

$$= P[T_{n+1} > t \mid T_1 = t_1, \ldots, T_n = t_n]\{1 - \lambda_t \epsilon\} + o(\epsilon) \tag{10.11}$$

for $0 < t_1 < t_2 < \cdots < t_n < t$ and $0 < \epsilon \ll 1$. Therefore, if

$P\{T_1 \in (t_1, t_1 + \epsilon), \ldots, T_n \in (t_n, t_n + \epsilon)\}$ is known, then one can determine that probability with n replaced by $n + 1$ from (10.11), by integration. By induction, this allows us to determine all the finite dimensional distributions of $\{J_t, t \geq 0\}$.

Example 10.2.10 [Poisson process] A simple particular case of Theorem 10.2.9 is that a process is Poisson with rate $\lambda > 0$ if and only if $\lambda_t \equiv \lambda$.

Example 10.2.11 [Output of multiclass M/M/1 queue] Customers of classes 1 and 2 arrive at a queue as independent Poisson processes with rates λ_1 and λ_2, respectively. The service times are independent and exponentially distributed with rate μ_i for $i = 1, 2$. The queue is FCFS and has a single server. Assume that $\lambda_i > 0$ for $i = 1, 2$ and that $\mu_1 \neq \mu_2$. (See Figure 10.2.)

Denote by J_t the number of departures in $[0, t]$, for $t \geq 0$. The claim is that $\{J_t, t \geq 0\}$ is **not a Poisson process,** under any initial distribution π_0 of the state of the queue. In particular, it follows that the queue is **not quasi-reversible** (which, incidently, is easier to prove, see Exercise 10.2).

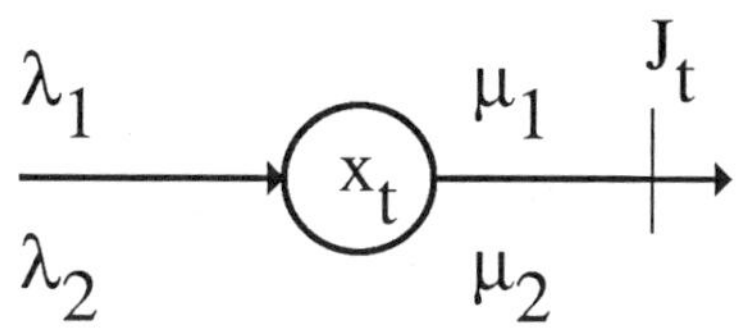

Figure 10.2. The two-class M/M/1 queue.

To prove this, first notice that the equations (10.4) imply that

$$\begin{aligned} \frac{d}{dt}\psi_t &= \psi_t\, Q_J + \lambda_t \psi_t, \quad t \neq T_n \\ \psi_t &= [\lambda_{t-}]^{-1} \psi_{t-}\, Q^J, \quad t = T_n \qquad (10.12) \\ \lambda_t &= \psi_t\, Q^J 1 . \end{aligned}$$

These are the equations satisfied by the normalized conditional distribution. These equations are nonlinear, except in the special case where J_t is Poisson. Indeed, this happens if and only if $\lambda_t \equiv a$, for some constant a (the rate of the Poisson process). Thus, if one assumes that J_t is Poisson with rate a, then the equations (10.12) become

$$\begin{aligned}
\frac{d}{dt}\psi_t(x) &= [a - \lambda - \mu(x)]\,\psi_t(x) + \lambda(x)\,\psi_t(x'), \quad t \neq T_n \\
\psi_t(x) &= a^{-1}[\mu_1\psi_{t-}(x1) + \mu_2\psi_{t-}(x2)], \quad t = T_n \qquad (10.13) \\
a &\equiv \mu_1\psi_t(\mathbf{X}1) + \mu_2\psi_t(\mathbf{X}2)
\end{aligned}$$

where $\lambda = \lambda_1 + \lambda_2$, the T_n are the jump times of J_t, and $\mathbf{X}$ is the set of possible states of the queue. A state x is a finite string (possibly empty) of 1s and 2s that indicates the configuration of customers in the queue, from last arrived to first arrived. (That is, the rightmost element of x is the class of the customer being served.) By x' one indicates the string x from which the leftmost element is deleted. The rate $\mu(x)$ [resp. $\lambda(x)$] is defined to be μ_i [resp. λ_i] if the rightmost [resp. leftmost] element of x is i $(i = 1, 2)$, and $\mu(x) = \lambda(x) = 0$ if x is the empty string. Finally, $\psi_t(\mathbf{A}) := \sum_{x \in \mathbf{A}} \psi_t(x)$ and $\mathbf{A}i := \{xi \mid x \in \mathbf{A}\}$.

The claim is that these equations are not compatible if $\mu_1 \neq \mu_2$. One starts by observing that (10.13) implies that, for $t \neq T_n$, (the subscript t is omitted)

$$\begin{aligned}
\frac{d}{dt}\psi(0) &= (a - \lambda)\,\psi(0) \\
\frac{d}{dt}\psi(\mathbf{X}1) &= (a - \lambda_1 - \mu_1)\,\psi(\mathbf{X}1) + \lambda_1\psi(\mathbf{X}2) + \lambda_1\psi(0) \qquad (10.14) \\
\frac{d}{dt}\psi(\mathbf{X}2) &= (a - \lambda_1 - \mu_2)\,\psi(\mathbf{X}2) + \lambda_1\psi(\mathbf{X}1) + \lambda_1\psi(0)\,.
\end{aligned}$$

Now, by (10.13), the quantities $\psi(0)$, $\psi(\mathbf{X}1)$, $\psi(\mathbf{X}2)$ must satisfy the following equations, for $t \neq T_n$:

$$\begin{aligned}
\psi(0) + \psi(\mathbf{X}1) + \psi(\mathbf{X}2) &= 1 \\
\mu_1\psi(\mathbf{X}1) + \mu_2\psi(\mathbf{X}2) &= a \qquad (10.15) \\
\mu_1\frac{d}{dt}\psi(\mathbf{X}1) + \mu_2\frac{d}{dt}\psi(\mathbf{X}2) &= 0\,.
\end{aligned}$$

Replacing the derivatives in the last equation by their values given in (10.14) shows that (10.15) correspond to three linear equations in the three unknown $\psi(0)$, $\psi(\mathbf{X}1)$, $\psi(\mathbf{X}2)$. Some algebra shows that this system is not degenerate, i.e, that it admits a unique solution. Thus, (10.14) implies that $[\psi(0), \psi(\mathbf{X}1), \psi(\mathbf{X}2)]$ is a constant vector for $t \neq T_n$.

The first equation (10.14) then implies that $\psi(0) = 0$. The second equation (10.13) shows therefore that $\psi_{t-}(1) = \psi_{t-}(2) = 0$ if

$t = T_n$. It then follows from the first equation (10.13) that $\psi_t(1) = \psi_t(2) = 0$ for all t. (For the T_n can take arbitrary values.)

Repeating the argument of the above paragraph then shows that $\psi_t(12) = \psi_t(21) = \psi_t(11) = \psi_t(22) = 0$ for all t. Continuing in this way shows that $\psi_t(x) = 0$ for all finite sequences x. But this is a contradiction, for it implies that $\psi_t(\mathbf{X}) = 0$, while we know that $\psi_t(\mathbf{X}) = 1$.

This concludes the proof that J_t is not Poisson for any π_0.

Problem 10.2.12 [Equivalence of point processes] Let $(\mathbf{X}, \pi, Q, \mathbf{J})$ define a point process $\{J_t, t \geq 0\}$ as in Definition 10.2.1. (The distribution π is that of x_0.) Let also $(\mathbf{X}', \pi', Q', \mathbf{J}')$ define another point process $\{J'_t, t \geq 0\}$ similarly.

When do $\{J_t, t \geq 0\}$ and $\{J'_t, t \geq 0\}$ have the **same distribution?** The answer is obtained by combining Theorems 10.2.9 and 10.2.2. This gives the following result.

Theorem 10.2.13 [Equivalence] The point processes $\{J_t, t \geq 0\}$ and $\{J'_t, t \geq 0\}$ have the same distribution (i.e., the same finite dimensional distributions) if and only if

$$\frac{\pi e^{Q_J t_1} Q^J e^{Q_J t_2} \ldots Q^J e^{Q_J t_k} Q^J 1}{\pi e^{Q_J t_1} Q^J e^{Q_J t_2} \ldots Q^J e^{Q_J t_k} 1}$$

and the corresponding expression for $\{J'_t, t \geq 0\}$ have the same value for all $k \geq 1$ and all $t_1, \ldots, t_k \in [0, \infty)$.

Variations 10.2.14 [Multiple observations] Many variations of the basic model of Definition 10.2.1 are possible. For instance, one may observe a collection of point processes that correspond to distinct sets $\mathbf{J}_c$ of transitions. (Of course, one has to assume that one observes which point process jumps for this problem to be different form that obtained by letting $\mathbf{J}$ be the union of the $\mathbf{J}_c$s.)

Another variation will be used in the next application. Say that there are two point processes $\{J_t^c, t \geq 0\}$ that correspond to the sets $\mathbf{J}_c$, $c = 1, 2$. It is assumed that $\mathbf{J}_1$ and $\mathbf{J}_2$ are disjoint subsets of $\{(i, j) \in \mathbf{X}^2 \mid i \neq j\}$. Assume that all the jump times $\{T_n^1, n \geq 1\}$ of $\{J_t^1, t \geq 0\}$ are known **in advance** and that the jumps $\{T_n^2, n \geq 1\}$ of $\{J_t^2, t \geq 0\}$ are observed as they occur. The relevant result is the following.

Theorem 10.2.15 [Nonlinear filtering equations] The conditional distribution

$$\psi_t(i) := P[x_t = i \mid J_s^1, s \geq 0, J_u^2, u \leq t]$$

is given by

$$\psi_t = \frac{\nu_t}{\nu_t 1}$$

where ν_t is determined by the equations

$$\begin{cases} \nu_0 = \pi_0 \\ \dfrac{d}{dt}\nu_t = \nu_t Q_0, \; T_n^1 \neq t \neq T_n^2 \quad \text{for } n \geq 1 \\ \nu_t = \nu_{t-} Q^1, \; t = T_n^1 \quad \text{for some } n \geq 1 \\ \nu_t = \nu_{t-} Q^2, \; t = T_n^2 \quad \text{for some } n \geq 1 \end{cases} \tag{10.16}$$

where the matrices Q_0, Q^1, Q^2 are defined by

$$\begin{cases} q_0(i,j) = q(i,j)\, 1\{(i,j) \notin \mathbf{J}_1 \cup \mathbf{J}_2\} + \lambda_1(i)\, 1\{i = j\} \\ q^1(i,j) = \dfrac{1}{\lambda_1(i)} q(i,j)\, 1\{(i,j) \in \mathbf{J}_1\} \\ q^2(i,j) = q(i,j)\, 1\{(i,j) \in \mathbf{J}_2\}\,. \end{cases}$$

In these expressions, $\lambda_1(i) := \sum_j q(i,j)\, 1\{(i,j) \in \mathbf{J}_1\}$.

The verification of this result can be done along the same lines as that of Theorem 10.2.2. The intuition behind the formulas is the following. Since the jump times of $\{J_t^1, t \geq 0\}$ are known in advance, one should set the rates of those jumps equal to zero in Q between those jump times. When a jump occurs in $\mathbf{J}_1$ when $x_t = i$, it is due to a given transition (i, j) with probability $[\lambda_1(i)]^{-1} q(i, j)$. The equations at the jumps T_n^2 are as in Theorem 10.2.2. (See Exercise 10.3.)

Example 10.2.16 [Interchangeability of M/M/1 queues] Consider the two M/M/1 queues in tandem illustrated in Figure 10.3.

Assume that the system is **initially empty,** and let $0 < a_1 < a_2 < a_3 < \cdots$ be an arbitrarily chosen sequence of arrival times at the first queue. Let J_t be the number of departures from the second queue in $[0, t]$ (for $t \geq 0$).

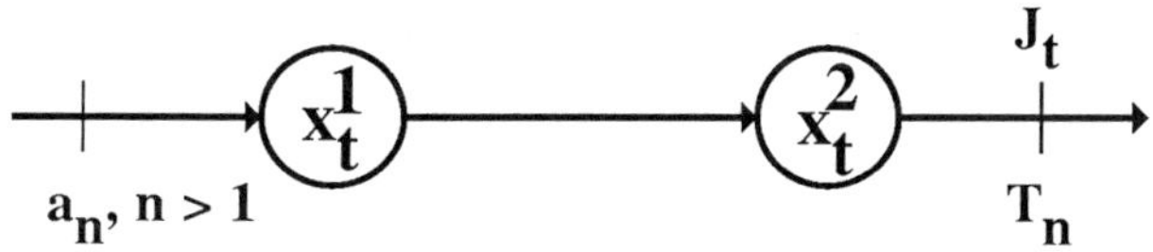

Figure 10.3. The M/M/1 queues in tandem.

The claim is that the law of that process is symmetric in μ_1 and μ_2. That is, the departure process **would be the same** (i.e., have the same law) if the two queues were **interchanged.**

To prove the result, it suffices to show that the stochastic intensity λ_t of J_t is symmetric in μ_1 and μ_2. This is done by a straightforward calculation, as explained below.

As in Theorem 10.2.15 one finds that

$$\begin{aligned} \frac{d}{dt}\nu_t &= \nu_t Q_J, \quad \text{if } t \neq a_n \text{ and } t \neq T_n \\ \nu_t &= \nu_{t-} Q^J, \quad \text{if } t = T_n \\ \nu_t &= \nu_{t-} A, \quad \text{if } t = a_n \end{aligned} \tag{10.17}$$

where

$$\begin{aligned} \nu Q_J(x^1, x^2) &= -[\mu_1 1\{x^1 > 0\} + \mu_2 1\{x^2 > 0\}]\, \nu(x^1, x^2) \\ &\quad + \mu_1 \nu(x^1 + 1, x^2 - 1) \\ \nu Q^J(x^1, x^2) &= \mu_2 \nu(x^1, x^2 + 1) \\ \nu A(x^1, x^2) &= \nu(x^1 - 1, x^2) . \end{aligned} \tag{10.18}$$

Also,

$$\lambda_t = \frac{\nu_t Q^J 1}{\nu_t 1} = \frac{\mu_2 \sum_{\{x \mid x^2 \geq 1\}} \nu_t(x)}{\sum_x \nu_t(x)} . \tag{10.19}$$

To show that λ_t is symmetric in μ_1 and μ_2, in view of (10.19), it suffices to show that, for all $0 \leq M \leq N$,

$$f_t(N, M) := \mu_2^M \nu_t\{A(N, M)\} := \mu_2^M \sum_{x \in A(N, M)} \nu_t(x) \tag{10.20}$$

with

$$A(N, M) := \{x \in \mathbf{X} \mid x^1 + x^2 = N, x^2 \geq M\} \tag{10.21}$$

is symmetric in μ_1 and μ_2. Indeed, choosing $M = 1$ [resp. $M = 0$] in (10.21) and summing over $N \geq M$ then shows that the numerator [resp. the denominator] of (10.20) is symmetric in μ_1 and μ_2.

At $t = 0$, (10.21) is symmetric since

$$f_0(N, M) = 1\{N = M = 0\} .$$

Now, (10.19) implies, by summing the equations (10.18) over $A(N, M)$, that

$$\frac{d}{dt} f_t(N, M) = -(\mu_1 + \mu_2) f_t(N, M) + \mu_1 \mu_2 f_t(N, M-1), \quad t \neq T_n$$

$$f_t(N, M) = f_{t-}(N+1, M+1), \quad t = T_n$$

$$f_t(N, M) = f_{t-}(N-1, M), \quad t = a_n$$

which proves the symmetry of the $f_t(N, M)$ for all $t \geq 0$ and concludes the proof of the interchangeability of the M/M/1 queues.

10.3. HYPOTHESIS TESTING

A common question that arises when observing a queueing system is whether it is **operating properly,** i.e., as it was intended to. This question can be formulated as a **hypothesis testing problem,** either with complete or partial observation. The solution to the problem involves the calculation of likelihood ratios.

Definition 10.3.1 [Likelihood ratio for Markov chains] Let $\mathbf{x} := \{x_t, \ t \geq 0\}$ be a possible realization of a Markov chain on $\mathbf{X}$. Assume that there are two possible rate matrices for $\mathbf{x}$: Q_0 and Q_1 and two possible initial distributions π_0 and π_1. Define

$$L_t = \frac{dP_1\{x_s, s \leq t\}}{dP_0\{x_s, s \leq t\}} \tag{10.22}$$

where

$$dP_0\{x_s, s \leq t\} = \pi_0(\xi_0) e^{-q_0(\xi_0) t_1} q_0(\xi_0, \xi_1) e^{-q_0(\xi_1)(t_2 - t_1)}$$
$$\times q_0(\xi_1, \xi_2) \ldots q_0(\xi_{n-1}, \xi_n) e^{-q_0(\xi_n)(t - t_n)} \tag{10.23}$$

if $\{x_s, s \leq t\}$ visits the states $\xi_0, \ldots, \xi_n$ successively and jumps at the times $t_1, \ldots, t_n$. The definition of dP_1 is similar.

Thus, L_t is the ratio of the densities of $\{x_s, s \leq t\}$ under (π_1, Q_1) and (π_0, Q_0), respectively. By P_0 one denotes the law of $\mathbf{x}$ under (π_0, Q_0), and similarly for P_1. The quantity L_t is called the **likelihood ratio** of P_1 over P_0 on $[0, t]$. It is a random variable that depends on the realization of $\mathbf{x}$ in $[0, t]$.

The relevance of this definition will be discussed below. Intuitively, $L_t(x_s, s \leq t)$ is large if $\{x_s, s \leq t\}$ is more likely under P_1 than under P_0, and vice versa. One should notice that L_t may be infinite if a given realization is possible under P_1 and not under P_0. Similarly, L_t may be equal to zero.

The next observation shows how to calculate L_t recursively.

Fact 10.3.2 The quantity L_t can be calculated by $L_t = \exp\{l_t\}$, where

$$dl_t = -[q_1(x_t) - q_0(x_t)]\,dt + \log\frac{q_1(x_{t-}, x_t)}{q_0(x_{t-}, x_t)}\,d\Delta_t .$$

In the expression above, Δ_t denotes the number of jumps of $\mathbf{x}$ in $[0, t]$ and $dl_t = l_{t+dt} - l_t$ if l_s does not jump at time t while $dl_t = l_t - l_{t-}$ if it jumps at t.

This fact is an immediate consequence of (10.22) with $l_t := \log L_t$.

Definition 10.3.3 [Bayesian Nonsequential Hypothesis Testing Problem] Say that the law of $\mathbf{x}$ is given by P_θ with probability $p(\theta)$ for $\theta \in \{0, 1\}$ for some given numbers $0 < p(0) = 1 - p(1) < 1$. A fixed $T > 0$ is given. On the basis of $\{x_s, s \leq T\}$ one has to choose a value $\delta \in \{0, 1\}$ so as to minimize

$$E\{L(\theta, \delta)\} \tag{10.24}$$

for a given function $L := \{0, 1\}^2 \to [0, \infty)$.

The **interpretation** is that $L(\theta, \delta)$ is the cost of choosing the value δ when θ is true. A typical choice of L is $L(\theta, \delta) = 1\{\theta \neq \delta\}$, in which case the problem is to maximize $P\{\theta = \delta\}$, i.e., the probability of correctly guessing θ.

It will be assumed that

$$L(0, 0) < L(0, 1) \text{ and } L(1, 1) < L(1, 0)\,, \tag{10.25}$$

which is consistent with the interpretation of L.

The problem is said to be **Bayesian** because the prior probabilities $p(0)$ and $p(1)$ of the **hypotheses** P_0 and P_1 are assumed known.

Theorem 10.3.4 [Bayes Decision Rule] The following decision rule, called the **Bayes Decision Rule,** is optimal for the problem of Definition 10.3.2:

$$\delta = 1\{L_T \geq \frac{p(0)\,[L(0,1) - L(0,0)]}{p(1)\,[L(1,0) - L(1,1)]}\}. \qquad (10.26)$$

Proof: One can write

$$E\{L(\theta,\delta)\} = E\{E[L(\theta,\delta)\,|\,x_0^T]\}$$

where $x_0^T := \{x_s, 0 \leq s \leq T\}$. Now, (the summations are over $(\theta_0, \delta_0) \in \{0,1\}^2$)

$$\begin{aligned} E[L(\theta,\delta)\,|\,x_0^T] &= \sum_{\theta_0,\delta_0} L(\theta_0,\delta_0)\, P[\theta = \theta_0, \delta = \delta_0\,|\,x_0^T] \\ &= \sum_{\theta_0,\delta_0} L(\theta_0,\delta_0)\, P[\theta = \theta_0\,|\,x_0^T]\, P[\delta = \delta_0\,|\,x_0^T] \\ &= \sum_{\theta_0,\delta_0} L(\theta_0,\delta_0)\, p(\theta_0)\, dP_{\theta_0}(x_0^T) \\ &\quad \times \{p(0)dP_0(x_0^T) + p(1)dP_1(x_0^T)\}^{-1} P[\delta = \delta_0\,|\,x_0^T]. \end{aligned}$$

Thus, $E[L(\theta,\delta)\,|\,x_0^T]$ is proportional to

$$\begin{aligned} &P[\delta = 0\,|\,x_0^T]\,\{p(0)L(0,0) + p(1)L(1,0)L_T\} \\ &\quad + P[\delta = 1\,|\,x_0^T]\,\{p(0)\,L(0,1) + p(1)\,L(1,1)L_T\}. \qquad (10.27) \end{aligned}$$

A decision rule corresponds to a value of

$$\alpha := P[\delta = 1\,|\,x_0^T] = 1 - P[\delta = 0\,|\,x_0^T].$$

The equation (10.27) shows that $E[L(\theta,\delta)\,|\,x_0^T]$ is linear in $\alpha \in [0,1]$. It is minimized by choosing $\alpha = 1$, i.e., $\delta = 1$, if x_0^T is such that the coefficient of α in (10.27) is nonpositive, and by choosing $\alpha = 0$, i.e., $\delta = 0$, otherwise. Some algebra shows that this corresponds to the decision rule (10.26).

Example 10.3.5 [M/M/1 queue] Let x_t be the queue length process of an initially empty M/M/1 queue. Under P_1 the M/M/1 queue has rates (λ_1, μ_1). Under P_0 the rates are (λ_0, μ_0). Assume that $p(0) = p(1) = 1/2$.

The objective is to minimize $P\{\delta \neq \theta\}$, where δ is to be decided on the basis of x_0^T.

The solution is provided by Theorem 10.3.4 and Fact 10.3.2. One first solves the equation of Fact 10.3.2. In this particular case, that equation reads

$$dl_t = -[\lambda_1 - \lambda_0 + (\mu_1 - \mu_0)\, 1\{x_t > 0\}]\, dt$$

$$+ \log \frac{\lambda_1 1\{x_t = x_{t-} + 1\} + \mu_1 1\{x_t = x_{t-} - 1\}}{\lambda_0 1\{x_t = x_{t-} + 1\} + \mu_0 1\{x_t = x_{t-} - 1\}} d\Delta_t \,. \quad (10.28)$$

The initial condition is $l_0 = 0$ (i.e., $L_0 = 1$).
One then uses the decision rule (10.26) which becomes

$$\delta = 1\{L_T \geq 1\}\,. \quad (10.29)$$

To be specific, let us say that $T = 10$ and that x_0^T is such that arrivals occur at times $\{1, 2, 4, 5, 8\}$ and that departures take places at times $\{3, 6, 7, 9\}$. Assume also that $(\lambda_0, \mu_0) = (1, 3)$ and $(\lambda_1, \mu_1) = (2, 4)$. Solving (3.6) gives

$$l_{10} = -10 - \int_0^{10} 1\{x_t > 0\}\, dt + A_{10} \log 2 + D_{10} \log \frac{4}{3}$$

where A_t [resp. D_t] is the number of arrivals [resp. departures] in $[0, t]$. Hence,

$$l_{10} = -19 + 5 \log 2 + 4 \log \frac{4}{3} < 0.$$

It follows that the optimal decision, given this particular realization, is

$$\delta = 0\,.$$

Our next topic is the **disorder problem,** i.e., the detection of the modification of the parameters of a system.

Example 10.3.6 [Disorder problem for a Poisson process] Let τ be an exponentially distributed random variable with rate α. Given τ, the process $\{N_t, t \geq 0\}$ is Poisson with rate λ_0 on $[0, \tau)$ and Poisson with rate λ_1 on $[\tau, \infty)$. One observes $\{N_t, t \geq 0\}$ and one wants to detect τ.

First observe that the determination of

$$p_t := P[\tau \leq t \mid N_s, 0 \leq s \leq t]$$

is a particular case of the problem discussed in Example 10.2.7, as is seen by letting $\beta = 0$ and $p_t = \phi_t(0)$, in the notation of that example.

Now, consider the following problem. Assume that at some time σ one declares that the rate of the Poisson process has switched to λ_1. Say that the cost of this declaration is

$$c(\sigma, \tau) = a(\tau - \sigma)^+ + b(\sigma - \tau)^+ \tag{10.30}$$

where a and b are given positive constants. That is, one is charged a units of cost per unit of time that the declaration was made too early and b units of cost per unit of time that the declaration was made too late. The problem is to minimize

$$E\{c(\sigma, \tau)\}$$

over all the stopping times σ of $\{N_t, t \geq 0\}$.

Proposition 10.3.7 [Optimality of threshold policy] The optimal policy is of the following form:

$$\sigma = \min\{t \geq 0 \mid p_t \leq A\}$$

for some **threshold** value A.

Thus, by using the equations of Example 10.2.7 one keeps updating the value of p_t on the basis of the observations. As soon as the likelihood that the process has not switched to rate λ_1 is smaller than some value A, one must declare that it has switched.

Proof: The problem can be solved by Dynamic Programming as follows. For $p \in (0, 1)$, let $V(p)$ be the minimum expected cost given that at time 0 the process has already switched to rate λ_1 with probability $1 - p$ and, with probability p, will remain at rate λ_0 for an exponentially distributed time τ with rate α.

By considering the evolution in $[0, \epsilon)$, one has the Dynamic Programming equations

$$V(p) = \min[pE\{a\tau\}, (1 - p)a\epsilon + E\{V(p_\epsilon)\}]. \tag{10.31}$$

Indeed, the first term in brackets is the average cost of choosing $\sigma = 0$ given p, and the second term is the minimum expected cost given that one decides to wait at least for ϵ time units before declaring σ. These equations show that one should choose $\sigma = 0$ if and only if $pE\{a\tau\} \leq V(p)$, and that one should wait otherwise.

To prove the proposition, it suffices to show that $V(p)$ is a **concave** function (i.e., such that $-V(p)$ is convex). Indeed, that will imply that the functions $V(p)$ and $pE\{a\tau\}$ are as indicated in Figure 10.4.

The concavity of $V(p)$ is proved as follows. For a given stopping time σ the cost $E\{c(\sigma, \tau)\}$ has the value

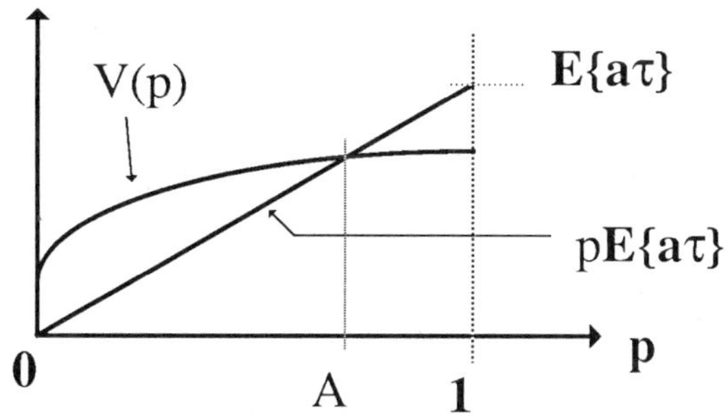

Figure 10.4. The value function.

$$K(p, \sigma) = (1 - p)E\{b\sigma\} + pE\{a(\tau - \sigma)^+\} .$$

Therefore,

$$V(p) = \min_{\sigma} K(p, \sigma)$$

is the minimum of a collection of linear functions in p. This implies that $V(p)$ is a concave function.

The threshold value A can be calculated by solving the equations (10.31).

Comments 10.3.8 It is not difficult to generalize the above example. One can consider a Markov chain $(\mathbf{x}, \pi, Q_0)$ with a rate matrix that switches to Q_1 after a random time τ with a phase-type distribution. Say that one observes a process $\{J_t, t \geq 0\}$ that counts the transitions of $\mathbf{x}$ in a given set $\mathbf{J}$ during $[0, t]$.

One can write the filtering equations for the distribution ν_t of the phase of τ at time t given $\{J_s, s \leq t\}$. If the problem is to minimize a cost function by choosing a stopping time σ as in Example 10.3.6, then one can write Dynamic Programming equations for the minimum cost $V(\nu_t)$ given the distribution ν_t. In principle, the equations can be solved to determine the optimal σ. The solution is generally difficult to obtain, and efficient numerical methods have to be used.

10.4. LARGE DEVIATIONS

This section deals with **rare events** in queueing networks. Specifically, we will develop a method for estimating the time until the population in a network reaches a given large value. This method leads to a procedure for **speeding up simulation experiments** that provide those estimates.

Model 10.4.1 [Network of GI/GI/1 queues] Consider the network in Figure 10.5. There are J queues and the routing is random.

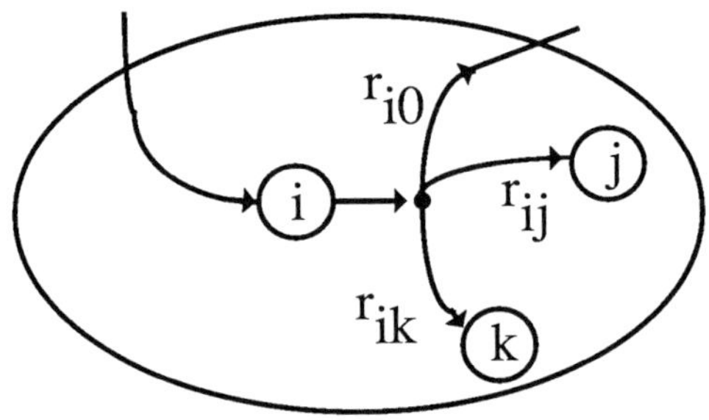

Figure 10.5. The network of GI/GI/1 queues.

The exogenous arrivals into the queues are independent renewal processes, and all the service times are independent and identically distributed in any given queue. A customer who leaves queue i is sent to queue j with probability r_{ij} for $j = 1, \ldots, J$ and leaves the network with probability $r_{i0} = 1 - \sum_{j=1}^{J} r_{ij}$.

Problem 10.4.2 [Time until large population] Choose N to be a large integer and define τ to be the first time that the population in the network reaches the value N. It is assumed that the network is initially empty and is stable. Our objective is to estimate the distribution of τ.

Let us start with an intuitive result.

Fact 10.4.3 [Approximately exponential] The distribution of τ is approximately **exponential.**

Justification: Call a **busy cycle** the evolution of the network from the time it ceases to be empty until it becomes empty again. Notice that the successive busy cycles are approximately i.i.d. A dependence is introduced between busy cycle by the residual times until the next arrivals. The busy cycles are i.i.d. if the arrivals are Poisson. We will assume that the busy cycles can be approximated as being i.i.d.

Denote by S_n the time when the nth busy cycle ends ($n \geq 1$) and let $T_n = S_n - S_{n-1}$, $n \geq 1$ with $S_0 := 0$. Let N be the index of the first busy cycle when the network population reaches the value N. (See Figure 10.6.)

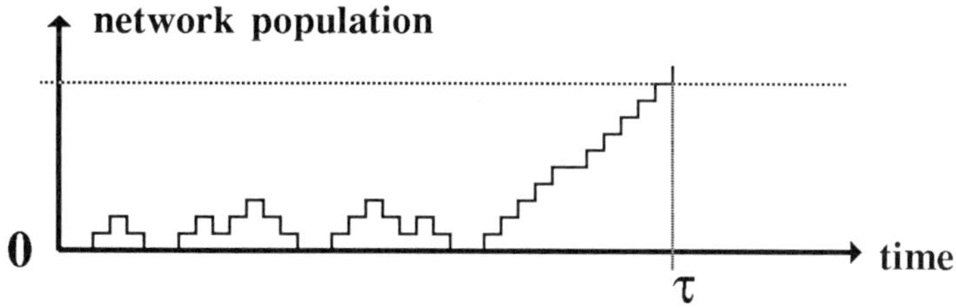

Figure 10.6. Typical evolution of the network population.

Call a busy cycle **successful** if the network population reaches N during that cycle, and unsuccessful otherwise. Notice also that unsuccessful cycles are approximately i.i.d. Now,

$$\tau = T_1 + \cdots + T_{N-1} + T_N$$

where N is such that $P\{N = n\} \approx (1 - \alpha)^{n-1}\alpha$ for $n \geq 1$ where α is the probability that a given busy cycle is successful. Hence

$$E\{e^{-s\tau}\} \approx \sum_{n=1}^{\infty} (1 - \alpha)^{n-1} \alpha E\{e^{-s(T_1 + \cdots + T_{n-1})} e^{-sT_n}\}$$

$$= \sum_{n=1}^{\infty} (1 - \alpha)^{n-1} \alpha \psi(s)^{n-1} \phi(s)$$

$$= \frac{\alpha \phi(s)}{1 - (1-\alpha)\psi(s)}$$

where $\psi(s)$ is the characteristic function of T_1 for an unsuccessful busy cycle and $\phi(s)$ is the characteristic function of T_1 for a successful busy cycle. Replacing s by αs, one finds after some algebra

$$E\{e^{-s\alpha\tau}\} = \frac{1}{1 - s\psi'(0)} + O(\alpha)$$

where $O(\alpha)$ denotes a function such that $\frac{O(\alpha)}{\alpha}$ converges to a nonzero finite number as $\alpha \to 0$.

This shows that when α is very small the random variable $\alpha\tau$ is almost exponentially distributed with mean $-\psi'(0)$, i.e., with the same mean as that of T_1 for an unsuccessful busy cycle. When α is very small, a busy cycle is very likely to be unsuccessful, so that the mean value of $\alpha\tau$ is equal to that of T_1. Thus,

$$P\{\alpha\tau \le t\} \approx 1 - e^{-\frac{t}{B}}, \quad t \ge 0$$

where B is the mean duration of an idle period plus a busy cycle.

It remains to estimate B and α. That is done below.

Problem 10.4.4. [Estimating B] If the network has a known invariant distribution, then one can calculate the average duration of a busy cycle. For instance, if the queues are M/M/1 with exogenous arrival rate γ_i and service rate μ_i in queue i $(1 \le i \le J)$, then $\pi(0) = \prod_{i=1}^{J} \rho_i$ where $\rho_i = \lambda_i^{-1}\mu_i$ and the λ_is are found by solving the flow conservation equations. (See Figure 10.7.)

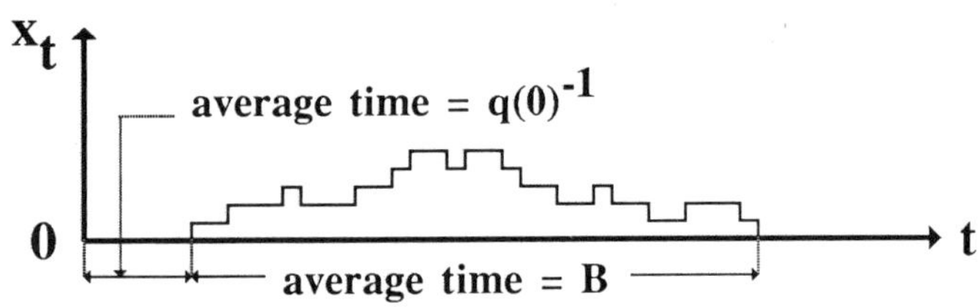

Figure 10.7. A typical evolution of the Jackson network.

To relate B to $\pi(0)$, one argues as follows. Say that the network just became empty at time $t = 0$. It will remain empty for an exponentially distributed time with mean γ^{-1} where $\gamma = \sum \gamma_i$. Then it will remain nonempty for an average of $B - \gamma^{-1}$ time units. Thus, the fraction of time the network remains empty is equal to $\pi(0) = \gamma^{-1} B^{-1}$. Hence

$$B = \frac{1}{\gamma\pi(0)}.$$

In general, the probability that the network is empty is not known and the value of B must be determined either numerically or by simulation. As we saw in Example 2.7.3, this may not be an easy task if the network is large.

Problem 10.4.5 [Estimating α] If N is small, then a direct simulation of the network can probably be used to estimate α. The method would be to generate n busy cycle and count the number of successful ones, say m. The ratio $\frac{m}{n}$ would then be an estimate of the probability α that a busy cycle is successful.

More specifically, if one denotes by Z_n the indicator that the nth busy cycle is successful, i.e., if Z_n takes the value 1 if the nth busy cycle is successful and the value 0 otherwise, then the random variables $\{Z_n, n \geq 1\}$ are i.i.d., so that

$$\tilde{\alpha}_n := \frac{Z_1 + \cdots + Z_n}{n} \to \alpha = P\{Z_1 = 1\}, \qquad (10.32)$$

with probability one, by the strong law of large numbers. Moreover, by the central limit theorem,

$$\sqrt{n}(\tilde{\alpha}_n - \alpha) = \frac{Z_1 + \cdots + Z_n - n\alpha}{\sqrt{n}} \to N(0, \sigma^2) \qquad (10.33)$$

in distribution, where $N(0, \sigma^2)$ denotes a Gaussian random variable with mean zero and variance σ^2. In this statement, σ^2 is the variance of each Z_n. That is,

$$\sigma^2 = E\{Z_n^2\} - \alpha^2 = \alpha(1 - \alpha).$$

This convergence result can be used to determine the number of cycles that should be generated in order to achieved a prescribed accuracy. The idea is as follows.

Say that an estimate $\tilde{\alpha}$ of α has a **level of confidence** $(a, b) \in (0, 1)^2$ if

$$P\{\frac{|\tilde{\alpha} - \alpha|}{\alpha} \leq a\} \geq b .$$

Using (10.33), one finds that

$$P\{\frac{|\tilde{\alpha}_n - \alpha|}{\alpha} \leq a\} \approx P\{|N(0, \sigma^2)| \leq a\alpha\sqrt{n}\}$$

$$= P\{|N(0, 1)| \leq \frac{a\alpha\sqrt{n}}{\sigma}\} .$$

Thus, if one defines $\gamma(b)$ such that

$$P\{\,|N(0,1)|\, \le \gamma(b)\} \;=\; b\,,$$

then one finds that $\tilde{\alpha}_n$ has a level of confidence (a, b) if n is such that $\frac{a\alpha\sqrt{n}}{\sigma} = \gamma(b)$, i.e., if

$$n \;=\; \frac{\gamma(b)^2}{a^2}\,\frac{\sigma^2}{\alpha^2}\,. \tag{10.34}$$

This calculation shows that for a given level (a, b) of confidence, the number of busy cycles that have to be generated is **proportional to the squared coefficient of variation** of Z_1, i.e, $\frac{\operatorname{var}(Z_1)}{(E\{Z_1\})^2}$. It is then highly desirable to find estimators with a small variance.

Formula (10.34), together with $\sigma^2 = \alpha(1-\alpha)$, shows that the number of busy cycles to be generated is of the order of

$$n \;=\; c\,.\,\alpha^{-1}\,, \tag{10.35}$$

which indicates that a direct simulation experiment is likely to be exceedingly slow.

In the remainder of this section we discuss one method for **speeding up** the simulation of α.

Comments 10.4.6 [Basic idea] The idea of the method is to simulate a network **modified** so that its population grows much faster and to translate the estimates measured for on the modified network in terms of the original one. The translation is done by using likelihood ratios.

In abstract terms, the method is as follows. Denote by ω a possible realization of the evolution of the queueing network. That is, ω is a possible trajectory in the state space. That evolution is random. Denote by $dP(\omega)$ the probability density of the evolution. Thus, $dP(\omega)$ is the probability that the actual evolution of the network is close to ω.

Let then ω be a realization of the network evolution. Assume that the network is empty and receives one customer at time 0. Let $Z(\omega)$ take the value 1 if the network population reaches N during the first busy cycle, and take the value 0 otherwise. Our objective is to estimate

$$\alpha \;=\; P\{Z(\omega) = 1\} \;=\; E\{Z(\omega)\} \tag{10.36}$$

where $E\{.\}$ denotes the expectation with respect to the probability distribution P.

Let $dP^*(\omega)$ be the probability density of the evolution for a modified version of the network. The specific modification is not important at this stage. Define

$$L(\omega) := \frac{dP(\omega)}{dP^*(\omega)} . \tag{10.37}$$

That is, L is the **likelihood ratio** of the distributions P and P^*. It follows that if one denotes by $E^*\{.\}$ the expectation with respect to P^*, then

$$E^*\{L(\omega)\, X(\omega)\} = E\{X(\omega)\} , \tag{10.38}$$

for any function X of the evolution of the network. This equality can be roughly justified by writing the left-hand side of (10.38) as $\sum_\omega \frac{dP(\omega)}{dP^*(\omega)} X(\omega)\, dP^*(\omega) = \sum_\omega X(\omega)\, dP(\omega)$ which agrees with the right-hand side of (10.38).

In particular, choosing $X = Z$, one has

$$E^*\{LZ\} = E\{Z\} = \alpha , \tag{10.39}$$

which shows that one could estimate α by simulating the network with distribution P^* and by measuring the value of $L(\omega)X(\omega)$. More precisely, if $L_n Z_n$ is the value measured on the nth busy cycle of the modified network, then the strong law of large numbers and (10.39) show that $n^{-1}(L_1 Z_1 + \cdots + L_n Z_n) \to \alpha$. Assume now that

$$L(\omega) \ll 1 \text{ whenever } Z(\omega) = 1 . \tag{10.40}$$

One then finds that

$$E^*\{[L(\omega)\, Z(\omega)]^2\} \ll E^*\{L(\omega)\, Z^2(\omega)\} = E\{Z^2(\omega)\} . \tag{10.41}$$

As a consequence, the variance of $L(\omega)Z(\omega)$ under P^* is much smaller than the variance of $Z(\omega)$ under P. This shows that estimating $L(\omega)Z(\omega)$ under P^* will be much more efficient (require fewer experiments) than estimating $Z(\omega)$ under P. It is, therefore, a better method for estimating α.

Optimal change of measure 10.4.7 The key idea of the method is to find how the network should be modified so that (10.40) holds. This relation means that the event $\{Z(\omega) = 1\}$ is much more likely under P^* than under P. Thus, one should find a distribution P^* under which the network is much more likely to become full before it becomes empty again.

The strategy consists in two steps: (1) one determines the **most likely** way that the network becomes full under its original evolution P; (2) one **modifies** the network to make that most likely way of becoming full exceedingly likely.

Before proceeding, we have to recall a basic result of the theory of large deviations.

Chernoff's Theorem 10.4.8 Let $\{\tau_n, n \geq 1\}$ be independent and identically distributed random variables with mean m. Then

$$\lim \frac{1}{n} \log P\{\frac{\tau_1 + \cdots + \tau_n}{n} \approx a\} = -h(a) \qquad (10.42)$$

where $h(.)$ is the **Cramer-Legendre transform** of the distribution of the τ_n, which is assumed to exist. That is

$$h(a) := \sup_{s \in \mathbb{R}} \{sa - E\{e^{s\tau_1}\}\} . \qquad (10.43)$$

The interpretation of this theorem is as follows. The weak law of large numbers tells us that the sample mean $(\tau_1 + \cdots + \tau_n)/n$ converges in distribution to the true mean m. Chernoff's theorem says that the probability of the sample mean being away from the true mean **vanishes exponentially fast** in the number of samples.

This is called a **large deviations** result because it provides an estimate for the probability that the sample mean will make a large deviation about its typical behavior. (Thus, $h(m) = 0$ and $h(a) > 0$ for $a \neq m$.)

Results of this type have been used extensively in Information Theory to prove Shannon's coding theorems. We will see how they can be applied to **quick simulation** methods.

Example 10.4.9 [Exponential distribution] Assume that τ_1 is exponentially distributed with rate λ. One can show that

$$h(a) = \lambda a - 1 - \log(\lambda a), \quad a > 0 . \qquad (10.44)$$

Example 10.4.10 [GI/GI/1 queue] Consider the queueing system in Figure 10.8. Customers arrive and have to be processed on a first-come–first-serve basis by a single server. Assume that the times between arrivals are independent and identically distributed and that so are the service times. Assume also that the arrival rate λ is less than the service rate μ. (The arrival rate λ is the reciprocal of the mean interarrival time, and similarly for the service rate.) Thus, the queue is "stable."

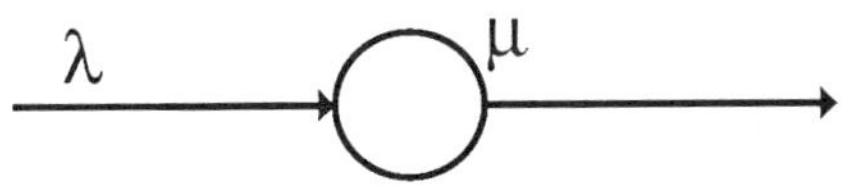

Figure 10.8. The queueing system of Example 10.4.10.

Denote by x_t the number of customers in the system at time t. We want to estimate the probability α that, starting with an empty system, the queue length will reach some large value N before going back to zero. That is, we are trying to estimate the probability that a given busy cycle will reach a queue length of N.

The reasoning for estimating α goes as follows. The queue length will reach the value N in one busy cycle if, on that busy cycle of some duration T, the queue **behaves** as a queue with an effective arrival rate λ' and with an effective service rate μ' such that $\lambda' > \mu'$ and $N = T(\lambda' - \mu')$, for some λ' and some μ'.

Now, the probability that the effective arrival rate over $[0, T]$ is λ' (instead of λ) can be estimated as the probability that the sample mean of the first $\lambda' T$ interarrival times is equal to $1/\lambda'$. According to Chernoff's Theorem, this probability is of the order of

$$\exp\{-\lambda' T h_\lambda(1/\lambda')\}$$

where h_λ is the Cramer-Legendre transform of the distribution of the interarrival times.

A similar estimate can be derived for the service rate. This shows that the probability that the queue behaves, for some T, some λ', and some μ', as a queue with arrival rate λ' and service rate μ' on the interval $[0, T]$ is of the order of the sum of

$$\exp\{-\lambda' T\, h_\lambda\,(1/\lambda') - \mu' T\, h_\mu(1/\mu')\}$$

where the sum is over all the $\lambda' > \mu'$ with $N = T(\lambda' - \mu')$. Substituting T by its value $N(\lambda' - \mu')^{-1}$, one finds that α is of the order of the sum of

$$\exp\{-N(\lambda' - \mu')^{-1}[\lambda' h_\lambda(1/\lambda') + \mu' h_\mu(1/\mu')]\}$$

where the sum is over $\lambda' > \mu'$.

If N is large, then this sum will be dominated by the term that corresponds to the **minimum coefficient** of $-N$ in the exponential. Thus, α will be of the order of

$$\exp\{-N(\lambda^* - \mu^*)^{-1}[\lambda^* h_\lambda(1/\lambda^*) + \mu^* h_\mu(1/\mu^*)]\}\,.$$

This shows also that the overwhelmingly most likely way in which the queue length can reach N during a busy cycle is for the queue to behave as a queue with arrival rate λ^* and service rate μ^* until it reaches N. These values λ^* and μ^* are those that achieve the minimum over $\lambda' > \mu'$ of

$$(\lambda' - \mu')^{-1}[\lambda' h_\lambda(1/\lambda') + \mu' h_\mu(1/\mu')] .$$

We will choose the probability P^* so that it corresponds to a queue with arrival rate λ^* and service rate μ^*. As explained in section 10.2.5, this will guarantee that (10.40) holds and this will lead to a much faster simulation.

Exponential change of measure 10.4.11 How should the **distribution** of the interarrival and service times so that the new rates are λ^* and μ^*? A convenient method for doing this is to use a change of measure of the following form. Denote by $F(.)$ the original distribution (of the interarrival times or of the service times). The new distribution is then defined by

$$dF^*(x) = A.e^{\theta x}\, dF(x) \tag{10.45}$$

where A is the normalizing constant. The value of the parameter θ is chosen so that the new rate agrees with the one selected by the discussion of 10.4.10. Notice that if F is exponential, then so is F^*.

A simple example 10.4.12 [The M/M/1 queue] Consider Example 10.4.10 and assume that the interarrival times and service times are exponentially distributed. One can then show that the values λ^* and μ^* are given by

$$\lambda^* = \mu \quad \text{and} \quad \mu^* = \lambda .$$

For instance, with $\lambda = 0.4$, $\mu = 0.6$, and $N = 30$, one finds that $\alpha = 2.6 \times 10^{-6}$. A direct simulation of the queue will require of the order of $\alpha^{-1} \approx 400{,}000$ busy cycles to be generated before obtaining satisfactory estimates of α (see (10.34)). The simulation of the queue with the parameters $\lambda^* = 0.6$ and $\mu^* = 0.4$ requires on the order of 2,000 busy cycles.

Example 10.4.13 [Queues in tandem] Let us explain one of the ideas involved in the extension of the method to networks in the case of a system of two queues in tandem. (See Figure 10.9.)

Say that one desires to estimate the probability α that a given busy cycle of the system reaches a total population of N. Arguing

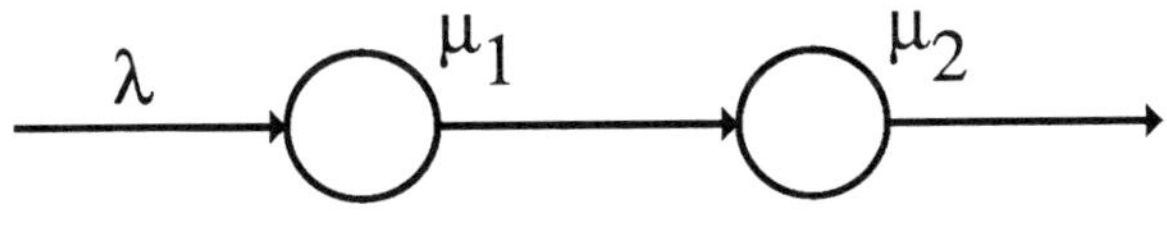

Figure 10.9. The two queues in tandem.

as before, one can express this probability as the sum over $\lambda', \mu'_1, \mu'_2, T$ of the probability that the system behaves as if the arrival rate were λ' and the service rate in queue i were μ'_i for $i = 1, 2$ over the time interval $[0, T]$. The set of parameters has to be determined with some care.

Consider the case $\lambda' > \mu'_1 > \mu'_2$. The two queues are building up and the "net" rate of increase of the total population is roughly equal to $\lambda' - \mu'_2$. Therefore, one should have $N = T(\lambda' - \mu'_2)$.

On the other hand, if $\lambda' > \mu'_2 > \mu'_1$, then the second queue is almost always empty and the net rate of increase is equal to $\lambda' - \mu'_1$. In this case, one should have $N = T(\lambda' - \mu'_1)$.

Again, by minimizing the appropriate exponent (involving the Cramer-Legendre transforms of the interarrival and service times), one finds the most likely effective rates $\lambda^*, \mu_1^*, \mu_2^*$. These rates are then used to change the measure and define the simulation experiment to be performed.

Simple example 10.4.14 [M/M/1 queues in tandem] In the case of M/M/1 queues in tandem, it turns out that the new parameters $(\lambda^*, \mu_1^*, \mu_2^*)$ are obtained from (λ, μ_1, μ_2) by interchanging λ and the smallest of μ_1 and μ_2. With $\lambda = 0.2, \mu_1 = 0.3, \mu_2 = 0.5$, and $N = 20$, one can calculate numerically $\alpha = 3.8 \times 10^{-4}$. One finds from (10.34) that about 2,600 busy cycles have to be generated. A simulation of the system with rates $\lambda^* = 0.3, \mu_1^* = 0.2$, and $\mu_2^* = 0.5$ shows that it suffices to generate 300 cycles to obtain a good estimate of α. It should be noticed that the modified simulation is substantially faster than the numerical evaluation of α based on the first step equations.

In order to extend this procedure to networks with routing, one needs a result similar to Chernoff's theorem for discrete random variables (the routing decisions). The result is as follows.

Sanov's Theorem 10.4.15 Let $\{Y_n, n \geq 1\}$ be i.i.d. random variables that take values in $\{1, 2, \ldots, J\}$ with $r_i = P\{Y_1 = i\}$ for $i = 1, \ldots, J$. Define $P_n(i) = n^{-1} \sum_{m=1}^{n} 1\{Y_m = i\}$ for $n \geq 1$ and $i = 1, \ldots, J$. That is, $P_n = \{P_n(i), 1 \leq i \leq J\}$ is the empirical distribution of the Y_ns after n observations. Then

$$\lim_{n \to \infty} \frac{1}{n} P\{P_n(i) \approx r'_i, \ 1 \leq i \leq J\} = -K(r, r')$$

where $K(r, r') = \sum_i r'_i \log(\frac{r'_i}{r_i})$. This will be used in the next example.

Example 10.4.16 [Queues in tandem with routing] Consider the network in Figure 10.10.

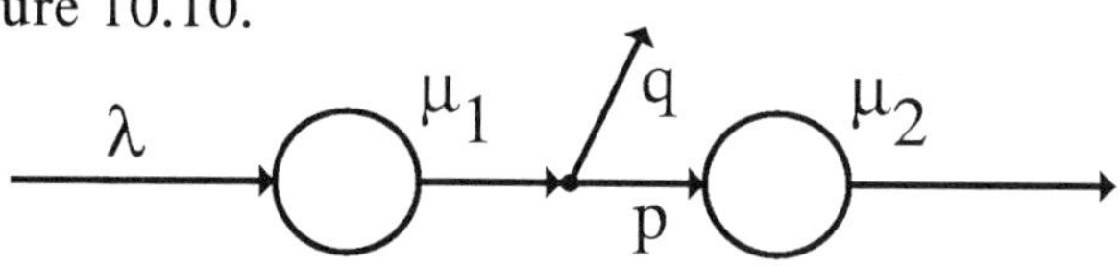

Figure 10.10. Two queues in tandem with routing.

The arrivals are renewal with rate λ and the service times are independent and are distributed as σ_i in queue i $(i = 1, 2)$. A customer who leaves queue 1 joins queue 2 with probability $p \in (0, 1)$ and leaves the network with probability $q = 1 - p$.

Fix a large number N. As before, we are interested in estimating the probability α that the population of the network reaches the value N in a given busy cycle.

One argues that if a busy cycle reaches the population N, it must be that the system is behaving during that busy cycle as a network with parameters $(\lambda', \mu'_1, \mu'_2, p')$ over some interval of time $[0, T]$.

For this to happen, it must be that the sample mean of the first $\lambda' T$ interarrival times is almost equal to $(\lambda')^{-1}$, that the sample mean of the first $\mu'_i T$ service times is almost equal to $(\mu'_i)^{-1}$ for $i = 1, 2$, and that a fraction p' of the customers leaving queue 1 in $[0, T]$ are sent to queue 2. Now, the number of customers leaving queue 1 in $[0, T]$ is roughly equal to $T\min\{\lambda', \mu'_1\}$ if the effective arrival rate is λ' and the effective service rate in queue 1 is μ'_1.

Arguing as before, one sees that the probability of such a behavior can be estimated as

$$\exp\{-T[\lambda' h_\lambda(\frac{1}{\lambda'}) + \sum_{i=1}^{2} \mu'_i \, h_{\mu_i}(\frac{1}{\mu'_i})$$

$$+ \min\{\lambda, \mu_1\} \, K(p, p')]\} \tag{10.46}$$

where $K(p, p') = p' \log(\frac{p'}{p}) + q' \log(\frac{q'}{q})$ with $q' := 1 - p'$.

It remains to relate T to N. If $\lambda' > \mu'_1$ and $\mu'_1 p' > \mu'_2$, then one can argue that both queues are "blowing up" and that the population increases at rate $\lambda' - \mu'_1$ in the first queue and at rate $\mu'_1 p' - \mu'_2$ in the second queue. Thus, one should have

$$N = T(\lambda' - \mu_1 + \mu'_1 p' - \mu'_2), \quad \text{if } \lambda > \mu'_1 \text{ and } \mu'_1 p' > \mu'_2 \,.$$

Similarly, one concludes that

$$N = T(\lambda' - \mu'_1), \quad \text{if } \lambda' > \mu'_1 \text{ and } \mu'_1 p' < \mu'_2$$

and

$$N = T(\mu'_1 p' - \mu'_2), \quad \text{if } \lambda' < \mu'_1 \text{ and } \mu'_1 p' > \mu'_2 \,.$$

One then writes the probability α as the sum of the probabilities (10.46) over the possible parameters, with the suitable substitutions of T in terms of N and those parameters. As before, if N is large, this sum of exponentials is almost equal to the exponential with the minimum coefficient of $-N$. This minimization then yields the parameters λ^*, μ_1^*, μ_2^*, p^* that are used in the simulation experiment.

Simple example 10.4.17 [M/M/1 queues with routing] In the case of M/M/1 queues with $\lambda = 0.2$, $\mu_1 = 0.6$, $\mu_2 = 0.2$, $p = 0.5$, and $N = 20$, one finds numerically that $\alpha = 2.3 \times 10^{-6}$. A direct simulation would require generating about 450,000 busy cycles. One finds, by performing the minimization explained in 10.4.16, that the modified parameters should be $\lambda^* = 0.3$, $\mu_1^* = 0.6$, $\mu_2^* = 0.1$, and $p^* = 0.33$. Also, a simulation of the modified system gives a good estimate of α after only 500 busy cycles.

10.5. EXERCISES

10.1. Nonlinear filtering Give a proof of Theorem 10.2.2 for a regular Markov chain (i.e., without assuming that it is uniform). *Hint:* Observe that the Kolmogorov equations hold for a killed Markov chain, since killing preserves regularity. (See Theorem 2.5.2.)

10.2. Poisson output Construct a queue that is not quasi-reversible but with a Poisson output.

10.3. Quasi-reversibility Give a simple proof that the multiclass queue of Example 10.2.11 is not quasi-reversible. *Hint:* Use the condition (10.8).

10.4. Known jumps Provide the details of the proof of Theorem 10.2.15.

10.5. Disorder problem Propose an algorithm for solving equations (10.31).

10.6. Disorder of Markov chains Derive the Dynamic Programming equations for solving the disorder problem of Comments 10.3.8.

10.7. Cramer-Legendre Transform Verify formula (10.44).

10.8. Quick simulation of an M/M/1 queue Derive the values of λ^* and μ^* given in Example 10.4.12.

10.9. Tandem M/M/1 queues Prove the result stated in Example 10.4.14.

10.6. REFERENCES AND COMMENTS

Section 10.2 Nonlinear filtering theory was first developed for diffusion processes. See, for instance, Liptser and Shiryayev (1975). Stochastic calculus is simpler for point processes than for diffusions, since the integrals are defined pathwise. See Brémaud (1981) and Varaiya, et al. (1981) for more details. The interchangeability of the M/M/1 queues is due to Weber (1979). The proof given here is from Anantharam (1985). See also Tsoucas, et al. (1986) for a proof based on a coupling argument.

Section 10.3 Hypothesis testing is discussed in a number of probability textbooks. See, e.g., Lehmann (1959) and Whetherill and Glazebrook (1986). See also Kumar and Varaiya (1986), and Wong and Hajek (1985). For the disorder problem, see Liptser and Shiryayev (1975), and Wan and Davis (1977).

Section 10.4 A nice introduction to large deviations is Varadhan (1984). Applications to statistical mechanics are discussed in Ellis (1985). The monograph by Freidlin and Wentzell (1984) explores large deviations of trajectories of dynamical systems. The paper by Cottrell, et al. (1983) discusses the quick simulation for a birth-and-death process. The results for networks are from Parekh, et al. (1986).

APPENDIX A

PROBABILITY

A.1. INTRODUCTION

The elementary probability theory, at the level of an upper division course, is assumed familiar. No measure theory is required. The next section reviews ideas that are used repeatedly in the text. Exercises allow the reader to check her/his understanding of the necessary concepts.

A.2. REVIEW

Notation A.2.1 In the following, $\mathbb{R}$ denotes the set of real numbers $(-\infty, +\infty)$, and Y, Y_n, Z are real-valued random variables. By $P(A)$ or $\text{Prob}\{A\}$ one denotes the probability of some event A. For a condition C, $1\{C\}$ takes the value 1 if C is satisfied, and takes the value 0 otherwise; thus, $1\{4 \leq 3\} = 0$. The abbreviation a.s. stands for **almost surely** and means that the corresponding statement holds with probability one. For instance, $Y_n \to Y$ a.s. as $n \to \infty$ if the probability that the sequence Y_n converges to Y as $n \to \infty$ is one. By $o(t)$ one denotes a function such that $\frac{o(t)}{t} \to 0$ as $t \to 0$ (e.g. t^2).

Conditional probability A.2.2 Let A and B be two events with $P(B) > 0$. The **conditional probability** of A given B is then defined by

$$P[A \mid B] := \frac{P(A \cap B)}{P(B)} .$$

Independence A.2.3 $\{Y_n,\ n \geq 0\}$ are **independent** if

$$\begin{aligned}\text{Prob}\{&Y_0 \leq y_0, Y_1 \leq y_1, \ldots, Y_n \leq y_n\} \\ &= \text{Prob}\{Y_0 \leq y_0\} \ldots \text{Prob}\{Y_n \leq y_n\} ,\end{aligned}$$

for all $n \geq 1$ and all $y_0, \ldots, y_n$ in $(-\infty, +\infty]$.

Expectation A.2.4 One defines the **expectation** of the random variable Z by $E\{Z\} = \int_{-\infty}^{+\infty} z\,dP(Z \leq z)$, provided that Z is integrable, i.e., that the integral does not yield $\infty - \infty$.

The **variance** of Z is defined by $\mathrm{var}(Z) = E\{(Z - E\{Z\})^2\}$.

By definition, $E\{(Y_1, \ldots, Y_n)\} = (E\{Y_1\}, \ldots, E\{Y_n\})$.

Conditional expectation A.2.5 Given two random variables Y and Z, the **conditional expectation** of Z given Y is denoted by $E[Z \mid Y]$ and is defined to be a random variable of the form $f(Y)$ that is such that $E\{g(Y)E[Z \mid Y]\} = E\{g(Y)Z\}$ for any function g such that $g(Y)Z$ is integrable.

The interpretation is that the random variable $E[Z \mid Y]$ is known when Y is known and that it has the same average value as Z on the events that are determined by Y. It follows, if $E\{Z^2\} < \infty$, that $E[Z \mid Y]$ is the function $f(Y)$ of Y that minimizes $E\{(Z - f(Y))^2\}$. It is therefore the **least mean squares estimate** of Z given Y.

Facts A.2.6 $E\{Z\} = E\{E[Z \mid Y]\}$; if Y and Z are independent, then $E[Z \mid Y] = E\{Z\}$; also, for $1 \leq m \leq n$,

$$E[E[Z \mid Y_1, \ldots, Y_n] \mid Y_1, \ldots, Y_m] = E[Z \mid Y_1, \ldots, Y_m].$$

Monotone Convergence Theorem A.2.7 If $0 \leq Y_n \leq Y_{n+1}$ a.s. for all $n \geq 1$, and if $Y_n \to Y$ a.s. when $n \to \infty$, then $E\{Y_n\} \to E\{Y\}$ as $n \to \infty$.

Dominated Convergence Theorem A.2.8 If $Y_n \to Y$ a.s. when $n \to \infty$, and if $|Y_n| \leq Z$ a.s. for all $n \geq 1$ with $E\{Z\} < \infty$, then $E\{Y_n\} \to E\{Y\}$ as $n \to \infty$.

Example A.2.9 [Interchanging limits and sums] Notice that

$$\lim_{n \to \infty} \sum_{i=1}^{\infty} \frac{1}{n} 1\{1 \leq i \leq n\} = 1 \neq \sum_{i=1}^{\infty} [\lim_{n \to \infty} \frac{1}{n} 1\{1 \leq i \leq n\}] = 0.$$

This example illustrates the necessity of some condition to allow the interchange of summation and limit operations. The monotone and dominated convergence theorems are the most useful sufficient conditions to justify that interchange.

Markov's inequality A.2.10 If $f : [0, \infty) \to [0, \infty)$ is a nondecreasing function, then the following inequality, known as **Markov's inequality,** holds:

$$\text{Prob}\{\,|\,Y\,|\, \geq a\} \;\leq\; \frac{E\{f(\,|\,Y\,|\,)\}}{f(a)} \text{ for any } \quad a > 0 \text{ with } f(a) > 0\,.$$

In particular, with $f(x) := x^2$,

$$\text{Prob}\{\,|\,Y\,|\, \geq a\} \;\leq\; \frac{E\{Y^2\}}{a^2},\; a > 0\,.$$

This last expression is **Chebyshev's inequality.**

Cauchy-Schwarz inequality A.2.11 For two random variables Y and Z with finite second moments $E\{X^2\}$ and $E\{Z^2\}$ one has the following **Cauchy-Schwarz inequality:**

$$|\,E\{YZ\}\,|^2 \;\leq\; E\{Y^2\}\,E\{Z^2\}\,.$$

Jensen's inequality A.2.12 If $f : \mathbb{R}^n \to \mathbb{R}$ is a **convex** function (i.e., $f(\theta x + (1 - \theta)y) \leq \theta\, f(x) + (1 - \theta) f(y)$ for all $x, y \in \mathbb{R}^n$ and $0 \leq \theta \leq 1$), then **Jensen's inequality** holds:

$$f(E\{(Y_1, \ldots, Y_n)\}) \;\leq\; E\{f(Y_1, \ldots, Y_n)\}\,.$$

Useful bounds A.2.13 One has

$$(a_1 + \cdots + a_n)^2 \;\leq\; n(a_1^2 + \cdots + a_n^2)$$

for all $n \geq 1$ and $a_1, \ldots, a_n$ in $\mathbb{R}$.
Also, for all $x \in \mathbb{R}$,

$$1 + x \leq e^x \text{ and } |\,e^{ix} - 1\,| \;\leq\; |\,x\,| \text{ with } i^2 = -1\,.$$

Countability A.2.14 A set $\mathbf{X}$ is **countable** if it has finitely many elements or if one can enumerate its elements as $\mathbf{X} = \{i_1, i_2, i_3, \ldots\}$. For instance, $\{1, 2, \ldots, 3359\}$, $\{0, 1, 2, \ldots\}$, and the rational numbers are countable. The interval $[0, 1]$ is not countable.

Strong Law of Large Numbers A.2.15 Let $Y_1, Y_2, \ldots$ be **i.i.d.** (independent and identically distributed) with

$$E\{Y_n\} \;=\; \mu \in (-\infty, +\infty)\,.$$

Then

$$n^{-1}(Y_1 + \cdots + Y_n) \rightarrow \mu \text{ a.s. as } n \rightarrow \infty .$$

Central Limit Theorem A.2.16 Let $Y_1, Y_2, \ldots$ be i.i.d. with $E\{Y_n\} = 0$ and $\text{var}(Y_n) = \sigma^2 < \infty$.

Define $Z_n = \frac{1}{\sigma\sqrt{n}}(Y_1 + \cdots + Y_n)$, $n \geq 1$. Then

$$\text{Prob}\{Z_n \leq z\} \rightarrow \text{Prob}\{Z \leq z\} := \frac{1}{\sqrt{2\pi}} \int_{-\infty}^{z} e^{-\frac{x^2}{2}} dx$$

for all $z \in R$ as $n \rightarrow \infty$.
This result says that the probability distribution of Z_n approaches (point-wise) that of a Gaussian random variable Z with zero mean and unit variance. One says that Z_n **converges weakly (or in distribution)** to an $N(0, 1)$ random variable, and one writes $Z_n \Rightarrow N(0, 1)$.

Conditional independence A.2.17 Let $\{X, Y, Z\}$ be three random variables. One says that X and Z are **conditionally independent given** Y if

$$E[f(X)\, g(Z) \mid Y] = E[f(X) \mid Y]\, E[g(Z) \mid Y]$$

for all $f(.)$ and $g(.)$ so that the expectations are defined.

Characteristic function A.2.18 Let X be a real-valued random variable. Its distribution is completely specified by its **characteristic function**

$$\phi(u) := E\{e^{iuX}\}$$

where $i^2 = -1$.

A.3. EXERCISES

A.1. Cauchy-Schwarz Prove the Cauchy-Schwarz inequality A.2.11. *Hint:* Use the fact that $E\{(Y + \lambda Z)^2\} \geq 0$ for all $\lambda \in \mathbb{R}$. A suitable choice of λ yields the result.

A.2. Bounds Prove the first inequality in A.2.13. *Hint:* Use A.2.11 with Z such that $\text{Prob}\{Z = a_i\} = \frac{1}{n}$ for $1 \leq i \leq n$ and $Y \equiv 1$.

A.3. Convergence Assume that $Y_n \to \infty$ a.s. as $n \to \infty$. Show that

$$E\{\frac{4Y_n^2 + 3Y_n + 5}{2Y_n^2 + 4Y_n + 7}\} \to 2 \text{ as } n \to \infty .$$

A.4. Convergence Find simple examples that show that the monotonicity is necessary in A.2.7 and that the domination is necessary in A.2.8.

A.5. Jensen's inequality Prove the inequality A.2.12 in the case where $Y_1, \ldots, Y_n$ take only finitely many possible values. *Hint:* This is an immediate consequence of the definition of convexity.

A.6. Characteristic function Assume that $E\{\,|Z|\,\} < \infty$ and define $\phi(u) = E\{e^{iuZ}\}$ for $u \in R$ $(i^2 = -1)$. Show that

$$\frac{1}{i}\frac{d}{du}\phi(0) = E\{Z\} .$$

Hint: Use dominated convergence and a bound in A.2.13.

A.7. Conditional expectation Prove the Facts A.2.6. *Hint:* Everything is immediate from Definition A.2.5.

A.8. Independence Find three random variables $\{X, Y, Z\}$ that are not independent but such that any two of them are independent.

A.9. Conditional independence Show that X and Z are conditionally independent given Y if and only if

$$E[g(Z) \mid X, Y] = E[g(Z) \mid Y]$$

for all function $g(.)$ so that the expectations are well defined.

A.10. Convex functions Let $f, g : \mathbb{R}^n \to \mathbb{R}$ be two convex functions. Show that $h(x) := \max\{f(x), g(x)\}$, $x \in \mathbb{R}$ defines a convex function.

A.4. REFERENCES AND COMMENTS

There are many excellent texts on probability theory. The book by Hoel, et al. (1971) is an easy introduction to the subject and should be a sufficient background for the present text.

It is difficult to be very precise without some concepts of measure theory. We hope that our decision to avoid the necessary measure theory will not create undue confusion.

Readers wishing for a higher-level discussion of probability and stochastic processes can find it in Breiman (1968), Feller (1971), Ross (1983a), Çinlar (1975), and Wong and Hajek (1985).

APPENDIX B

QUEUES

B.1. INTRODUCTION

It turns out that the standard methods of queueing theory play a very small role in the text. The reason is that few of those methods extend to networks. However, the intuition and the terminology are most useful. This appendix summarizes a few items of interest.

The **work-conserving disciplines** and the fact that **determinism minimizes delays** are discussed in section B.2. Some basic results on the **M/GI/1 queue** are presented in section B.3. The **GI/M/1 queue** is analyzed in section B.4.

B.2. GENERAL QUEUEING CONCEPTS

Model B.2.1 Consider the situation depicted in Figure B.1.

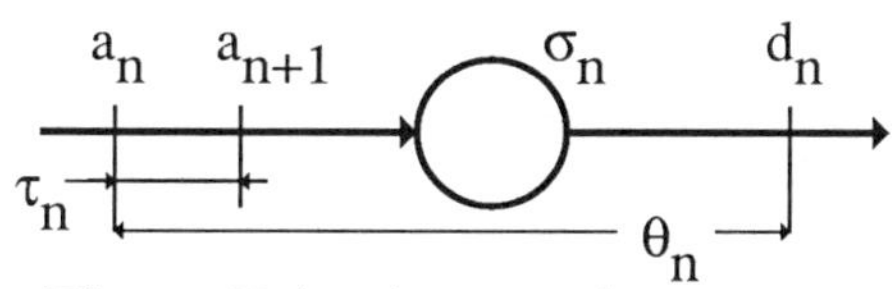

Figure B.1. A queueing system.

Customers arrive at the (possibly random) successive times $0 \leq a_1 < a_2 < \cdots$ into the initially empty queueing system. Denote by σ_n, d_n, and θ_n the service time, the departure time, and the sojourn time of the customer who arrives at time a_n, for $n \geq 1$. Thus, $\theta_n = d_n - a_n$.

Service discipline B.2.2 The **service discipline** determines which customer is served at any given time. It will be assumed in this section that service is provided at a total rate of one unit per unit of time, as long as the system is nonempty. Such systems are said to be **work-conserving.**

For instance, **first-come–first-served (FCFS)** and **last-come–first-served (LCFS)** that serve, at any given time, the first [resp. last] arrived customer are work-conserving. In the LCFS queue, it is implicitly assumed that the service of a customer will be interrupted (preempted) by that of a subsequent arrival and that the amount of partial service received is not lost when the service is resumed. (The proper name of that discipline is LCFS-preemptive-resume; we will simply call it LCFS.)

The **processor-sharing (PS)** discipline in which the unit service rate is shared equally among all the customers present in the queue is also work-conserving.

Work-to-be-done B.2.3 Define the right-continuous process $\{w_t, t \geq 0\}$ by

$$\begin{aligned} w_t &= 0 \quad \text{for } t \in [0, a_1) \\ w_t &= \{w_{a_n} - (t - a_n)\}^+ \quad \text{for } a_n < t < a_{n+1}, \; n \geq 1 \\ w_{a_n} &= w_{a_n -} + \sigma_n, \quad n \geq 1 . \end{aligned}$$

In these expressions, $\{x\}^+ := \max\{0, x\}$ for $x \in \mathbb{R}$. This definition is illustrated in Figure B.2.

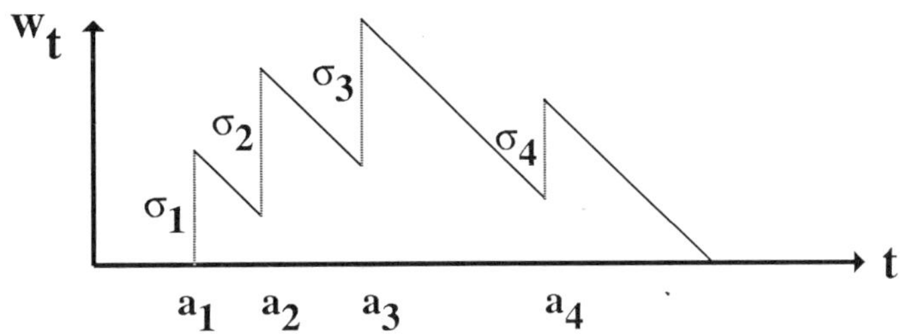

Figure B.2. The work-to-be-done process.

The interpretation is that w_t is the total service time still required at time t by the customers present at time t. Notice that the work-to-be-done does not depend on the specific work-conserving discipline of the queue.

The intervals of time when w_t is nonzero are the **busy periods** of the system, and the others are the **idle periods.** Thus, the busy and idle periods do not depend on the work-conserving priority.

Departure and sojourn times B.2.4 The departure and sojourn times of the customers depend on the service discipline. For instance, it should be intuitively clear that the discipline that always serves the customer with the smallest [resp. largest] residual service time minimizes [resp. maximizes] the number of customers present in the system at all times over all the work-conserving policies. (See Proposition 8.3.6.)

FCFS queue B.2.5 Assume that the customers are served in their order of arrivals. For $n \geq 1$, the $(n+1)$st customer starts service at time $t = \max\{d_n, a_{n+1}\}$. Therefore,

$$\begin{aligned} d_1 &= a_1 + \sigma_1 \\ d_{n+1} &= \max\{d_n, a_{n+1}\} + \sigma_{n+1}, \quad n \geq 1 . \end{aligned}$$

This shows by induction on n that

$$\begin{aligned} d_n = \max\{&a_1 + \sigma_1 + \cdots + \sigma_n , \\ &a_2 + \sigma_2 + \cdots + \sigma_n, \ldots, a_n + \sigma_n\}, \quad n \geq 1 . \end{aligned}$$

As a consequence,

$$\begin{aligned} \theta_n &= d_n - a_n \\ &= \max\{a_1 - a_n + \cdots + \sigma_n, \ldots, \\ &\qquad a_{n-1} - a_n + \sigma_{n-1} + \sigma_n, \sigma_n\} \\ &= \max\{(\sigma_1 - \tau_1) + \cdots + (\sigma_{n-1} - \tau_{n-1}) + \sigma_n, \ldots, \\ &\qquad (\sigma_{n-1} - \tau_{n-1}) + \sigma_n, \sigma_n\} \end{aligned} \tag{B.1}$$

where $\tau_n := a_{n+1} - a_n$ is the nth interarrival time.

Observe that θ_n is convex in $(\tau_1, \ldots, \tau_{n-1}, \sigma_1, \ldots, \sigma_n)$. Indeed, linear functions are convex (see A.2.12), and the maximum of convex functions is a convex function (see Exercise A.10). Also, this function is nonincreasing in the τ_ms and nondecreasing in the σ_ms.

Remark B.2.6 [Fluctuations] Notice that in an initially empty queue with $0 < \sigma_n < \tau_n$ for all $n \geq 1$, every customer finds the queue empty upon his arrival and leaves before the next arrival. In that case, the sojourn times are equal to the service times: there is no time spent waiting for the service to start.

If the interarrival times τ_n or the service times σ_n fluctuate in such a way that more arrivals may occur than can be served during some time interval, then the queue builds up. This gives rise to queueing delays that increase the sojourn times.

This suggests that fluctuations induce congestion. This is made precise next.

Convexity and congestion B.2.7 Assume that in the FCFS queue the σ_ns are i.i.d. and are independent of the τ_ns. Define

$$\psi_n[\tau_1, \ldots, \tau_{n-1}] := E[\theta_n \mid \tau_1, \ldots, \tau_{n-1}], \quad n \geq 1 . \tag{B.2}$$

It follows from the discussion of B.2.5 that ψ_n is convex in $(\tau_1, \ldots, \tau_{n-1})$ and nonincreasing in each τ_m. Indeed, ψ_n is a positive linear combination of the convex nonincreasing functions

θ_n (with weights given by the density of the σ_ms in the expectation).

Fix $m \geq 2$. For any $k \geq m$ one has from (B.1)

$$\theta_k \geq \max\{(\sigma_{k-m+1} - \tau_{k-m+1}) + \cdots + (\sigma_{k-m} - \tau_{k-1}) + \sigma_k, \ldots, (\sigma_{k-1} - \tau_{k-1}) + \sigma_k, \sigma_k\}.$$

By taking expectation and using the definition (B.2) one finds that

$$E[\theta_k \mid \tau_1, \ldots, \tau_{k-1}] \geq \psi_m[\tau_{k-m+1}, \tau_{k-m+1}, \ldots, \tau_{k-1}].$$

Therefore,

$$\frac{1}{n-m} \sum_{k=m+1}^{n} E\{\theta_k\} \geq \frac{1}{n-m} \sum_{k=m+1}^{n} E\{\psi_m[\tau_{k-m+1}, \ldots, \tau_{k-1}]\}$$

$$\geq \psi_m[\frac{1}{n-m} \sum_{k=m+1}^{n} E\{(\tau_{k-m+1}, \ldots, \tau_{k-1})\}]$$

$$= \psi_m[\frac{1}{n-m} E\{(a_{n-m+2}, \ldots, a_n - a_m)\}]$$

$$\geq \psi_m[\frac{1}{n-m} E\{(a_n, \ldots, a_n)\}]. \tag{B.3}$$

The second inequality follows from Jensen's inequality A.2.12 applied to the expectation $\frac{1}{n-m} \sum_{k=m+1}^{n} E\{.\}$ of the convex function $\psi_m[.]$. The last inequality follows from the fact that ψ_m is nonincreasing in its arguments.

Assume now that

$$\limsup_{n \to \infty} \frac{1}{n} E\{a_n\} \leq \lambda^{-1}, \tag{B.4}$$

for some $\lambda > 0$. (Recall that $\limsup_{n \to \infty} x_n := \lim_{n \to \infty} \sup_{k \geq n} x_k$. This limit is always well defined, since $\sup_{k \geq n} x_k$ is a nonincreasing sequence. One defines $\liminf x_n$ similarly and one can check that $\limsup x_n \geq \liminf x_n$ and that the equality holds if and only if $\lim x_n$ exists.)

The condition (B.4) means that the long term average arrival rate is at least λ. Then (B.3) implies that

$$\liminf_{n \to \infty} \frac{1}{n} \sum_{k=m+1}^{n} E\{\theta_k\} \geq \psi_m[\lambda^{-1}, \ldots, \lambda^{-1}]. \tag{B.5}$$

As m tends to infinity, the right-hand side of (B.5) converges to the limiting sojourn time of a customer in the FCFS queue when all the interarrival times are equal to λ^{-1}. Thus, this result shows that the average delay of the customers in an FCFS queue with i.i.d. service times and with an arrival process with rate at least equal to λ is minimized when all the interarrival times are equal to λ^{-1}. This is a precise statement of the fact that **determinism minimizes congestion.**

B.3. M/GI/1 QUEUE

This section relies on some notions about the Poisson process (2.11) and discrete-time Markov chains (2.9.1 and 2.9.2).

Definition B.3.1 The M/GI/1 queue is an FCFS queue with a single server; the arrivals are a Poisson process, and the service times $\sigma_1, \sigma_2, \ldots$ are independent and identically distributed. Let λ be the arrival rate and define

$$F(t) = P\{\sigma_n \leq t\}, \quad t \geq 0$$

$$F^*(s) = E\{e^{-s\sigma_n}\} = \int_0^\infty e^{-st}\, dF(t)\,.$$

Embedded Markov chain B.3.2 For $n \geq 1$ let x_n denote the queue length just after the nth service completion. It follows from the memoryless property of the Poisson process that $\mathbf{x} = \{x_n, n \geq 0\}$ is a Markov chain. Also,

$$x_{n+1} = (x_n - 1)^+ + \xi_n, \; n \geq 0 \tag{B.6}$$

where ξ_n is the number of arrivals during the $(n + 1)$st service.

Pollaczek-Khintchine formulas B.3.3 Assume that $\rho := \dfrac{\lambda}{\mu} < 1$, where μ^{-1} is the average service time. Denote by $\pi = \{\pi(i), i \geq 0\}$ the invariant distribution of $\mathbf{x}$. Then

$$\sum_{i=0}^{\infty} z^i \pi(i) = \frac{(1 - \rho)(1 - z)F^*(\lambda - \lambda z)}{F^*(\lambda - \lambda z) - z}\,. \tag{B.7}$$

In particular, the average value of x_n under π is given by

$$\sum_{i=0}^{\infty} i\, \pi(i) = \rho + \frac{\rho^2 + \lambda^2\sigma^2}{2(1 - \rho)} \tag{B.8}$$

where σ^2 is the variance of the service times.

The formulas (B.7)–(B.8) are called the **Pollaczek-Khintchine formulas.**

Proof: Let $A(z) = \sum_{i=0}^{\infty} z^i \pi(i)$ and

$$B(z) = \sum_{i=0}^{\infty} z^i P\{\xi_n = i\}$$

$$= \sum_{i=0}^{\infty} z^i \int_0^{\infty} \frac{(\lambda t)^i}{i!} e^{-\lambda t} dF(t) = F^*(\lambda - \lambda z).$$

From (B.6) one has

$$A(z) = E_\pi\{z^{x_{n+1}}\} = E_\pi\{z^{(x_n - 1)^+ + \xi_n}\}$$

$$= E_\pi\{z^{(x_n - 1)^+}\} B(z) = [\pi(0) + \sum_{i=0}^{\infty} \pi(i) z^{i-1}] B(z)$$

$$= [\pi(0) + z^{-1}(A(z) - \pi(0))] B(z).$$

Hence,

$$A(z) [B(z) - z] = \pi(0) (1 - z) B(z). \tag{B.9}$$

Differentiating (B.9) with respect to z gives successively

$$A'(z) [B(z) - z] + A(z) [B'(z) - 1]$$
$$= \pi(0) (1 - z) B'(z) - \pi(0) B(z) \tag{B.10}$$

and

$$A^{(2)}(z) [B(z) - z] + 2A'(z) [B'(z) - 1] + A(z) B^{(2)}(z)$$
$$= \pi(0) (1 - z) B^{(2)}(z) - 2\pi(0) B'(z) \tag{B.11}$$

where A' and $A^{(2)}$ denote the first and second derivative of A, respectively, and similarly for B.

In order to evaluate these expressions at $z = 1$ one observes that

$$A(1) = B(1) = 1, A'(1) = E_\pi\{x_n\}, B'(1) = E\{\xi_n\} = \rho$$

and

$$
\begin{aligned}
B^{(2)}(1) &= \frac{d^2}{dz^2} F^*(\lambda - \lambda z)_{z=1} \\
&= \lambda^2 F^{*(2)}(0) = \lambda^2 E\{\sigma_n^2\} \\
&= \lambda^2[\mathrm{var}(\sigma_n) + (E\{\sigma_n\})^2] \\
&= \lambda^2\sigma^2 + \rho^2 .
\end{aligned}
$$

Substituting these values into (B.10) and (B.11) with $z = 1$ gives (B.7)–(B.8).

B.4. GI/M/1 QUEUE

Definition B.4.1 The GI/M/1 queue is a single server FCFS queue with i.i.d. interarrival times τ_n and i.i.d. service times σ_n that are exponentially distributed with mean μ^{-1}.

Let

$$
\begin{aligned}
H(t) &= P\{\tau_n \le t\}, \quad t \ge 0 \\
H^*(s) &= E\{e^{-s\tau_n}\} = \int_0^\infty e^{-st} dH(t) .
\end{aligned}
$$

Embedded Markov chain B.4.2 For $n \ge 0$ let x_n be the queue length just before the $(n + 1)$st arrival. Then $\mathbf{x} = \{x_n, n \ge 0\}$ is seen to be an irreducible aperiodic Markov chain on $\mathbf{X} = \{0, 1, 2, \ldots\}$. Its transition matrix P is given by

$$
P = \begin{bmatrix}
\beta_0 & \alpha_0 & 0 & 0 & . & . \\
\beta_1 & \alpha_1 & \alpha_0 & 0 & . & . \\
\beta_2 & \alpha_2 & \alpha_1 & 0 & . & . \\
. & . & . & . & . & . \\
. & . & . & . & . & .
\end{bmatrix}
$$

where α_i is the probability that exactly i service completions take place during one interarrival time and $\beta_i = 1 - \sum_{j=0}^{i} \alpha_j$. That is,

$$\alpha_i = \int_0^\infty \frac{(\mu t)^i}{i!} e^{-\mu t} dH(t) .$$

A simple verification shows that the balance equations $\pi = \pi P$ are satisfied by

$$\pi(i) = (1 - \nu)\nu^i, \quad i \geq 0 , \tag{B.13}$$

where

$$\nu = \sum_{i=0}^\infty \nu^i \alpha_i = \sum_{i=0}^\infty \nu^i \int_0^\infty \frac{(\mu t)^i}{i!} e^{-\mu t} dH(t) ,$$

i.e.,

$$\nu = H^*(\mu - \mu\nu) . \tag{B.14}$$

Since the Markov chain $\mathbf{x}$ has at most one invariant distribution, it must be that (B.13) has at most one solution in (0, 1). One can see that it has one if $\lambda < \mu$ where $\lambda^{-1} = E\{\tau_n\}$. This is a simple consequence of Pakes' Lemma (see Exercise 2.15).

B.5. EXERCISES

B.1. Determinism State and prove the results corresponding to those discussed in B.2.7 for the service times.

B.2. Regularity Assume that the τ_n and the σ_n in the FCFS queue of B.2.7 are two sequences of i.i.d. random variables. Show that among the deterministic strategies that accept at least a fraction N^{-1} of customers ($N \geq 2$ is a given integer), the one that accepts exactly every Nth customer minimizes the long-term average expected sojourn time per accepted customer.

Hint: Let $1 = l_1 < l_2 < \cdots$ denote the accepted customers. One assumes $\limsup \frac{l_n}{n} \leq N$. As in B.2.7 one finds that

$$\frac{1}{n - m} \sum_{k=m+1}^{n} E\{\theta_{l_k}\} \geq \psi_m[\frac{\lambda^{-1}}{n - m}(l_{n-m+2} - l_1, \ldots, l_n - l_m)] .$$

B.3. M/D/1 queue Calculate the variance of the sojourn time in an M/D/1 queue (i.e., an M/GI/1 queue in which all the service times are equal to a common constant).

B.4. GI/M/1 queue Verify the invariant distribution of **x** for a GI/M/1 queue.

B.5. D/M/1 and M/D/1 queues Compare the average queue lengths in these two queues.

B.6. Sojourn time in a GI/M/1 queue Show the the sojourn time of a typical customer in a GI/M/1 queue is exponentially distributed with mean $\mu^{-1}(1 - \nu)^{-1}$. *Hint:* Proceed by analogy with the M/M/1 queue (see Exercise 4.2).

B.6. REFERENCES AND COMMENTS

For expositions of queueing theory, see Cooper (1981), Cohen (1982), Kleinrock (1975), Gross and Harris (1974), and Newell (1971). A nice presentation of coupling arguments can be found in the highly recommended text by Asmussen (1987). Our discussion of B.2.7 follows closely Hajek (1983a) and Humblet (1982).

REFERENCES

Agrawala, A. K., Coffman, E. G. Jr., Garey, M. R., and **Tripathi, S. K.** [1983], "A stochastic optimization algorithm minimizing expected flow times on uniform processors." *IEEE Trans. Comp.,* to appear.

Albin, S. L. [1984], "Approximating a point process by a renewal process, II: superposition arrival processes to queues." *Oper. Res.,* **32,** 1133-1162.

Aldous, D. and **Schepp, L.** [1987], "The least variable phase type distribution is Erlang." Preprint.

Altiok, T. [1982], "Approximate analysis of exponential tandem queues with blocking." *European J. Oper. Res.,* **11,** 390-398.

Anantharam, V. [1985], "Probabilistic proof of the interchangeability of M/M/1 queues in series." *Stoch. Proc. Appl.,* to appear.

Anantharam, V. [1986], "Optimal experimentation in decision making under uncertainty." Ph. D. thesis, Department of Electrical Engineering and Computer Sciences, University of California at Berkeley.

Anantharam, V. [1987a], "Relaxation time of closed Jackson networks." Preprint, School of Electrical Engineering, Cornell University.

Anantharam, V. [1987b], "The optimal buffer allocation problem." Preprint, School of Electrical Engineering, Cornell University.

Asmussen, S. [1987], *Applied Probability and Queues.* John Wiley & Sons, New York.

Baccelli, F. and Brémaud, P. [1987], *Palm Probabilities and Stationary Queues.* Lecture Notes in Statistics, **41,** Springer-Verlag.

Baras, J. S., Ma, D. J., and Makowsky, A. M. [1985], "K competing queues with linear costs and geometric service requirements: the μc-rule is always optimal." *Syst. Control Letters,* **6,** 173-180.

Barbour, A. D. [1976], "Networks of queues and the method of stages." *Adv. Appl. Prob.,* **8,** 584-591.

Barbour, A. D. and Schassberger, R. [1981], "Insensitive average residence times in generalized semi-Markov processes." *Adv. Appl. Prob.,* **13,** 720-735.

Bard, Y. [1980], "Estimation of state probabilities using the maximum entropy principle." *IBM J. Res. Dev.,* **24,** 563-569.

Bartlett, M. S. [1984], "Recent investigations involving stochastic population models." *Adv. Appl. Prob.,* **16,** 449-470.

Bartroli, M. and Stidham, S., Jr. [1987], "Towards a unified theory of structure of optimal policies for control of networks of queues." Preprint, Dept. of Oper. Res., Univ. of North Carolina, Chapel Hill.

Baskett, F., Chandy, K. M., Muntz, R. R., and Palacios, F. [1975], "Open, closed, and mixed networks of queues with different classes of customers." *J.A.C.M.,* **22,** 248-260.

Beneš, V. E. [1963], *General Stochastic Processes in the Theory of Queues.* Addison-Wesley, Reading, Mass.

Beneš, V. E. [1965], *Mathematical Theory of Connecting Networks and Telephone Traffic.* Academic Press, New York.

Berbee, H. C. P. [1979], "Random walks with stationary increments and renewal theory." Ph.D. thesis, Mathematisch Centrum, Amsterdam.

Berman, M. and **Westcott, M.** [1983], "On queueing systems with renewal departure processes." *Adv. Appl. Prob.,* **15,** 657-673.

Berry, D. A. and **Fristedt, B.** [1985], *Bandit Problems – Sequential Allocations of Experiments.* Chapman and Hall, London, New York.

Bertsekas, D. and **Gallager, R.** [1987], *Data Networks.* Prentice-Hall, Englewood Cliffs, New Jersey.

Bertsekas, D. [1987], "Dynamic Programming: Deterministic and Stochastic Models." Prentice-Hall, Englewood Cliffs, New Jersey.

Beutler, F. J. [1983], "A note on multivariate Poisson flows on stochastic processes." *J. Adv. Prob.,* **15,** 219-220.

Bhat, U. N. [1969], "Sixty years of queueing theory." *Mgmt Sci.,* **15B,** 280-294.

Bhat, U. N. [1986], "Finite capacity assembly-like queues." *Queueing Systems,* **1,** 85-101.

Bhat, U. N. [1987], "Statistical issues in queueing theory: a survey." To appear in *Queueing Systems.*

Billingsley, P. [1968], *Convergence of Probability Measures.* John Wiley & Sons, New York.

Bitran, G. R. and **Tirupati, D.** [1986], "Multiproduct queueing networks with deterministic routing: decomposition approach and the notion of interference." Report WP # 1764-86, Massachusetts Institute of Technology.

Blanc, J. P. C. [1984], "Asymptotic analysis of a queueing system with a two-dimensional state space." *J. Appl. Prob.,* **21,** 870-886.

Blanc, J. P. C. and **van Doorn, E. A.** [1984], "Relaxation times for queueing systems." Report OS-R8407, Centre for Math. and Comp. Science, Amsterdam, Netherlands.

Borovkov, A. A. [1967], "Boundary-value problems for random walks and large deviations in function space." *Theory of Prob. and Appl.,* **4,** 575-595.

Borovkov, A. A. [1976], *Stochastic Processes in Queueing Theory.* Springer-Verlag, New York.

Borovkov, A. A. [1984], *Asymptotic Methods in Queueing Theory.* John Wiley & Sons, New York.

Boxma, O. J. [1979a], "On a tandem queueing model with identical service times at both counters, I." *Adv. Appl. Prob.,* **11,** 616-643.

Boxma, O. J. [1979b], "On a tandem queueing model with identical service times at both counters, II." *Adv. Appl. Prob.,* **11,** 644-659.

Boxma, O. J. [1983], "The cyclic queue with one general and one exponential server." *Adv. Appl. Prob.,* **15,** 857-873.

Boxma, O. J. and **Konheim, A. G.** [1981], "Approximate analysis of exponential queueing systems with blocking." *Acta Informatica,* **15,** 19-66.

Brémaud, P. [1981], *Point Processes and Queues.* Springer-Verlag, New York.

Breiman, L. [1968], *Probability.* Addison-Wesley.

Brown, T. C. and **Pollett, P. K.** [1982], "Some distributional approximations in Markovian queueing networks." *Adv. Appl. Prob.,* **14,** 654-671.

Bruno, J., Downey, P., and **Frederickson, G. N.** [1981], "Sequencing tasks with exponential service times to minimize the expected flowtime or makespan." *J. Assoc. Comp. Mach.,* **28,** 100-113.

Burke, P. J. [1956], "The output of a queueing system." *Oper. Res.,* **4,** 699-704.

Burke, P. J. [1968], "The output of a stationary M/M/s queueing system." *Ann. Math. Stat.,* **39,** 1144-1152.

Burke, P. J. [1969], "The dependence of delays in tandem M/M/s queues." *Oper. Res.,* **17,** 754-755.

Burke, P. J. [1976], "Proof of a conjecture on the interarrival time distribution in an M/M/1 queue with feedback." *IEEE Trans. Comm.,* **COM-24,** 575-576.

Burman, D. Y. [1981], "Insensitivity in queueing systems." *Adv. Appl. Prob.,* **13,** 846-859.

Burman, D. Y. and **Smith D. R.** [1983], "A light traffic theorem for multi-server queues." *Math. Oper. Res.,* **8,** 15-25.

Burman, D. Y. and **Smith D. R.** [1986], "An asymptotic analysis of a queueing system with Markov-modulated arrivals." *Oper. Res.,* **34,** 105-119.

Burman, D. Y., Lehoczy, J. P., and **Lim, Y.** [1984], "Insensitivity of blocking probabilities in a circuit-switched network." *J. Appl. Prob.,* **21,** 853-859.

Buyukkoc, C., Varaiya, P., and **Walrand, J.** [1985], "The $c\mu$ rule revisited." *Adv. Appl. Prob.,* **17,** 237-238.

Buzacott, J. A. and **Yao, D. D.** [1986], "On queueing network models of flexible manufacturing systems." *Queueing Systems,* **1,** 5-27.

Buzen, J. P. [1973], "Computational algorithm for closed networks with exponential servers." *Comm. Assoc. Comput. Mach.,* **16,** 527-531.

Callaert, H. [1974], "On the rate of convergence in birth-and-death processes." *Bull. Soc. Math. Belg.,* **26,** 173-184.

Calo, S. B. [1982], "The cumulative waiting time in GI/(G/1) ^{M} queueing systems." IBM Thomas J. Watson Research Center, Yorktown Heights, New York.

Chandy, K. M., Herzog, U., and **Woo, L. S.** [1975], "Parametric analysis of queueing networks." *IBM J. of Research and Development,* **19,** 43-49.

Chandy, K. M., Howard, J. H., and **Towsley, D. F.** [1977], "Product form and local balance in queueing networks." *J.A.C.M.,* **24,** 250-263.

Chandy, K. M. and **Reynolds, P. F.** [1977], "Scheduling partially ordered tasks with probabilistic execution times." *Proceedings 5th Symposium on Operating Systems, A.C.M.,* 167-172.

Chung, K. L. [1967], *Markov Chains with Stationary Transition Probabilities.* 2nd ed. Springer-Verlag, New York.

Cinlar, E. [1975], Stochastic Processes. Prentice-Hall, Englewood Cliffs, New Jersey.

Coffman, E. G. and **Reiman, M. I.** [1984], "Diffusion approximations for computer/communications systems." In *Mathematical computer performance and reliability* (eds. G. Iazeolla, P. J. Courtois, and A. Hordijk) Elsevier (North-Holland).

Cohen, J. W. [1957], "The generalized Engset formulae." *Philips Telecom. Review,* **18,** 158-170.

Cohen, J. W. [1982], *The Single Server Queue (2nd Ed.],* North-Holland, Amsterdam.

Cohen, J. W. and **Boxma, O. J.** [1983], *Boundary Value Problems in Queueing System Analysis.* North-Holland.

Cohen, G., Dubois, D., Quadrat, J. P., and **Viot, M.** [1985], "A linear-system-theoretic view of discrete-event processes and its use for performance evaluation in manufacturing." *IEEE Trans. Auto. Control,* **AC-30,** 210-220.

Conway, R. W., Maxwell, W. L., and **Miller, L. W.** [1967], *The Theory of Scheduling.* Addison-Wesley, Reading, MA.

Cooper, R. B. [1981], *Introduction to Queueing Theory, 2d ed.* North-Holland, New York.

Cottrell, M., Fort, J., and **Malgouyres, G.** [1983], "Large deviations and rare events in the study of stochastic algorithms." *IEEE Trans. Aut. Control,* **AC-28,** 907-920.

Courtois, P. J. [1977], *Decomposability.* Academic Press, New York.

Cox, D. R. and **Smith, W. L.** [1961], *Queues.* Methunen, London.

Crabill, T., Gross, D., and **Magazine, M.** [1977], "A classified bibliography of research on optimal design and control of queues." *Oper. Res.,* **25,** 219-232.

Daduna, H. [1982], "Passage times for overtake-free paths in Gordon-Newell networks." *Adv. Appl. Prob.,* **14,** 672-686.

Daduna, H. [1986], "Cycle times in two-stage closed queueing networks: applications to multiprogrammed computer systems with virtual memory." *Oper. Res.,* **34,** 281-288.

Daley, D. J. [1976], "Queueing output processes." *Adv. Appl. Prob.,* **8,** 395-415.

de Smit, J. H. A. [1972], "The time dependent behaviour of the queue length process in the system M/M/s." *CORE* paper no. 7217, Univ. Louvain.

Descloux, A. [1967], "On the validity of the particular subscriber's point of view." *Proc. Fifth Intern. Teletraffic Congress,* New York.

Disney, R. L. [1975], "Random flows in queueing networks: a review and a critique." *Trans. A.I.E.E.,* **7,** 268-288.

Disney, R. L., Farrell, R. L., and **de Morais, P. R.** [1973], "A characterization of M/G/1 queues with renewal departures." *Management Sci.,* **20,** 1222-28.

Disney, R. L. and **Kiessler, P. C.** [1987], *Traffic Processes in Queueing Networks: A Markov Renewal Approach.* The Johns Hopkins University Press, Baltimore, MD.

Disney, R. L. and **König, D.** [1985], "Queueing networks: a survey of their random processes." *SIAM Review,* **27,** 335-403.

Disney, R. L., König, D., and **Schmidt, V.** [1983], "Stationary queue length and waiting time distributions in single-server feedback queues." *Adv. Appl. Prob.,* **16,** 437-446.

Doob, J. L. [1953], *Stochastic Processes.* John Wiley & Sons, New York.

Ellis, R. S. [1985], *Entropy, Large Deviations, and Statistical Mechanics.* Springer-Verlag.

Ephremides, A., Varaiya, P., and **Walrand, J.** [1980], "A simple dynamic routing problem." *IEEE Trans. Aut. Control,* **AC-25,** 690-693.

Fayolle, G., Iasnogorodski, R., and **Mitrani, I.** [1983], "The distribution of the sojourn times in a queueing network with overtaking: reduction to a boundary value problem." In *Performance '83,* (eds. A. K. Agrawala and S. K. Tripathi). North-Holland, Amsterdam.

Feller, W. [1971], *An Introduction to Probability Theory and Its Applications.* Vol. I & II, Wiley.

Fischer, M. J. [1977], "A queueing analysis of an integrated telecommunications system with priorities." *INFOR,* **15,** 277-288.

Flores, C. [1985], "Diffusion approximations for computer communications networks." In *Computer Communications,* Proc. of Symposia in Applied Mathematics (Ed. B. Gopinath), A.M.S., Providence, RI.

Foley, R. D. and **Disney, R. L.** [1983], "Queues with delayed feedback." *Adv. Appl. Prob.,* **15,** 162-182.

Franken, P., König, D., Arndt, U., and **Schmidt, V.** [1982], *Queues and point processes.* John Wiley & Sons, New York.

Freedman, D. [1971], *Markov Chains.* Holden-Day, San Francisco.

Freidlin, M. I. and **Wentzell, A. D.** [1984], *Random Perturbations of Dynamical Systems.* Springer-Verlag.

Gelenbe, E. [1975], "On approximate computer system models." *J.A.C.M.,* **22,** 261-269.

Gelenbe, E. and **Mitrani, I.** [1980], *Analysis and Synthesis of Computer Systems.* Academic Press, New York.

Gelenbe, E. and **Muntz, R. R.** [1976], "Probabilistic models of computer systems-part I (exact results)." *Acta Informatica,* **7,** *35-60.*

Gelenbe, E. and **Pujolle, G.** [1982], *Introduction aux Réseaux de Files d' Attente.* Eyrolles.

Gittins, J. C. and **Jones, D. M.** [1974], "A dynamic allocation index for the sequential design of experiments." In *Progress in Statistics* (eds. J. Gani, et al.), North-Holland, Amsterdam.

Gittins, J. C. [1979], "Bandit processes and dynamic allocation indices." *J. Roy. Stat. Soc.,* **B41,** *148-164.*

Glazebrook, K. D. [1979], "Scheduling tasks with exponential service times on parallel processors." *J. Appl. Prob.,* **16,** 685-689.

Glazebrook, K. D. [1983], "Optimal strategies for families of alternative bandit processes." *IEEE Trans. Aut. Control,* **AC-28,** 858-861.

Gnedenko, B. V. and **Kovalenko, I. N.** [1968], *Introduction to Queueing Theory.* Israel Program for Scientific Translations, Jerusalem.

Gordon, W. J. and **Newell, G. F.** [1967], "Closed queueing systems with exponential servers." *Operations Research,* **15,** 254-265.

Greenberg, B. S. [1986], "Queueing systems with returning customers and the order of tandem queues." Ph. D. thesis, Department of Industrial Engineering and Operations Research, University of California at Berkeley.

Gross, D. and **Harris, C. M.** [1974], *Fundamentals of Queueing Theory.* John Wiley & Sons, New York.

Hajek, B. [1983a], "The proof of a folk theorem on queueing delays with applications to routing in networks." *J.A.C.M.,* **30,** 834-851.

Hajek, B. [1983b], "Extremal splitting of point processes." Presented at the *Conference Dedicated to the Memory of Jack Kieffer and Jacob Wolfowitz,* Cornell

Univ. (To appear in *Math. Oper. Res.*)

Hajek, B. [1984], "Optimal control of two interacting service stations." *IEEE Trans. Aut. Control,* **AC-29,** 491-499.

Halfin, S. [1985], "The shortest queue problem." *J. Appl. Prob.,* **22,** 865-878.

Harrison, J. M. [1975a], "A priority queue with discounted linear costs." *Oper. Res.,* **23,** 260-269.

Harrison, J. M. [1975b], "Dynamic scheduling of a multi-class queue: discount optimality." *Oper. Res.,* **23,** 270-282.

Harrison, J. M. [1985], *Brownian Motion and Stochastic Flow Systems.* John Wiley & Sons, New York.

Harrison, J. M. [1986], "Brownian models of queueing networks with heterogeneous customers." Presented at the *IMA Workshop on Stochastic Differential Systems,* Minneapolis, MI, June 9-19.

Harrison, J. M. and **Reiman, M. I.** [1981], "On the distribution of a multidimensional reflected Brownian motion." *SIAM J. Appl. Math.,* **41,** 345-361.

Harrison, J. M. and **Williams, R.** [1987], "Brownian models of open queueing networks with homogeneous customer populations," *to appear.*

Harrison, P. G. [1984], "A note on cycle time in tree-like queueing networks." *Adv. Appl. Prob.,* **16,** 216-219.

Heidelberger, P. and **Lavenberg, S. S.** [1984], "Computer performance evaluation methodology." *IEEE Trans. Comp.,* **C-33,** 1195-1220.

Helm, W. E. and **Waldmann, K.-H.** [1984], "Optimal control of arrivals to multiserver queues in a random environment." *J. Appl. Prob.,* **21,** 602-615.

Heyman, D. P. and **Sobel, M. J.** [1982], *Stochastic Models in Operations Research, Vol. I.* McGraw-Hill, New York.

Heyman, D. P. and **Stidham, S. Jr.** [1980], "A note on the relation between customer and time averages in queues." *Oper. Res.,* **15,** 1109-1116.

Hoel, P. G., Port, S. C., and **Stone, C. J.** [1971], *Introduction to Probability Theory.* Houghton Mifflin Company, Boston.

Hordijk, A. [1984], "Insensitivity for stochastic networks." In *Mathematical computer performance and reliability* (eds. G. Iazeolla, P. J. Courtois. and **A. Hordijk) Elsevier (North-Holland).**

Hordijk, A. and **van Dijk, N.** [1982], "Stationary probabilities for networks of queues." In *Applied Probability-Computer Science, the Interface: Proc. of the ORSA-TIMS Boca Raton Symposium* (eds. R. Disney and T. Ott), Birkhauser, Boston, Cambridge, Mass.

Hordijk, A. and **van Dijk, N.** [1983], "Networks of queues, part I: job-local-balance and the adjoint process; part 2: general routing and service characteristics." *Proceedings International Seminar on Modelling and Performance Evaluation Methodology,* INRIA, Vol. I, 79-135.

Howard, R. A. [1960], *Dynamic Programming and Markov Processes.* MIT Press, Cambridge, Mass.

Hu, T. C. [1961], "Parallel sequencing and assembly line problems." *Oper. Res.,* **9,** 841-848.

Humblet, P. [1982], "Determinism minimizes waiting times in queues." *Tech. Report, LIDS-P-1207,* Mass. Inst, Technology, Cambridge, MA.

Iglehart, D. L. [1965], "Limiting diffusion approximations for the many server queue and the repairman problem." *J. Appl. Prob.,* **2,** 429-441.

Iglehart, D. L. and **Whitt, W.** [1970a], "Multiple channel queues in heavy traffic, I." *Adv. Appl. Prob.,* **2,** 150-177.

Iglehart, D. L. and **Whitt, W.** [1970b], "Multiple channel queues in heavy traffic, II: sequences, networks, and batches." *Adv. Appl. Prob.,* **2,** 355-369.

Jackson, J. R. [1957], "Networks of waiting lines." *Oper. Res.,* **5,** 518-521.

Jackson, J. R. [1963], "Jobshop-like queueing systems." *Mgmt. Sci.,* **10,** 131-142.

Jackson, R. R. P. [1954], "Queueing systems with phase-type service." *Operat. Res. Quart.,* **5,** 109-120.

Jackson, R. R. P. and **Aspden, P.** [1980], "A transient solution to the multistage Poisson queueing system with infinite server." *Oper. Res.,* **28,** 618-622.

Jansen, U. and **König, D.** [1980], "Insensitivity and steady-state probabilities in product form queueing networks." *Elektronische Informationsverarbeitung und Kybernetik,* **16,** 385-397.

Jansen, U. [1984], "Conditional expected sojourn times in insensitive queueing systems and networks." *Adv. Appl. prob.* **16,** 906-919.

Johnson, D. P. [1983], "Diffusion approximations for optimal filtering of jump processes and for queueing networks." Ph.D. thesis, Dept. Math., Univ. Wisconsin, Madison.

Kanter, M [1985], "Lower bounds for the probability of overload in certain queueing networks." *J. Appl. Prob.,* **22,** 429-436.

Karlin, S. and **McGregor, J. L.** [1957], "The differential equations of birth-and-death processes and the Stieltjes moment problem." *Trans. Amer. Math. Soc.,* **85,** 489-546.

Katehakis, M. N. and **Veinott, A. F. Jr.** [1985], "The multi-armed bandit problem: decomposition and computation." Tech. Report No. 41, Dept. of Oper. Res., Stanford University.

Kelly, F. P. [1975], "Networks of queues with customers of different types." *J. Appl. Prob.,* **12,** 542-554.

Kelly, F. P. [1976a], "Networks of queues." *Adv. Appl. Prob.,* **8,** 416-432.

Kelly, F. P. [1976b], "The departure process from a queueing system." *Math. Proc. Camb. Phil. Soc.,* **80,** 283-285.

Kelly, F. P. [1979], *Reversibility and Stochastic Networks.* John Wiley & Sons, New York.

Kelly, F. P. [1982a], "Networks of quasi-reversible nodes." In *Applied Probability-Computer Science, the Interface: Proc. of the ORSA-TIMS Boca Raton Symposium* (eds. R. Disney and T. Ott), Birkhauser, Boston, Cambridge, Mass.

Kelly, F. P. [1982b], "The throughput of a series of buffers." *Adv. Appl. Prob.,* **14,** 633-653.

Kelly, F. P. [1985], "Stochastic models of computer communication systems." *J. Roy. Statist. Soc. B.*

Kelly, F. P. [1986a], "Blocking probabilities in large circuit-switched networks." *Adv. Appl. Prob.,* **18.**

Kelly, F. P. [1986b], "Segregating the input to a series of buffers." Preprint, Statistical Laboratory, Cambridge University.

Kelly, F. P. and **Pollett, P. K.** [1983], "Sojourn times in closed queueing networks." *Adv. Appl. Prob.,* **15,** 638-656.

Kiessler, P. C. [1987], "A simple proof of the equivalence of input and output intervals in Jackson networks." *Oper. Res.,* to appear.

Kingman, J. F. C. [1963], "The exponential decay of Markov transition probabilities." *Proc. London Math. Soc.,* **13,** 337-358.

Kingman, J. F. C. [1969], "Markov population processes." *J. Appl. Prob.,* **6,** 1-18.

Kingman, J. F. C. [1982], "Queue disciplines in heavy traffic." *Math. Oper. Res.,* **7,** 262-271.

Kleinrock, L. [1975], *Queueing Systems, Vol. I: Theory.* John Wiley & Sons, New York.

Kleinrock, L. [1976], *Queueing Systems, Vol. II: Computer Applications.* John Wiley & Sons, New York.

Klimov, G. P. [1974], "Time sharing service systems I." *Th. Prob. Appl.,* **19,** 532-551.

Knessl, C., Matkowsky, B. J., Schuss, Z., and **Tier, C.** [1985], "An asymptotic theory for Markov jump processes." *SIAM J. Appl. Math.,* **46,** 1006-1028.

Ko, S-J. [1985], "Scheduling with precedence constraints." *Ph.D. thesis,* Department of Electrical Engineering and Computer Sciences, University of California at Berkeley.

Kobayashi, H. [1974], "Application of the diffusion approximation to queueing networks I: equilibrium queue distributions." *J.A.C.M.,* **21,** 316-328.

Kobayashi, H. and **Konheim, A. G.** [1977], "Queueing models for computer communications systems analysis." *IEEE Trans. Comm.,* **COM-25,** 2-29.

Koenigsberg, E. [1958], "Cyclic queues." *Oper. res. Quart.,* **9,** 22-35.

Koenigsberg, E. [1982], "Twenty-five years of cyclic queues and closed queue networks: a review." *J. Oper. Res. Soc.,* **33,** 605-619.

Kolmogorov, A. [1936], "Zur Theorie der Markoffschen Ketten." *Mathematische Annalen,* **112,** 155-160.

König, D., Miyazawa, M., and **Schmidt, V.** [1983], "On the identification of Poisson arrivals in queues with Conditions pour que, en regime stationaire, coinciding time-stationary and customer-stationary state distributions." *J. Appl. Prob.,* **20,** 860-871.

Kovalenko, I. N. [1962], "Conditions pour que, en regime stationaire, la distribution soit independante des lois des durees des conversations." *Oper. Res.,* **24,** 516-529.

Kuehn, P. J. [1979], "Approximate analysis of general queueing networks by decomposition." *IEEE Trans. Comm.,* **COM-27,** 113-126.

Kumar, P. R. [1985], "A survey of some results in stochastic adaptive control." *SIAM J. on Control and Optim.,* **23(3),** 329-380.

Kumar, P. R. and **Varaiya, P. P.** [1986], *Stochastic Systems: Estimation, Identification, and Adaptive Control.* Prentice-Hall, Englewood Cliffs, New Jersey.

Labetoulle, J., Pujolle, G., and **Soula, C.** [1981], "Stationary distribution of flows in Jackson networks." *Math. Oper. Res.,* **6,** 173-185.

Lai, T. L. and **Robbins, H.** [1985], "Asymptotically efficient adaptive allocation rules." *Adv. Appl. Math.,* **6,** 4-22.

Langaris, C. and **Conolly, B.** [1984], "On the waiting time of a two stage queueing system with blocking." *J. Appl. Prob.,* **21,** 628-638.

Lavenberg, S. S. [1975], "Stability and maximum departure rate of certain open queueing networks having finite capacity constraints." *IBM Research Report RJ1625.*

Lavenberg, S. S. (Ed.) [1983], *Computer Performance Modeling Handbook.* Academic Press, New York.

Lavenberg, S. S. and **Reiser, M.** [1980], "Stationary state probabilities at arrival instants for closed queueing networks with multiple types of customers." *J. Appl. Prob.,* **17,** 1048-1061.

Lazar, A. A. [1983], "Optimal control of a class of queueing networks in equilibrium." *IEEE Trans. Aut. Control,* **AC-28,** 1001-1007.

Lee, H. L. and **Cohen, M. A.** [1983], "A note on the convexity of performance measures of M/M/c queueing systems." *J. Appl. Prob.,* **20,** 920-923.

Lehmann [1959], *Testing Statistical Hypotheses.* John Wiley & Sons, New York.

Lehtonen, T. [1986], "On the ordering of tandem queues with exponential servers." *J. Appl. Prob.,* **23,** 215-229.

Leitnaker, M. G. and **Purdue, P.** [1985], "Compartmental models with transfer delays: a semi-Markov approach." *J. Appl. Prob.,* **22,** 570-582.

Lemoine, A. J. [1977], "Networks of queues-a survey of equilibrium analysis." *Mgmt Sci.,* **24,** 464-481.

Lemoine, A. J. [1978], "Networks of queues-a survey of weak convergence results." *Mgmt Sci.,* **24,** 1175-1193.

Lemoine, A. J. [1979], "On the total sojourn time in networks of queues." *Mgmt Sci.,* **25,** 1034-1035.

Liggett, T, M. [1985], *Interacting Particle Systems.* Springer-Verlag, New York.

Lin, W. and **Kumar, P. R.** [1984], "Optimal control of a queueing system with two heterogeneous servers." *IEEE Trans. Aut. Control,* **AC-29,** 696-705.

Lindvall, T. [1977], "A probabilistic proof of Blackwell's renewal theorem." *Ann. Probab,* **5,** 482-485.

Liptser, R. S. and **Shiryayev, A. N.** [1975], *Statistics of Random Processes.* Springer-Verlag, New York.

Little, J. D. C. [1961], "A proof of the queueing formula $L = \lambda W$." *Oper. Res.,* **9,** 383-387.

Loynes, R. M. [1962], "The stability of a queue with non-independent inter-arrival times and service times." *Proc. Cambridge Phil. Soc.,* **58,** 497-520.

Loynes, R. M. [1964], "The stability of a system of queues in series." *Proc. Cambridge Phil. Soc.,* **60,** 569-574.

Lu, F. V. and **Serfozo, R. F.** [1985], "M/M/1 queueing decision processes with monotone hysteretic optimal policies." *Oper. Res.,* **32,** 1116-1132.

McKenna, J. and **Mitra, D.** [1982], "Integral representations and asymptotic expansions for closed Markovian queueing networks: normal usage." *Bell Syst. Tech. J.,* **61,** 661-683.

McKenna, J., Mitra, D., and **Ramakrishnan, K. G.** [1981], "A class of closed Markovian queueing networks: integral representations, asymptotic expansions, and generalizations." *Bell Syst. Tech. J.,* **60,** 599-641.

Mandelbaum, A. [1985], "Discrete multi-parameter processes and multi-armed bandits." *Z. Wahrsch. verw. Gebiete,* **65.**

Marie, R. A. [1979], "An approximate analytical method for general queueing networks." *IEEE Trans. Software Eng.*, **SE-5**, 530-538.

Massey, W. [1984a], "Open networks of queues: their algebraic structure and estimating their transient behavior." *Adv. Appl. Prob.*, **16**, 176-201.

Massey, W. [1984b], "An operator-analytic approach to the Jackson network." *J. Appl. Prob.*, **21**, 379-393.

Matthes, K. [1962], "Zur Theorie der Bedienungsprozesses." *Trans. 3rd Prague Conf. Inform. Theory.*

Meilijson, I. and **Weiss, G.** [1977], "Multiple feedbacks at a single server service station." *Stoch. Proc. and Appl.*, **195-205.**

Melamed, B. [1979], "Characterization of Poisson traffic streams in Jackson queueing networks." *Adv. Appl. Prob.*, **11**, 422-438.

Melamed, B. [1982a], "On Markov jump processes imbedded at jump epochs and their queueing-theoretic applications." *Math. Oper. Res.*, **7**, 111-128.

Melamed, B. [1982b], "Sojourn times in queueing networks." *Math. Oper. Res.*, **7**, 233-244.

Mitrani, I. [1984], "Fixed-point approximations for distributed systems." In *Mathematical computer performance and reliability* (eds. G. Iazeolla, P. J. Courtois, and A. Hordijk) Elsevier (North-Holland).

Miyazawa, M. [1983], "The derivation of invariance relations in complex queueing systems with stationary inputs." *Adv. Appl. Prob.*, **15**, 874-885.

Miyazawa, M. [1985], "The intensity conservation law for queues with randomly changed service rate." *J. Appl. Prob.*, **22**, 836-843.

Morrison, J. A. and **Mitra, D.** [1985], "Heavy-usage asymptotic expansions for the waiting time in closed processor-sharing systems with multiple classes." *Adv. Appl. Prob.*, **17**, 163-185.

Muntz, R. R. [1972], "Poisson departure processes and queueing networks." *IBM Research Report RC 4145.* IBM Thomas J. Watson Research Center, Yorktown Heights, New York.

Nain, P. and **Ross, K. W.** [1986], "Optimal multiplexing of heterogeneous traffic with hard constraint." *Performance Evaluation Review*, **14.**

Nash, P. and **Weber, R. R.** [1982], "Dominant strategies in stochastic allocation and scheduling problems." In *Deterministic and Stochastic Scheduling.* Dempster, et al. (eds), 343-354.

Neuts, M. F. [1981], *Matrix-Geometric Solutions in Stochastic Models.* Johns Hopkins, Baltimore, Md.

Neveu, J. [1977], "Processus ponctuels." *Ecole d'´eté de Probabilités de St. Flour VI, Lecture Notes in Mathematics*, **598**, 249-447, Springer-Verlag.

Neveu, J. [1983], "Construction de files d'attentes stationaires." *Lect. Notes Control and Inform. Sc.*, **60**, Springer-Verlag, 31-41.

Newell, G. [1971], *Applications of Queueing Theory.* Chapman and Hall, London.

Odoni, A. R. and **Roth, E.** [1983], "An empirical investigation of the transient behavior of stationary queueing systems." *Oper. Res.*, **31**, 432-455.

Pakes, A. G. [1969], "Some conditions of ergodicity and recurrence of Markov chains." *Oper. Res.*, **17**, 1058-1061.

Papadimitriou, C. and **Tsitsiklis, J.** [1985], "On the stochastic scheduling of a tree." *Proc. 23rd Allerton Conference*, 63-66.

Parekh, S. and **Walrand, J.** [1986], "The quick simulation method for excessive backlogs in networks of queues." Department of Electrical Engineering and Computer Sciences, E.R.L. Memo, University of California at Berkeley. To appear in *IEEE Trans. on Aut. Control.*

Peterson, W. P. [1985], "Diffusion approximations for networks of queues with multiple customer types." Ph.D. thesis, Dept. Oper. Res., Stanford University.

Pinedo, M. [1982], "On the optimal order of stations in tandem queues." In *Applied Probability-Computer Science, the Interface: Proc. of the ORSA-TIMS Boca Raton Symposium* (eds. R. Disney and T. Ott), Birkhauser, Boston, Cambridge, Mass.

Pinedo, M. and **Weiss, G.** [1979], "Scheduling stochastic tasks on two parallel processors." *Naval Res. Logist. Quart.,* **26,** 527-535.

Pinedo, M. and **Weiss, G.** [1984], "Scheduling exponential tasks with precedence constraints and different rates." Preprint.

Prabhu, N. U. [1980], *Stochastic Storage Processes.* Springer-Verlag, New York.

Reich, E. [1957], "Waiting times when queues are in tandem." *Ann. Math. Stat.,* **28,** 527-530.

Reich. E. [1963], "Notes on queues in tandem." *Ann. Math. Stat.,* **34,** 338-341.

Reiman, M. I. [1982], "The heavy traffic approximation for sojourn times in Jackson networks." In *Applied Probability-Computer Science, the Interface: Proc. of the ORSA-TIMS Boca Raton Symposium* (eds. R. Disney and T. Ott), Birkhauser, Boston, Cambridge, Mass.

Reiman, M. I. [1984], "Open queueing networks in heavy traffic." *Math. Oper. Res.,* **9,** 441-458.

Reiman, M. I. and **Simon, B.** [1985a], "Open queueing systems in light traffic." To appear in *Math. of Oper. Res.*

Reiman, M. I. and **Simon, B.** [1985b], "An interpolation approximation for queueing systems with Poisson input." To appear in *Math. of Oper. Res.*

Reiman, M. I. and **Simon, B.** [1986], "Light traffic limits of sojourn time distributions in Markovian queueing networks." Preprint.

Reiser, M. [1977], "Numerical methods in separable queueing networks." In *Algorithmic Methods in Probability,* (Ed. M. Neuts), TIMS Studies in Management Sciences, Vol. 7, North-Holland, Amsterdam.

Reiser, M. [1981], "Mean value analysis and convolutional method for queue-dependent servers in closed queueing networks." *Perf. Eval.,* **1,** 7-18.

Reiser, M. and **Lavenberg, S. S.** [1980], "Mean value analysis of closed multichain queueing networks." *J.A.C.M.,* **27,** 313-322.

Rényi, A. [1967], "Remarks on the Poisson process." In *Lecture Notes in Mathematics,* Vol 31, Springer-Verlag, New York, 280-286.

Righter, R. [1986], "A resource allocation problem in a random environment." Submitted to *Operations Research.* **1,** 189-202.

Righter, R. [1987], "Job scheduling to minimize weighted flow time on uniform processors." To appear in *Syst. and Control Letters.*

Righter, R. [1987], "The stochastic sequential assignment problem with random deadlines." *Probability in the Engineering and Informational Sciences,*

Rolski, T. [1981], *Stationary Random Processes Associated With Point Processes.* Lec. Note Series, Springer-Verlag, New York.

Rosberg, Z. [1980], "A positive recurrence criterion associated with multidimensional queueing processes." *J. Appl. Prob.*, **17**, 790-801.

Rosberg, Z. [1981], "A note on the ergodicity of Markov chains." *J. Appl. Prob.*, **18**, 112-121.

Rosberg, Z. [1983], "Optimal decentralized control in a multiaccess channel with partial information." *IEEE Trans. Aut. Control,* **AC-28**, 187-193.

Rosberg, Z. [1984], "Deterministic routing to buffered channels." Comp. Sci. Rep., Technion, Israel.

Rosberg, Z., Varaiya, P., and **Walrand, J.** [1982], "Optimal control of service in tandem queues." *IEEE Trans. Aut. Control,* **AC-27**, 600-609.

Ross, S. M. [1983a], *Stochastic Processes.* John Wiley & Sons, New York.

Ross, S. M. [1983b], *Introduction to Stochastic Dynamic Programming.* Academic Press, New York.

Ross, S. [1987], "Approximating transition probabilities and mean occupation times in continuous-time Markov chains." *Probability in the Engineering and Informational Sciences,* **1**, 251-264.

Rumswicz, N. [1987], "Insensitivity of GSMPs with age-dependent routing." Preprint, University of Adelaide, Australia.

Sacks, J. [1960], "Ergodicity of queues in series." *Ann. Math. Stat.*, **31**, 579-588.

Salza, S. and **Lavenberg, S. S.** [1981], "Approximating response time distributions in closed queueing network models of computer performance." In *Performance '81,* (Ed. F. J. Klystra). North-Holland, Amsterdam.

Sanders, B. A. [1985], "A private good/public good decomposition for optimal flow control of an M/M/1 queue." *IEEE Trans. Aut. Control.*

Schassberger, R. [1977], "Insensitivity of steady-state distributions of generalized semi-Markov processes, Part I." *Ann. Prob.*, **5**, 87-99.

Schassberger, R. [1978a], "Insensitivity of steady-state distributions of generalized semi-Markov processes, Part II." *Ann. Prob.*, **6**, 85-93.

Schassberger, R. [1978b], "Insensitivity of steady-state distributions of generalized semi-Markov processes with speeds." *Adv. Appl. Prob.*, **10**, 836-851.

Schassberger, R. [1978c], "The insensitivity of stationary probabilities in networks of queues." *Adv. Appl. Prob.*, **10**, 906-912.

Schassberger, R. and **Daduna, H.** [1983], "The time for a round trip in a cycle of exponential queues." *J.A.C.M.*, **30**, 146-150.

Schrage, L. [1968], "A proof of the optimality of the shortest remaining processing time discipline." *Oper. Res.*, **16**, 687-690.

Schwartz, M. [1977], *Computer Communication Network Design and Analysis.* Prentice-Hall, Englewood Cliffs, New Jersey.

Schwartz, M. [1987], *Telecommunication Networks: Protocols, Modeling and Analysis.* Addison-Wesley.

Schweitzer, P. [1979], "Approximate analysis of multiclass closed networks of queues." In *Proc. Int. Conf. Stoch. Control and Optimization,* Amsterdam.

Serfozo, R. [1979], "An equivalence between continuous and discrete time Markov decision processes." *Oper. Res.*, **27**, 616-620.

Serfozo, R. [1981], "Optimal control of random walks, birth and death processes, and queues." *Adv. Appl. Prob.*, **13**, 61-83.

Sevcik, K. C. and **Mitrani, I.** [1981], "The distribution of queueing network states at input and output instants." *J.A.C.M.*, **28,** 358-371.

Shantikumar, J. G. and **Buzacott, J. A.** [1981], "Open queueing network models of dynamic job shops." *Int. J. of Prod. Res.*, **19,** 255-266.

Shantikumar, J. G. and **Yao, D. D.** [1986], "The effect of increasing service rates in a closed queueing network." *J. Appl. Prob.*, **23,** 474-483.

Simon, B. and **Foley, R. D.** [1979], "Some results on sojourn times in acyclic Jackson networks." *Mgmt Sci.*, **25,** 1027-1034.

Stidham, S. Jr. [1970], "On the optimality of the single-server queueing system." *Oper. Res.*, **18,** 708-732.

Stidham, S. Jr. [1974], "A last word on $L = \lambda W$." *Oper. Res.*, **22,** 417-421.

Stidham, S. Jr. [1985], "Optimal control of admission to a queueing system." *IEEE Trans. Auto. Control,* **AC-30,** 705-713.

Stidham, S. Jr., and **Prabhu, N. U.** [1974], "Optimal control of queueing systems." In *Mathematical Methods in Queueing Theory* (Lect. Notes in Economics and Math. Syst., Vol. 98), Springer-Verlag, New York.

Stidham, S. Jr. and **Weber, R. R.** [1983], "Monotonic and optimal policies for control of queues with undiscounted cost." To appear.

Stoyan, D. [1983], *Comparison Methods for Queues and Other Stochastic Models.* (Edited by D. J. Daley). John Wiley & Sons, New York.

Strauch, R. E. [1970], "When a queue looks the same to an arriving customer as to an observer." *Mgmt Sci.*, **17,** 140-141.

Szpankowski, W. [1985], "Some sufficient conditions for non-ergodicity of Markov chains." *J. Appl. Prob.*, **22,** 138-147.

Takacs, L. [1963], "A single server queue with feedback." *Bell Syst, Tech, J.*, **42,** 505-519.

Tcha, D. and **Pliska, S. R.** [1977], "Optimal control of single-server queueing networks and multiclass M/G/1 queues with feedback." *Oper.·Res.*, **25,** 248-258.

Tembe, S. W. and **Wolff, R. R.** [1974], "The optimal order of service in tandem queues." *Oper. Res.*, **22,** 824-832.

Tijms, H. C. [1986], *Stochastic Modelling and Analysis: A Computational Approach.* John Wiley & Sons, New York.

Towsley, D. F. [1980], "Queueing network models with state-dependent routing." *J.A.C.M.*, **27,** 323-337.

Tsoucas, P. and **Walrand, J.** [1984a], "A note on the processor sharing queue in a quasireversible network." *J. Appl. Prob.*

Tsoucas, P. and **Walrand, J.** [1984b], "A note on stochastic bounds for queueing networks." *Adv. Appl. Prob.*, **16,** 926-928.

Tsoucas, P. and **Walrand, J.** [1986], "On the interchangeability and comparison of M/M/1 queues in tandem." To appear in *Adv. Appl. Prob.*

van der Wal, J. [1986], "Monotonicity of the throughput of a closed exponential network in the number of jobs." *Proceedings of IEEE Conference on Decision and Control, Dec. 1986, Athens.*

van Dijk, N. M., Tsoucas, P., and **Walrand, J.** [1987], "Simple bounds for the call congestion of multiserver queues." To appear in *Probability in Engineering and Information Systems.*

Varadhan, S. R. S. [1984], *Large Deviations and Applications.* SIAM, Philadelphia.

Varaiya, P. and **Walrand, J.** [1980a], "Sojourn times and the overtaking condition in Jacksonian networks." *Adv. Appl. Prob.,* **12,** 1000-1018.

Varaiya, P. and **Walrand, J.** [1980b], "Interconnection of Markov chains and quasi-reversible queueing networks." *Stoc. Proc. Appl.,* **10,** 209-219.

Varaiya, P. and **Walrand, J.** [1981], "Flows in queueing networks: a martingale approach." *Math. Oper. Res.,* **6,** 387-404.

Varaiya, P., Walrand, J., and **Buyukkoc, C.** [1985], "Extensions of the multiarmed bandit problem: the discounted case." *IEEE Trans. Aut. Control,* **AC-30,** 426-439.

Walrand, J. [1983], "A probabilistic look at networks of quasi-reversible queues." *IEEE Trans. Info. Theory,* **IT-29,** 825-831.

Wan, C. B. and **Davis, M. H. A.** [1977], "The general point process disorder problem." *IEEE Trans. Info. Theory,* **IT-23,** *538-540.*

Weber, R. R. [1979], "The interchangeability of M/M/1 queues in series." *J. Appl. Prob.,* **16,** 690-695.

Weber, R. R. [1980], "Optimal organization of multiserver systems." Ph. D. thesis, Cambridge University.

Weber, R. R. [1982a], "Multiserver stochastic scheduling." In *Deterministic and Stochastic Scheduling* (eds. Dempster, et al.), Reidel, Dordrecht, Holland.

Weber, R. R. [1982b], "Scheduling stochastic jobs on parallel machines to minimize makespan or flowtime." In *Applied Probability-Computer Science, the Interface: Proc. of the ORSA-TIMS Boca Raton Symposium* (eds. R. Disney and T. Ott), Birkhauser, Boston, Cambridge, Mass.

Weber, R. R. and **Stidham, S. Jr.** [1987], "Control of service rates in networks of queues." *Adv. Appl. Prob.,* **19,** 202-218.

Weber, R. R., Varaiya, P. P., and **Walrand, J.** [1986], "Scheduling jobs with stochastically ordered processing times on parallel machines to minimize expected flowtime." *J. Appl. Prob.,* **23,** 841-847.

Weiss, G. [1982], "Multiserver stochastic scheduling." In *Deterministic and Stochastic Scheduling* (eds. Dempster, et al.), Reidel, Dordrecht, Holland.

Weiss, G. and **Pinedo, M.** [1979], "Scheduling tasks with exponential service times on non-identical processors to minimize various cost functions." *J. Appl. Prob.,* **17,** 187-202.

Wetherill, G. B and **Glazebrook, K. D.** [1986], *Sequential Methods in Statistics,* (3rd edition). Chapman and Hall, London.

Whitt, W. [1981], "Comparing point processes and queues." *Adv. Appl. Prob.,* **13,** 207-220.

Whitt, W. [1982], "Approximating a point process by a renewal process: two basic methods." *Oper. Res.,* **50,** 125-147.

Whitt, W. [1983], "The queueing network analyzer." *Bell Syst. Tech. J.,* **62,** 2779-2815.

Whitt, W. [1984], "Open and closed models for networks of queues." *AT&T Bell Labs Tech. J.,* **63,** 1911-1979.

Whittle, P. [1967], "Nonlinear migration processes." *Bull. Int. Inst. Statist.,* **42,** 642-647.

Whittle, P. [1968], "Equilibrium distributions for an open migration process." *J. Appl. Prob.,* **5,** 567-571.

Whittle, P. [1980], "Multi-armed bandits and the Gittins index." *J. Roy. Stat. Soc.*, **B42,** 143-149.

Whittle, P. [1981], "Arm-acquiring bandits." *Ann. Prob.*, **9,** 284-292.

Whittle, P. [1982], *Optimization Over Time: Dynamic Programming and Stochastic Control, Vol. I.* John Wiley & Sons, New York.

Whittle, P. [1983a], *Optimization Over Time: Dynamic Programming and Stochastic Control, Vol. II.* John Wiley & Sons, New York.

Whittle, P. [1983b], "Relaxed Markov processes." *Adv. Appl. Prob.*, **15,** 769-782.

Whittle, P. [1985], "Partial balance and insensitivity." *J. Appl. Prob.*, **22,** 168-176.

Whittle, P. [1986a], *Systems in Stochastic Equilibrium.* John Wiley & Sons, New York.

Whittle, P. [1986b], "Partial balance, insensitivity and weak coupling." *Adv. Appl. Prob.*, **18,** 706-723.

Williams, D. [1979], *Diffusions, Markov Processes, & Martingales.* John Wiley & Sons, New York.

Wolff, R. W. [1970], "Work conserving priorities." *J. Appl. Prob.*, **7,** 327-337.

Wolff, R. W. [1982a], "Poisson arrivals see time averages." *Oper. Res.*, **30.**

Wolff, R. W. [1982b], "Tandem queues with dependent service in light traffic." *Oper. Res.*, **30,** 619-635.

Wolff, R. W. and **Wrightson, C. W.** [1976], "An extension of Erlang's formula." *J. Appl. Prob.*, **13,** 628-632.

Wolfson, B. [1984], "Some moment results for certain tandem and multiple-server queues." *J. Appl. Prob.*, **21,** 901-910.

Wong, E. and **Hajek, B.** [1985], *Stochastic Processes in Engineering Systems.* Springer-Verlag, New York.

INDEX